HAZARDOUS AND INDUSTRIAL SOLID WASTE TESTING: FOURTH SYMPOSIUM

A symposium
sponsored by
ASTM Committee D-34
on Waste Disposal
Arlington, VA, 2–4 May 1984

ASTM SPECIAL TECHNICAL PUBLICATION 886
James K. Petros, Jr., Union Carbide Corp.,
William J. Lacy, U.S. Environmental
Protection Agency, and
Richard A. Conway, Union Carbide Corp.,
editors

ASTM Publication Code Number (PCN)
04-886000-16

 1916 Race Street, Philadelphia, PA 19103

Library of Congress Cataloging-in-Publication Data

Hazardous and industrial solid waste testing.

(ASTM special technical publication; 886)
"A symposium sponsored by ASTM Committee D-34 on
Waste Disposal, Arlington, Va., 2-4 May 1984."
"ASTM publication code number (PCN) 04-886000-16."
Includes bibliographies and index.
1. Hazardous wastes—Testing—Congresses.
2. Factory and trade waste—Testing—Congresses.
I. Petros, James K.   II. Lacy, W. J. (William J.)
III. Conway, Richard A.   IV. ASTM Committee D-34 on
Waste Disposal.   V. ASTM Symposium on Testing of
Hazardous Solid Waste (4th : 1983 : Arlington, Va.)
VI. Series.
TD811.5.H3825   1986   628.5′028′7   85-28730
ISBN 0-8031-0430-8

NOTE

The Society is not responsible, as a body,
for the statements and opinions
advanced in this publication.

Printed in Baltimore, MD
February 1986

Foreword

The fourth symposium on Hazardous and Industrial Solid Waste Testing was held on 2–4 May 1984 in Arlington, VA. This symposium was the fourth in a series of annual symposia on hazardous waste sponsored by ASTM Committee D-34 on Waste Disposal.

Chairing the symposium were James K. Petros, Jr., Union Carbide Corp., and William J. Lacy, U.S. Environmental Protection Agency, both of whom also served as editors of this publication. Richard A. Conway, Union Carbide Corp., also served as an editor.

Related
ASTM Publications

Hazardous and Industrial Waste Management and Testing: Third Symposium, STP 851 (1984), 04-851000-16

Hazardous and Industrial Solid Waste Testing: Second Symposium, STP 805 (1983), 04-805000-16

Hazardous Solid Waste Testing: First Conference, STP 760 (1982), 04-760000-16

Permeability and Groundwater Contaminant Transport, STP 746 (1981), 04-746000-38

Aquatic Toxicology and Hazard Assessment: Sixth Symposium, STP 802 (1983), 04-802000-16

A Note of Appreciation
to Reviewers

The quality of the papers that appear in this publication reflects not only the obvious efforts of the authors but also the unheralded, though essential, work of the reviewers. On behalf of ASTM we acknowledge with appreciation their dedication to high professional standards and their sacrifice of time and effort.

ASTM Committee on Publications

ASTM Editorial Staff

Contents

INCINERATION

SUMMARY

INDEXES

Introduction

One of the major national problems facing the United States and other nations as well is the need for safe handling and disposal of hazardous solid waste. The Resource Conservation and Recovery Act (RCRA) mandates that the U.S. Environmental Protection Agency (EPA) promulgate and enforce regulations relating to the proper management of hazardous wastes. Neither the EPA nor any other organization can be expected to accomplish this major task without significant external input. Toward that end, Committee D-34 on Waste Disposal has sponsored a series of symposia. These symposia have brought together recognized experts from all sectors—federal, state, and municipal governments, industry, commerce, private laboratories, consultants, legislators, academicians, and knowledgeable citizens—for the express purpose of sharing technical knowledge in the field of hazardous waste testing.

ASTM has worked diligently for years toward learning more about the problems of waste disposal and developing test methods to solve these problems in a scientifically sound, economical manner with a minimum risk to the public health. Committee D-34 has within its ranks a considerable array of talent with which to accomplish this monumental task.

The first ASTM symposium on Testing of Hazardous Solid Wastes was held in Fort Lauderdale, Florida, on 14–15 Jan. 1981, under the chairmanship of Richard A. Conway of Union Carbide Corp. and B. Charles Malloy of Jones, Malloy, and Associates.

As stated in the resulting publication,[1] the purposes of the symposium were to:

- Present and discuss new knowledge on testing.
- Stimulate new technology by building on reported work.
- Provide bases for new and improved ASTM methods.
- Bring more science into an area dominated by regulations.
- Develop a series of special technical publications (STP) on testing of solid wastes.

Papers for the first symposium were selected in five general areas:

- Laboratory extraction and leaching procedures.
- Large-scale leaching tests versus laboratory tests.

[1]*Hazardous Solid Waste Testing: First Conference, ASTM STP 760*, American Society for Testing and Materials, Philadelphia, 1982.

- Analysis of residues, extracts, solids, and groundwaters.
- Evaluation of land disposal sites and materials.
- Risk assessment approaches.

The 300 scientists and engineers who attended thoroughly discussed the presentation made in these areas. The resulting STP[1] addresses the suitability of the land disposal option for various types of residues.

The second ASTM Committee D-34 symposium of this series was held at Lake Buena Vista, Florida, 28–29 Jan. 1982, under the cochairmanship of Richard Conway and William P. Gulledge of the Chemical Manufacturers Association. In the resulting STP,[2] papers were selected in the following categories:

- Sampling considerations.
- Batch extraction.
- Column leaching and transport.
- Analytical techniques.
- Linear testing and closure,
- Biological tests.

The third symposium, held in Philadelphia 7–10 March 1983 and cochaired by Larry P. Jackson of the University of Wyoming Research Center and Carlo Merli of the University of Rome, was a departure from the previous two in that it had many presentations from 13 countries and it addressed the overall problem of waste management as well as solid waste testing. This international conference was a follow-up to an international conference in Rome in 1981 led by B. C. Malloy, L. P. Jackson, C. Merli, and W. J. Lacy.

The principle objective of the third symposium was to accelerate technology transfer that could lead to future standards. Papers were selected in four areas:

- Sampling and analysis of wastes and waste disposal sites.
- Amelioration of wastes in the disposal environment.
- Waste as a resource.
- National perspectives in waste management.

These papers[3] became the basis for relating waste behavior in the disposal environment. This topic was expanded into the broader activity of risk assessment, which is also of interest to other ASTM committees. Risk assessment and risk management of hazardous waste have been the subjects of a series of recent books and articles, highlighting the shift to this area of technical concern.

---

[2]*Hazardous and Industrial Solid Waste Testing: Second Symposium, ASTM STP 805,* American Society for Testing and Materials, Philadelphia, 1983.

[3]*Hazardous and Industrial Waste Management and Testing: Third Symposium, ASTM STP 851,* American Society for Testing and Materials, Philadelphia, 1984.

The fourth ASTM D-34 symposium on Hazardous and Industrial Solid Waste Testing, the basis for the present STP, was held in Arlington, Virginia, on 2-4 May 1984. The chairpersons were James K. Petros, Union Carbide Corp. and William J. Lacy, U.S. EPA.

The keynote speakers were Norman H. Nosenchuck, director of the Solid Waste Department, New York State Department of Environmental Conservation, and Norbert B. Schomaker, director of Solid and Hazardous Waste Research, U.S. EPA.

The subjects discussed at this symposium and contained in this STP cover the following categories:

- Thermal treatment.
- Hazard degrees, health effects, and risk assessment.
- Waste characterization (chemical, physical, biological).
- Extraction of toxicants (batch and continuous).
- Evaluation of extracts and leachates.
- Landfill/landfarm simulation.
- Sampling of residues and sites (soil, groundwater, air).

In addition, a special session on risk assessment including panel discussions was held under the leadership of William Gulledge of the Chemical Manufacturers Association. Also held in conjunction with the standards development meetings of ASTM Committee D-34 was a Workshop on Ocean Disposal, chaired by Rosalie T. Matthews of Matthews Consulting and Construction, Inc.

The organizing committee for this symposium worked diligently in soliciting abstract submittals, in selecting promising presentations, and in chairing sessions. The committee was composed of the following people:

D. R. Bowlus	W. J. Lacy
D. Buskirk	D. J. Lorenzen
J. P. Chu	B. C. Malloy
R. A. Conway	C. L. Perket
D. Friedman	J. K. Petros
C. Glover	A. R. Rohlik
W. P. Gulledge	N. B. Schomaker
K. Jackson	W. C. Webster
L. P. Jackson	

As in any endeavor such as this, the technical quality of this publication has largely resulted from the dedicated and diligent efforts of the scientific reviewers of the technical papers. Considerable support staff efforts also were needed to assure the success of this symposium. The editors gratefully acknowledge the valuable assistance of this key group, Kathy Greene and Don Viall of ASTM, and Betty Maisonneuve of the U.S. EPA. We hope that the technical papers in this STP will be as valuable an aid to industry; federal, state, and local governments; and the scientific and engineering community in general as were the previous ones.

Future symposia in this series should address other issues, including the following:

- The relationship between laboratory characterizations and field findings.
- Stabilization, soil attenuation/migration.
- Improved liner and landfill closure site testing.
- More accurate testing of physical/chemical parameters.
- Deep well injection and resource recovery.
- Legal issues related to assessment and testing.
- Quality assurance.

Symposia addressing these issues were held in 1985 in Colorado Springs, Colorado, and Alexandria, Egypt, and in 1986 in New Orleans.

William J. Lacy

U.S. Environmental Protection Agency, Washington, DC 20460; symposium cochairman and coeditor.

James K. Petros, Jr.

Union Carbide Corp., South Charleston, WV 25303; symposium cochairman and coeditor.

Richard A. Conway

Union Carbide Corp., South Charleston, WV 25303; coeditor.

Analysis and Characterization of Wastes

John H. Martin, Jr.,[1] *and Raymond C. Loehr*[2]

Determination of the Oil Content of Soils

REFERENCE: Martin, J. H., Jr., and Loehr, R. C., "**Determination of the Oil Content of Soils,**" *Hazardous and Industrial Solid Waste Testing: Fourth Symposium, ASTM STP 886*, J. K. Petros, Jr., W. J. Lacy, and R. A. Conway, Eds., American Society for Testing and Materials, Philadelphia, 1986, pp. 7-14.

ABSTRACT: The precision and accuracy of acidification followed by Soxhlet extraction with trichlorotrifluoroethane to estimate the oil and grease content of oil-contaminated soils was determined. The amounts of oil and grease were determined gravimetrically after solvent extraction. Stock oil-soil mixtures prepared with vegetable oil, No. 2 fuel oil, No. 6 fuel oil, motor oil, and an oily waste were analyzed. The coefficient of variation for all oil-soil mixtures analyzed never exceeded 4%. The percentage of recovery varied from 68% for No. 2 fuel oil to 102% for motor oil. The results indicate that this method is precise and results in reasonable recoveries when used to measure the oils tested.

KEY WORDS: oil and grease, soil, Soxhlet extraction, trichlorotrifluoroethane, precision, accuracy, hazardous wastes

Land treatment has been used for the treatment and disposal of oily wastes by petroleum refineries for many decades. About 100 petroleum industry land treatment facilities exist in the United States, with additional facilities at refineries in Canada and Europe [1].

Land treatment is a managed technology that involves the controlled application of a waste into the upper soil zone, the zone of incorporation. The successful performance of land treatment depends on an understanding of the site and waste characteristics that affect the transformation, degradation, and immobilization of the wastes that are applied to the site. Petroleum industry wastes contain a large oil and grease fraction. Therefore, an ability to measure the oil and grease content of both the wastes and the soil in the zone of incorporation is important for determining appropriate waste application

---

[1] Research associate, Department of Agricultural Engineering, Cornell University, Ithaca, NY 14853.

[2] Professor of civil engineering, University of Texas, Austin, TX 78712; formerly professor of agricultural engineering and environmental engineering, Cornell University, Ithaca, NY 14853.

rates and for understanding the performance of the site in terms of oil and grease degradation.

All of the analytical methods for oil and grease that are now in use or have been considered for use are based on the fact that the oil and grease are insoluble in water, but are soluble in organic solvents. Usually, the sample is acidified to convert soaps to fatty acids before the oil and grease is extracted. In the determination of oil and grease, an absolute quantity of a specific substance is not measured. Rather, groups of substances with similar physical characteristics are determined quantitatively on the basis of their common solubility in the solvent that is used. Unlike some elements or compounds, oils and greases are defined by the method used for their determination.

The solvents that have been used to determine oil and grease include petroleum ether, hexane, benzene, chloroform and methanol, and carbon tetrachloride [2]. Both dry and wet extraction procedures have been used. The solvent now commonly used in the water pollution control field is trichlorotrifluoroethane [3]. This solvent represents less of a hazard in the laboratory than many of the solvents noted earlier, since trichlorotrifluoroethane is not flammable or explosive and has no known toxic properties [4]. The Soxhlet extraction method with trichlorotrifluoroethane used to extract oil and grease from sludge samples (American Public Health Association [APHA] Method 503D, Extraction Method for Sludge Samples) appears to be appropriate for petroleum wastes and soils and was evaluated in this study.

As the method states, the results are empirical and duplicate results can be obtained only by strict adherence to all details. The rate and time of extraction in the Soxhlet apparatus must be exactly as directed because of the varying solubilities of different materials. In addition, the length of time required for evaporating the solvent and cooling the extracted material cannot be varied. There may be a gradual increase in weight, presumably caused by absorption of oxygen or a gradual loss of weight due to volatilization [3]. Compounds volatilized at or below 70°C will be lost during the evaporating process.

A tentative infrared detection method that does permit the measurement of many relatively volatile hydrocarbons also is available (APHA Method 503B, Partition-Infrared Method [Tentative]). This evaluation focuses on APHA 503D because it is an approved rather than tentative method.

Objectives

The objective of this study was to evaluate the precision and accuracy of a modified form of the Soxhlet extraction method for sludge samples (APHA 503D) for gravimetric determination of the oil and grease content of soils.

Experimental Approach

Soil samples containing known amounts of several types of oil, about 5% on a dry weight basis, were analyzed by using a modified form of the oil and grease extraction method for sludge samples (APHA 503D). For each oil-soil mixture, eight samples were analyzed. The oils used were vegetable oil, No. 2 fuel oil, No. 6 fuel oil, Society of Automotive Engineers (SAE) 30 nondetergent motor oil, and an oily industrial waste. Eight samples of the soil without the oils were analyzed as blanks.

In addition, a second series of five oil-soil mixtures containing 1 to 5% of the oily industrial waste on a dry weight basis were analyzed. For each oil concentration, two samples were analyzed.

Materials

Each stock mixture of oil and soil was prepared by adding the amount of oil, on a weight basis, necessary to produce the desired concentration of oil in 300 g of soil. After thorough mixing, these stock mixtures were stored at room temperature (20 to 22°C) in screw-cap glass bottles until analyzed. When each oil-soil mixture was prepared, the dry matter content of the soil was determined using the American Society of Agronomy's Method for Gravimetry with Oven Drying (7-2.2) so that the oil content of the stock mixtures could be expressed in milligrams of oil per gram of moisture-free soil. Coefficients of variation for dry matter determinations never exceeded 1%, with two replicates per sample.

The soil used for this study was a heavy silt loam identified as Rhinebeck silt loam [5]. This soil is a deep, somewhat poorly drained, fine textured soil that formed in clayey, calcareous lake deposits. Soils of the Rhinebeck series are moderately acidic to neutral.

To eliminate soil characteristics as a variable, all of the soil used in this study was acquired at the beginning of the study and processed in the following manner. First, the soil was hand-sorted to remove stones and vegetative matter. It then was mixed thoroughly for about 30 min in a portable concrete mixer and allowed to air-dry at room temperature (20°C) until it was friable. When dry, the soil was pulverized and passed through a 1.19-mm (No. 16) U.S. standard sieve and stored at room temperature in a closed container until used.

The oils used in this study were vegetable oil (partially hydrogenated soybean oil), No. 2 fuel oil, No.6 fuel oil, SAE 30 nondetergent motor oil, and an oily industrial waste. The oily waste was of unknown origin but was from a lagoon that had been used for the storage of oil refinery sludges. The characteristics of this oily waste are summarized in Table 1.

Analytical Method

The method used in this study to determine the oil and grease content of soils and oil-soil mixtures was a modified form of APHA 503D. The procedure was as follows:

1. About 15 g of the soil or oil-soil mixture to be analyzed was placed in a preweighed porcelain evaporating dish, 120 mL or larger, that contained a glass stirring rod. The dish was weighed again with the glass rod and the soil to determine the weight of the added soil or mixture.
2. Any soluble metallic soaps present were hydrolyzed by acidification with concentrated hydrochloric acid to pH 2.0 or lower. To facilitate acidification, enough distilled water to produce a smooth paste-like mixture was added before the acid.
3. After the acidified soil was mixed thoroughly with the stirring rod, the soil was dried by adding $MgSO_4 \cdot H_2O$ that was prepared by drying $MgSO_4 \cdot 7H_2O$ overnight at 150°C.
4. The acidified soil and $MgSO_4 \cdot H_2O$ were mixed with the stirring rod, then placed in a desiccator and allowed to cool.
5. After the dish with its contents and the stirring rod was weighed, the contents of the dish were transferred to a porcelain mortar and ground.
6. A subsample of the material from Step 5, containing no more than 150 mg of oil, was placed in a tared cellulose extraction thimble and weighed. The thimble was then filled with small glass beads and placed in a Soxhlet extraction tube.
7. Using a preweighed extraction flask that had been dried at 103°C and cooled in a desiccator, the oil in the sample was extracted with 75 mL of reagent grade 1,1,2-trichloro-1,2,2-trifluoroethane for 4 h at a rate of 20 cycle/h.
8. After extraction, the extraction flask was placed in a 70°C water bath for 1 h to evaporate the solvent. At the end of the hour, air was drawn through the flask for 1 min using a vacuum pump to remove any remaining vapors.
9. Finally, the flask was weighed again after being cooled in a desiccator for 30 min to determine the weight of extracted oil and grease.

Although this procedure is almost identical to APHA 503D, it differs in one important respect. The APHA procedure requires the quantitative transfer of

TABLE 1—*Characteristics of oily waste used in study.*

Parameter, % Wet Basis	$\bar{X} \pm SD^a$	Method
Water	48.8 ± 0.7	ASTM D 95[b]
Ash	16.1 ± 0.1	ASTM D 482[c]
Oil and grease	26.2 ± 0.3	as described in this paper

[a] Mean ± standard deviation, $n = 8$.
[b] Test for water in petroleum products and bituminous materials by distillation.
[c] Test for ash from petroleum products.

the mixture of acidified sample and $MgSO_4 \cdot H_2O$ first to a mortar for grinding and then to an extraction thimble. The procedure used in this study eliminates possible errors associated with such transfers. It also permits the use of a larger, more representative original sample, which is an advantage when the oil and grease concentration limits the size of the sample what can be extracted. The subsample extracted contains portions of the original soil and the water and chemicals that were added.

Calculations

The concentration of oil and grease in each soil sample was expressed in milligrams of oil and grease per gram of moisture-free soil. The calculations were as follows:

$$\text{oil and grease (mg/g)} = \frac{\text{extraction flask weight gain (mg)}}{\text{soil extracted (g)} \times \text{dry matter fraction}} \quad (1)$$

where

$$\begin{array}{c} \text{soil} \\ \text{extracted} \\ \text{(g)} \end{array} = \begin{array}{c} \text{subsample} \\ \text{extracted} \\ \text{(g)} \end{array} \times \frac{\text{soil sample (g)}}{\text{soil sample} + H_2O + HCl + MgSO_4 \cdot H_2O \text{ (g)}}$$

$$(2)$$

Results and Discussion

The results of this evaluation are summarized in Tables 2, 3, and 4. The precision (coefficient of variation) of this method was excellent (Tables 2 and

TABLE 2—*Precision of Soxhlet extraction with trichlorotrifluoroethane for measuring the oil and grease content of soil.*

	Oil, mg per g moisture-free soil			
Oil	Range	$\bar{X} \pm SD^a$	CI^b	CV, %[c]
Blank[a]	0.1–0.3	0.2 ± 0.1	0.1–0.3	50.0
Vegetable oil	41.1–45.8	43.2 ± 1.6	42.0–44.6	3.7
Fuel oil, No. 2	33.6–35.1	34.5 ± 0.6	34.0–35.0	1.7
Fuel oil, No. 6	41.9–45.0	43.9 ± 0.9	43.1–44.7	2.0
Motor oil, SAE 30	50.8–57.1	54.7 ± 2.2	52.9–56.7	4.0
Oily waste	47.0–53.1	50.2 ± 2.0	48.5–51.9	4.0

[a] Mean ± standard deviation, $n = 8$.
[b] 95% confidence interval estimate, $t_{0.05}$.
[c] Coefficient of variance.
[d] Air-dried soil.

TABLE 3—*Precision and accuracy as functions of concentration for oily waste-soil mixtures.*

Oil, mg per g moisture-free soil		Coefficient of Variance, %	Recovered, %
Stock Mixture	Recovered[a]		
10.5	8.8 ± 0.1	1.1	83.8
21.0	18.0 ± 0.4	2.2	85.7
31.4	29.4 ± 0.1	0.3	93.6
41.9	40.5 ± 0.6	1.5	96.7
52.2	50.0 ± 0.6	1.2	95.8

[a] Blank corrected mean ± standard deviation, $n = 2$.

TABLE 4—*Accuracy of measuring the oil content of soil—Soxhlet extraction with trichlorotrifluoroethane.*

Oil	Oil, mg per g moisture-free soil		Recovered, %
	Stock Mixture	Recovered[a]	
Blank[b]	0	0.2 ± 0.1	...
Vegetable oil	50.6	43.0 ± 1.6	85.4
Fuel oil, No. 2	50.7	34.3 ± 0.6	68.0
Fuel oil, No. 6	52.9	43.7 ± 0.9	83.0
Motor oil, SAE 30	53.5	54.5 ± 2.2	102.2
Oily waste	52.2	50.0 ± 2.0	96.2

[a] Blank corrected mean ± standard deviation, $n = 8$.
[b] Air-dried soil.

3). The accuracy (percent recovery) varied with type of oil (Table 4) and with the concentration of the oily waste (Table 3).

Precision

As shown in Tables 2 and 3, the coefficient of variation for all of the oil-soil mixtures analyzed never exceed 4%. There was little variation in precision as the concentration of oily waste varied (Table 3).

The reason for the observed variability in the results for the blank air-dried soil samples is unclear, but it has been observed by the authors in other investigations. This coefficient of variation (50%) is large only in relative terms because the oil content of these samples was so low. Routine analysis of the residue content of the reagent grade trichlorotrifluoroethane used in this study indicated that solvent contamination was not responsible for the observed variability. Residue after evaporation never exceeded 0.1 mg per 100 mL.

One possible explanation for this variability could be the loss of random quantities of fibers from the extraction thimbles during extraction. Filtering of the solvent after extraction, as allowed in Method 503D, would eliminate this possible source of variability.

Accuracy

The accuracy of this method varied with the oil type (Table 4). With the exception of motor oil, the quantities added were underestimated.

The reasons for these differences in accuracy were not determined. It is known that extraction efficiency varies with oil composition and with the constituents being extracted. Only those substances that are soluble in the solvent used for extraction can be determined quantitatively [2,3].

The percentage of recovery for No. 2 fuel oil was the lowest among all of the oils used in this study (Table 4). It is possible that losses of volatile constituents could have occurred during the preparation of the stock oil-soil mixtures as well as during sample acidification and drying. McGill and Rowell [6] have reported recoveries of about 70% for untopped crude oils (volatiles not removed) versus about 95% recovery for crude oils that had been topped at 21°C for three days. The solvent used was methylene chloride.

As shown in Table 3, the percentage of recovery increased as the concentration of oil in the oily waste-soil mixtures increased. Regression analysis of the

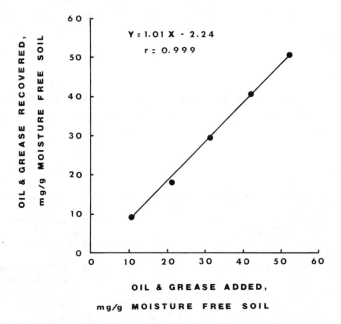

FIG. 1—*Regression analysis of the analytical results for oily waste–soil mixtures ranging from 1 to 5% oil and grease.*

TABLE 5—*Summary of results of measuring the oil content of soil using Soxhlet extraction with trichlorotrifluoroethane.*

Oil	Coefficient of Variation, %	Recovery, %
Blank[a]	50.0	...
Vegetable oil	3.7	85.4
Fuel oil, No. 2	1.7	68.0
Fuel oil, No. 6	2.0	83.0
Motor oil, SAE 30	4.0	102.2
Oily waste	4.0	96.2

[a] Air-dried soil.

analytical results indicated that the magnitude of the underestimate was constant (Fig. 1). This indicates that the observed difference was not a function of concentration but of an unknown constant error. The difference between the intercept of the regression line (-2.24 mg oil and grease per gram of moisture-free soil) and the origin was tested statistically [7] and was not found to be significant ($t < .05$).

Summary

The results of this study, summarized in Table 5, indicate that the modified method identified in this paper, which consists of acidification followed by Soxhlet extraction with trichlorotrifluoroethane, is precise and results in reasonable recoveries when used to measure the oil and grease content of contaminated soils.

Acknowledgments

This study was conducted as part of Cooperative Agreement CR-809285 between Cornell University and the Robert S. Kerr Environmental Research Laboratory of the U.S. Environmental Protection Agency.

References

[1] American Petroleum Institute, *Land Treatment Practices in the Petroleum Industry*, Washington, DC, 1983.
[2] Loehr, R. C. and Higgins, G. C. "Comparison of Lipid Extraction Methods," *International Journal of Air and Water Pollution*, Vol. 9, No.1/2, Feb. 1965, pp. 55-67.
[3] *Standard Methods For The Examination of Water and Wastewater*, 15th ed., American Public Health Association, Washington, DC, 1980.
[4] *The Merck Index*, 9th ed., M. Windholz, Ed., Merck & Co., Inc., Rahway, NJ, 1976.
[5] Soil Conservation Service, *Soil Survey, Tompkins County, New York*, Series 1961, No. 25, U.S. Department of Agriculture, Washington, DC, 1965.
[6] McGill, W. B. and Rowell, M. J., "Determination of Oil Content of Oil Contaminated Soil," *Science of the Total Environment*, Vol, 14, No. 3, April 1980, pp. 245-253.
[7] Snedecor, G. W. and Cochran, W. G., *Statistical Methods*, 7th ed., Iowa State University Press, Ames, 1980.

C. W. Francis,[1] *M. P. Maskarinec,*[1] *and J. C. Goyert*[2]

A Laboratory Extraction Method to Simulate Codisposal of Solid Wastes in Municipal Waste Landfills

REFERENCE: Francis, C. W., Maskarinec, M. P., and Goyert, J. C., "**A Laboratory Extraction Method to Simulate Codisposal of Solid Wastes in Municipal Waste Landfills**," *Hazardous and Industrial Solid Waste Testing: Fourth Symposium, ASTM STP 886,* J. K. Petros, Jr., W. J. Lacy, and R. A. Conway, Eds., American Society for Testing and Materials, Philadelphia, 1986, pp. 15-35.

ABSTRACT: The objective of this research was to develop a laboratory extraction method for solid wastes that simulates concentrations of inorganic and organic constituents in leachates resulting from codisposal with municipal wastes in landfills. A scientific rationale and a data base are presented supporting the use of such an extraction method to characterize the potential hazards of leachates from a solid waste codisposed with municipal waste in a landfill containing about a 95:5 ratio of municipal and industrial waste, respectively. Two field lysimeters, each containing approximately 1.5 Mg of assorted municipal wastes, were used to generate a municipal waste leachate (MWL) that in turn was used to leach four industrial wastes under anoxic simulated codisposal conditions.

Leachates from the four industrial wastes were monitored for concentrations of inorganic and organic constituents over 79 days at a liquid/solid ratio of MWL and industrial waste of approximately 20:1, similar to the liquid/solid ratio currently being used in the extraction procedure to determine toxicity under the Resource Conservation and Recovery Act. Air-tight Tedlar® bags were used to collect leachate to avoid loss of volatile organic compounds and to maintain an anoxic environment. Data from the field lysimeter test facility were used to identify concentrations of "target" constituents (those constituents in the leachates of the industrial wastes that were observed at concentrations statistically higher than those observed in the control leachates). This data base was used as a model to evaluate several laboratory extraction methods that might be used to simulate the target concentrations.

KEY WORDS: hazardous wastes, solid wastes, codisposal, leachates, laboratory extraction tests, lysimeters, industrial wastes, toxicity, mobility, municipal waste leachate

[1]Group leader, Environmental Sciences Division, and research scientist, Analytical Chemistry Division, respectively, Oak Ridge National Laboratory, Oak Ridge, TN 37830.
[2]Systems analyst, Science Applications, Inc., Oak Ridge, TN 37830.

15

To evaluate the potential threat to groundwater posed by improper disposal of an industrial waste in a sanitary landfill containing municipal waste, a laboratory extraction procedure is needed that models the concentrations of organic and inorganic constituents in the leachate of that waste. Currently, under the Resource Conservation and Recovery Act (RCRA), such mobility is determined by an extraction procedure (EP) toxicity test [1]. The EP is a 24-h batch-type laboratory extraction procedure that uses acetic acid to acidify the liquid-solid suspension (20:1 ratio) to a pH of 5. The intent of the EP is to simulate the leaching action of acetic acid, the dominant carboxylic acid in municipal waste leachate (MWL). Hazardous waste identification thresholds for the EP toxicity characteristics are based on the concentrations of eight elements (arsenic, barium, cadmium, chromium, lead, mercury, selenium, and silver), four pesticides (endrin, lindane, methoxychlor, and toxaphene), and two herbicides (2,4-D and 2,4,5-TP silvex) for which National Interim Primary Drinking Water Standards (NIPDWS) have been established.

The EP has a number of limitations. The most important ones are: (1) its ability to simulate a real-world disposal environment has not been tested, and (2) solid waste leachates containing toxic constituents other than those listed above are not included in the current criteria. In terms of applying biological testing to EP extracts, the EP is limited because the acetic acid used in the procedure has been shown to interfere with certain aquatic toxicity and phytotoxicity testing protocols [2,3].

Objectives

The objective of the present study was to develop an experimental data base from which a laboratory extraction procedure (referred to as EP-III) can be selected to simulate the leaching of an industrial waste codisposed with municipal wastes in a landfill. The intended characteristics of EP-III include the following:

1. The method should be capable of simulating leaching in a landfill containing municipal and industrial wastes in proportions of about 95 and 5% by weight, respectively.
2. It should be compatible with biological toxicity tests (such as mutagenic, aquatic, and phytotoxic tests).
3. It should be relatively inexpensive to conduct in terms of time, equipment, and personnel.

Strategy

The strategy used for the development of EP-III was as follows:

1. Large-scale field lysimeters filled with municipal waste were used to generate MWL.

2. This MWL was then used to leach selected industrial wastes under simulated landfill conditions.
3. The concentrations of inorganic and organic constituents observed in the industrial waste leachates (in excess of the control MWL) were plotted relative to liquid-to-solid ratios of the leachate (that is, the volume of leachate divided by the weight of the waste).
4. A variety of laboratory methods (combinations of extraction procedures and media) were used to produce extracts.
5. The preferred laboratory extraction method was determined by comparing the concentrations of the inorganic and organic constituents in the laboratory extracts to five sets of target concentrations established to simulate various leaching scenarios.

Methods and Materials

Industrial Wastes

Four industrial wastes were used in the field and laboratory studies. One waste, the heavy ends and column bottoms from the production of trichloroethylene and perchloroethylene (henceforth called dichloroethylene still bottoms), was predominantly organic. Two wastes, one a paint production sludge and the other a mixture of petroleum refining incinerator ash and American Petroleum Institute (API) separator sludge, contained both inorganic and organic hazardous constituents. The fourth waste, an electroplating wastewater-treatment sludge, was primarily inorganic in character. These wastes were selected in collaboration with the U.S. Environmental Protection Agency Office of Solid Waste.

Sawdust was added to the dichloroethylene still bottoms and the paint sludge because these wastes contained considerable free liquid. Sawdust, clays, and other sorbing-type materials are often added to liquid wastes for solidification, stabilization, and fixation purposes. Thus admixtures of sawdust with dichloroethylene still bottoms and paint sludge were used in the field and laboratory studies. Air-dried sawdust was added to the wastes in amounts large enough to sorb the free liquid component of the waste effectively. Laboratory experiments were conducted to ascertain the desired proportion of each admixture component. The criterion used was addition of the liquid component until free liquid emerged from the bottom of the laboratory test column. All wastes were rolled on a drum-roller before mixing. Mixing of the admixtures (for each column) was conducted in 19-L buckets lined with polytetrafluoroethylene (PTFE).

The physicochemical characteristics of the wastes (and the admixtures made from the wastes and sawdust) and inorganic elemental concentrations (except dichloroethylene still bottoms) are presented in Table 1. The solid content was determined after drying for 24 h at 90°C. The porosity represents

TABLE 1—*Physicochemical characteristics of the wastes used in the field and laboratory studies, concentrations in milligrams per kilogram.*

Inorganic Element	API Sludge-Incinerator Ash Mixture[a]	Dichloroethylene Still Bottoms[a]	Paint Sludge[a]	Electroplating Waste
Sodium				
ICP[b]	9.8E3	. . .	<0.25E3	1.0E4
NAA[c]	9.6E3	. . .	0.16E3	1.0E4
Magnesium				
ICP	1.1E4	. . .	1.0E4	0.6E3
NAA	1.3E4	. . .	1.1E3	0.55E3
Potassium				
ICP	3.4E3	. . .	0.49E3	<0.78E3
NAA	2.94E3	. . .	0.36E3	
Calcium				
ICP	8.9E4	. . .	2.0E3	3.0E3
NAA	9.5E4	. . .	1.95E3	2.64E3
Chromium				
ICP	3.9E3	. . .	0.8E3	1.9%
NAA	3.9E3	. . .	0.68E3	1.4%
Manganese				
ICP	0.41E3	. . .	0.037E3	0.19E3
NAA	0.41E3	. . .	0.034E3	0.15E3
Boron (ICP)	<15	. . .	70	610
Barium (ICP)	190	. . .	3 100	64
Cadmium (ICP)	2.9	. . .	2.2	<11
Iron (ICP)	2 100	. . .	6 900	22 000
Molybdenum (ICP)	82	. . .	83	630
Nickel (ICP)	40	. . .	<22	14 000
Lead (ICP)	1 700	. . .	1 500	590
Strontium (ICP)	260	. . .	22	<19
Titanium (ICP)	930	. . .	3 700	22
Vanadium (ICP)	290	. . .	<3	<18
Zinc (ICP)	1 800	. . .	4 100	36 000
Solids, %	82.8	72.9	42.3	30.3
Porosity, %	18.4	33.0	58.8	71.1

[a]Wastes as admixtures, as described in text.
[b]Analyses determined by inductive couple plasma (ICP).
[c]Analyses determined by neutron activation analyses (NAA).

the quantity of interstitial water held by the waste under simulated field conditions. Analyses for organic compounds in the API sludge-incinerator ash mixture, dichloroethylene still bottoms, and paint sludge revealed a variety of organic compounds (Table 2). The method used to determine inorganic elemental content and to isolate and measure the organic compounds in the wastes is presented by Francis et al [4].

TABLE 2—*Concentrations of organic compounds in wastes.*

Waste and Organic Compound	Concentration, mg/kg[a]
API sludge–incinerator ash[b]	
Toluene	6.1 ± 0.3
Xylene	1.0 ± 0.2
Naphthalene	20.5 ± 1.2
Dichloroethylene still bottoms[b]	
trans-1,2-Dichloroethylene	425 ± 122
1,2-Dichloroethane	1 360 ± 117
1,1,1-Trichloroethane	630 ± 49
1,1-Dichloropropane	62.7 ± 2.5
Trichloroethylene	100 ± 19
Hexachlorobutadiene	22 400 ± 26 000
Paint waste[b]	
Ethoxyethanol	5 290 ± 192
Toluene	298 ± 51
Ethoxyethyl acetate	1 790 ± 196
Xylenes	24 530 ± 448
C_3-benzenes	1 920 ± 34
Electroplating waste No target organic compounds	. . .

[a]Average of three analytical replicates.
[b]Wastes as admixtures as described in text.

Chemical Analyses of Leachates

To ensure that the total quantity of inorganic and organic constituents leached from the wastes was detected, analyses were made on unfiltered leachate samples. Inorganic elemental analyses were also conducted on filtered leachates. Filtration was carried out using 0.4-μm (nominal pore size) polycarbonate filters (Nuclepore) and vacuum filtration. Samples of filtered and unfiltered leachate were preserved by acidifying to pH 2. Organic analyses of leachates were determined on unfiltered leachates. All inductively coupled plasma (ICP) inorganic analyses of lysimeter leachates, unless otherwise specified, are average concentrations determined on filtered and unfiltered leachates. On the other hand, all atomic absorption spectroscopy inorganic analyses, to determine elemental NIPDWS concentrations that were below ICP detection limits, were determined on filtered leachates (0.4-μm polycarbonate Nuclepore filters).

Field Lysimeters

About 1.5 Mg of municipal waste obtained from the city of Oak Ridge, Tennessee, was packed in each of the two large-scale "lysimeters" (concrete cylinders lined with epoxy resin, 1.8 m in diameter and 3.6 m in height). The municipal waste consisted of residential waste (collected from households) and commercial waste (collected from the downtown shopping area and fast-food restaurants). The lysimeters were filled on 26 February 1982, and approximately 46 cm of clay loam soil was added to the top of the waste to prevent moisture and vapor loss. To provide moisture for microbial degradation of the waste, about 750 L of water was added to each lysimeter within five days of filling.

On 3 June 1982, pumping of distilled water to each of the lysimeters began at a flow rate of 26.5 L/day. Leachate from the lysimeters was diverted to an overflow collector until 6 August 1982 (after 64 days elapsed time). On this date, the leaching of the selected industrial wastes began. Operation of the lysimeters from 3 June to 6 August 1982 was intended to allow sufficient time for a steady state to be reached between the leachate flow and decomposition rate of the municipal waste. The objective was to generate an MWL that did not vary significantly in its physicochemical characteristics over the duration in which it was to be used as a leaching medium. Major changes in MWL characteristics would likely result in significant changes in leaching of the various constituents from the industrial wastes. Hydraulic residence time for each of the lysimeters was an estimated 30 to 40 days.

Municipal waste leachate from each of the two lysimeters was pumped to two columns containing each of the industrial wastes. Thus the experiment included four industrial waste columns for each waste, making a total of 16 test columns (two lysimeters times four wastes times two replicates). Four additional columns, two containing sand only and two others containing sand and sawdust, were used as controls. The columns containing these treatments were randomly assigned positions on the support rack in front of the lysimeters (Fig. 1).

Leachate from each of the lysimeters was collected in 30-L glass carboys that served as sumps for two recycling pumps (10 to 15 mL/min) discharging to two glass manifolds. The glass manifolds were used to divert MWL to columns containing the industrial wastes and to support the electrodes necessary to monitor pH and Eh (the redox potential using a combination glass pH and silver/silver chloride reference electrode and a platinum electrode). Flow at 0.8 mL/min of MWL to each of the industrial waste columns was accomplished by using two 15-channel peristaltic pumps. Borosilicate glass columns (38.7 cm inside diameter [ID] by 30.5 cm in height) used to contain the industrial wastes were supported on a wooden frame in front of the lysimeters (Fig. 1). The columns were covered with a 2-cm-thick Plexiglas® plate whose underside was covered with an overlay made of PTFE to protect against (1) decomposition of the Plexiglas by organic solvents and (2) contamination of the resulting leachate with organic compounds solubilized from the plexiglass plate. Wastes (3.6 kg) were layered between 5 and 7 cm of sand. The sand was used to prevent entrainment of the waste by the leachate and to distribute the

FIG. 1—*Overview of field lysimeter facility.*

MWL uniformly across the surface of the waste. A thin layer of glass wool was placed between the top layer of sand and the waste to further facilitate lateral dispersion of the MWL. Leachate was sampled twice weekly the first 30 days and once a week for the remainder of the experiment (79 days of leaching). The intent was to leach the wastes until a liquid-to-solid ratio of 20:1 was obtained (similar to the ratio now being used in the EP). There is no implication in the experimental design that such leaching simulates MWL generated over the lifetime of a landfill.

Leachate was collected from the industrial waste columns in 15-L Tedlar® bags. Tedlar bags, made from polyvinyl fluoride, are known to preserve the integrity of air samples collected for volatile organic compounds [5]. Any excess leachate remaining after sampling was discarded, and the same Tedlar bag was used for collection of the next sample. Using these Tedlar bags allowed collection of volatile organic compounds and maintenance of an anoxic environment to minimize oxidation of reduced iron and formation of secondary precipitation products. Depending on the wastes, aliquots were taken for analyses for volatile and nonvolatile organic compounds and inorganic constituents.

Laboratory Extractions

The two procedures, upflow-column and rotary-batch, were conducted with four extraction media: (1) 0.1 M sodium acetate pH 5 buffer, (2) carbonic acid, (3) deionized distilled water, and (4) fresh MWL at four liquid-to-solid ratios (2.5, 5, 10, and 20:1) with two replications. Two ancillary extraction methods were also included: (1) the RCRA-approved EP and (2) a bisequential extraction method. The bisequential method was developed to extract high concentrations of readily soluble constituents and acid-soluble metals from predominantly alkaline wastes. The method consisted of a rotary-batch procedure at an initial liquid-to-solid ratio of 5:1, followed by adding another 100 g of waste to the filtered extract and extracting for another 18 h for a final liquid-to-solid ratio of 2.5:1. The extraction medium chosen was sodium acetate pH 5 buffer. For the electroplating waste, a preliminary experiment revealed that it was necessary to use a concentration of 0.25 M acetate to bring the pH of the final extract to values less than 7.5. For the other three wastes, 0.1 M sodium acetate pH 5 buffer was used.

Rotary-batch extractions were made by using a rotary extractor (equipped with six 2-L borosilicate glass vessels and rotating at 30 rpm) over an 18-h period (overnight). One hundred grams of waste was used for all batch extractions except the 2.5:1 liquid-to-solid ratio, in which 200 g of waste was used with 500 mL of extracting medium to obtain a sufficient quantity of leachate for inorganic and organic analyses.

Upflow-column extractions were conducted by using glass chromatography columns (2.5 cm ID and 45 cm in length) equipped with PTFE-tipped adjustable plungers to accommodate wastes of varying densities. A flow rate of about 1.4 mL/min was used to reach a liquid-to-solid ratio of 20:1 in 24 h for a 100-g sample of waste [4]. The electroplating waste was mixed with 100 g of sand (50:50 mixture) to avoid excessive pressure; however, mixing with sand was not required for the other wastes. Aliquots were taken at specific intervals of time to make liquid-to-solid ratios of 2.5, 5, 10, and 20:1. Thus, unlike the batch extractions, a single-column extraction was conducted with aliquots taken to correspond to the appropriate liquid-to-solid ratio (aliquots were taken at 250, 250, 500, and 1000 mL).

All extracts were filtered through an 0.4-μm polycarbonate filter (Nuclepore). The column extracts were filtered via an in-line filter (47 mm ID) fitted with special PTFE gaskets. The rotary-batch extracts were filtered by using a Millipore pressure filtration apparatus and 142-mm diameter filter disks at <510 kPa (75 psi). For waste samples that contained volatile organic compounds, the column effluents were collected in 5-L Tedlar bags directly connected to the in-line filter. Fresh MWL from the lysimeters was collected in a Tedlar bag immediately before use as an extractant to maintain its anoxic status. For the upflow-column extractions, the MWL was pumped directly from the Tedlar bag. For the rotary-batch extractions, fresh anoxic MWL was added directly to the preweighed waste and then immediately extracted. No attempt was made to maintain an anoxic environment in the batch extractions other than adding the fresh MWL. The carbonic acid solution was produced by bubbling carbon dioxide gas (99.8% purity) through a fritted glass dispersion tube in deionized distilled water for 18 h before using it as an extracting medium. The solution was added directly to the preweighed wastes in the extraction flasks; the extraction flasks were immediately closed and extraction was initiated. As with the rotary-batch extractions with MWL, no attempt was made to maintain an anoxic environment during extraction. In the upflow-column extractions, however, the supply of influent solution was continually purged with carbon dioxide gas throughout the extraction period. Thus upflow-column extractions by the MWL and carbonic acid extracting media were maintained in an anoxic status, while the rotary-batch extractions were anoxic only to the degree of the poising capacity of the MWL and carbonic acid. The rationale for the selection of these extraction procedures and media is detailed by Francis et al [4].

Results and Discussion

Municipal Waste Leachate

Leachate from each of the two lysimeters was monitored for pH, Eh, and organic and inorganic constituents during the time in which MWL was used as a leaching medium (6 August to 20 October 1982). Values for pH ranged from 3.5 to 6.0, and values for Eh ranged from −300 to 50 mV [4].

Total organic carbon (TOC) content of the MWL decreased from approximately 3 g/L at the beginning of the leaching period to 1.6 to 2 g/L at the end of the experiment. Carboxylic acid concentrations in MWL from the two lysimeters followed a similar trend and made up about 72% of the TOC over the 79 days [4].

Concentrations of inorganic cations and anions in lysimeter leachate also decreased with time. The major changes over the duration of the experiment were as follows: (1) potassium—130 to 50 mg/L, (2) sodium—250 to 115 mg/L, (3) calcium—410 to 310 mg/L, (4) chlorine—190 to 100 mg/L, and (5) sulfate—110 to <40 mg/L.

Comparisons of lysimeter leachate characteristics [4] with those generated by other lysimeter test facilities indicated that the Oak Ridge National Laboratory (ORNL) leachate was generally a "weaker" MWL than the MWL generated by other lysimeter test facilities. The most likely reason for the lower concentrations of TOC and inorganic elemental constituents was the faster flow of water to the lysimeters. The ORNL flow rate was such that sufficient leachate could be generated to leach the industrial wastes (packed to simulate a 95:5 municipal-to-industrial waste scenario) to the desired 20:1 liquid-to-solid ratio over the experimental time table of 80 to 100 days. The ORNL leachate more closely resembled the characteristics of natural leachates collected from municipal waste landfills than leachates generated in lysimeter test facilities [4]. As noted previously by Fungaroli and Steiner [6], it appears that leachates from lysimeter test facilities are characteristically more concentrated with respect to both carboxylic acids and inorganic constituents than are MWLs measured under field conditions. The wide range in MWL composition from landfills often results from dilution by water from outside sources. For example, transitory breachments into a landfill by surface runoff or shallow-land groundwater result in major changes in leachate composition. These breachments are the major forces in contaminant dispersion from the landfill. Consequently, there is no such entity as a "typical" MWL.

Waste Leachates

The objective of the field lysimeter experiment was to determine the concentrations of inorganic and organic constituents in waste leachates over a leaching interval equivalent to a 20:1 liquid-to-solid ratio. The intent was to target inorganic and organic constituents and determine their concentration over time so that a laboratory extraction method could be developed to simulate selected segments of that leaching pattern.

Volumes of leachate collected from the 20 waste columns averaged slightly less than the intended 20:1 liquid-to-solid ratio (18.5 ± 1.5). Over the 79 days of operation, the average flow rate (0.6 mL/min) to each column was approximately 25% slower than the tubing specifications (0.8 mL/min); however, the low coefficient of variability (approximately 10%) among the 16 columns was considerably better than anticipated.

Inorganic ICP analyses indicated that leachates from the industrial wastes contained several elements in excess of those concentrations observed in the leachates from the sand and sand-sawdust controls. The elements included (1) calcium, chromium, potassium, sodium, molybdenum, and strontium from the API sludge–incinerator ash mixture; (2) barium and zinc from the paint sludge; and (3) boron, barium, potassium, manganese, sodium, nickel, strontium, and zinc from the electroplating waste. Many of these elements are not typically considered to be toxic (for example, calcium, sodium, potassium, manganese, and strontium). However, they do represent matrix elements of the waste and their dissolution and leaching by MWL are indicators of weathering of the wastes in a codisposal environment. A laboratory extraction procedure that simulates their leachate concentration is useful because it predicts the stability of that waste in a codisposal environment, and for that reason all of the above listed elements were considered targets for the comparison of laboratory extraction methods.

Organic compounds found to be leached from the industrial wastes in excess of those observed in control MWL over the 79 days of leaching were (1) naphthalene from the API sludge–incinerator ash mixture; (2) dichloroethane, trichloroethane, trichloroethylene, and hexachlorobutadiene from the dichloroethylene still bottoms; and (3) ethoxyethanol, ethoxyethyl acetate, toluene, and xylenes from the paint sludge. These organic compounds and the inorganic elements listed above constitute the target constituents that are used to make comparisons among laboratory extraction methods to best simulate lysimeter leachate concentrations.

Laboratory Extractions

A detailed description of the comparative influence of procedure, media, and liquid-to-solid ratios on extraction of the target constituents from the four wastes is presented in Francis et al [4]. Taking into consideration only the main effects, a comparison of the batch-rotary and upflow-column procedures across all four media and the four liquid-to-solid ratios indicated that the rotary-batch extraction procedure removed higher concentrations for 21 of the 25 target constituents. Likewise, in comparing media across the two procedures and the four liquid-to-solid ratios, the acetate and the ORNL/MWL media consistently extracted higher concentrations of the target constituents than carbonic acid or deionized distilled water. This relationship was more pronounced for inorganic constituents than for organic constituents. As expected, increasing the liquid-to-solid ratio of the extraction procedure generally decreased the concentration of the target constituents.

Target Concentrations

Target concentrations are the concentrations of the organic and inorganic target constituents observed in the lysimeter leachates that the EP-III will model. The concentrations of organic and inorganic target constituents in each of the laboratory extractions were compared to the lysimeter leachate target concentrations.

The criteria used to establish the target concentrations in the lysimeter leachates depend on the intended use of the solid waste extraction procedure. Several alternatives were considered: (1) maximum concentration observed in lysimeter leachate over 79 days; (2) the average concentration integrated over some predetermined leaching interval bracketing (or containing) the maximum observed leachate concentration; and (3) the average concentration integrated over some predetermined leaching interval beginning the first day of leaching, irrespective of the maximum concentration. The first alternative reflects the most aggressive and the most conservative extraction method with respect to classifying a waste as potentially toxic (that is, it would tend to extract higher concentrations of toxic constituents and thus classify more wastes as toxic than would extraction methods based on either of the other two alternatives). The concern in adopting such a criterion is that for most wastes, the maximum leaching concentration emanating from a landfill will be short-lived temporally and spatially, in that maximum concentrations most often appear shortly after leaching begins (see Fig. 2, *left*). Thus this first alternative would produce an extraction method that represents landfill leachate concentrations over a very short leaching interval.

The second alternative is a modification of the first one in that an average concentration is determined over a leaching interval that encompasses the period of time (in terms of liquid-to-solid ratios) during which the maximum concentration in the lysimeter leachate was observed. An extraction method that accurately models target values obtained by using this alternative would be less aggressive than the first alternative and consequently more moderate with respect to identifying wastes likely to contaminate subsurface environments adjacent to a landfill. The intent is to select a leaching interval that encompasses the maximum concentration at its midpoint. As an example, the leaching curve for zinc in the electroplating waste (Fig. 2, *right*) shows a maximum leachate concentration occurring at an average liquid-to-solid ratio of 14:1. If a liquid-to-solid ratio of 8:1 were assumed to be an appropriate leaching interval, the target concentration based on the second alternative would involve determining the average zinc concentration between a 10:1 and 18:1 liquid-to-solid ratio. The important issue here, as with the third alternative, is selecting the appropriate leaching interval.

The third alternative involves integrating the concentration of a constituent as it is leached from a waste over a selected leaching interval beginning on Day 1. It basically represents the total quantity of a constituent leached from

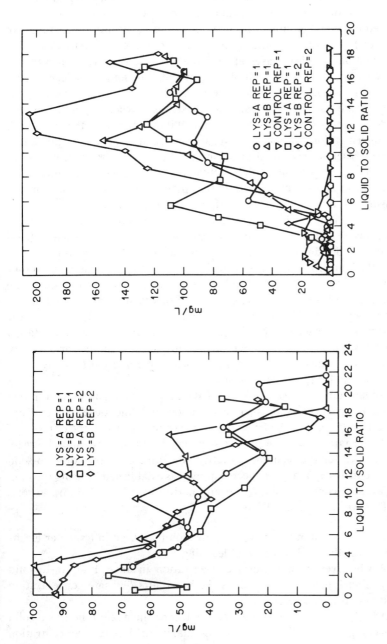

FIG. 2—*Leaching of (left) trichloroethane from dichloroethylene still bottoms and (right) zinc from electroplating wastes.*

a waste over a selected leaching time, divided by the volume of leachate. This differs from the second alternative in that it is not based on maximum leachate concentrations. Because peak concentrations from the lysimeter would be integrated over background concentrations (those concentrations similar to the controls), the third alternative would reflect the least aggressive extraction method and be the least conservative of the three alternatives. For example, for those wastes with readily soluble constituents, the longer the leaching interval over which the target concentration is integrated, the lower the resulting target value. On the other hand, for those wastes whose dissolution of the target constituents increases upon continued interaction with the MWL, the target concentration will increase over the longer interval. For wastes whose target constituents are limited by solubility in the MWL, the target concentration will be independent of the leaching interval selected. This alternative would result in an extraction method that predicts average leachate concentrations over a leaching interval starting from the first day.

The primary issue in the use of the latter two alternatives is the selection of the appropriate leaching interval. For example, the quantity of leachate (generated from groundwater that infiltrates the site or rainwater that percolates through the landfill) that moves through a landfill is dependent on numerous geologic, meteorologic, and hydrologic factors as well as the engineering method used in filling and closing the landfill. Ideally, the selection of the leaching interval should approximate or reflect these factors in modeling the movement of the leachate. The intent of this research is to establish target concentrations using all three alternatives and then to rank the various laboratory extraction methods, depending on which method most closely simulates the particular target concentration.

Five sets of target concentrations for the 25 inorganic and organic target constituents have been established, based upon guidelines developed by Kimmell and Friedman [7]. The first set was determined by using the criterion of the first alternative. Three of the five sets were established by using the criterion developed for the second alternative over three different liquid-to-solid ratios. The fifth set of target concentrations was established by using the integrated average concentration (third alternative) over an 8:1 liquid-to-solid ratio. The five sets are as follows:

1. MLC—maximum leachate concentration measured in lysimeter leachates over the 79-day period of leaching.
2. AMC8—average maximum concentration in an 8:1 liquid-to-solid leaching interval that brackets the maximum leachate concentration measured in lysimeter leachate.
3. AMC20—average maximum concentration in a 20:1 liquid-to-solid leaching interval that brackets the maximum leachate concentration measured in lysimeter leachate.
4. AMC40—average maximum concentration in a 40:1 liquid-to-solid

leaching interval that brackets the maximum leachate concentration measured in lysimeter leachate.

5. IAC8—integrated average concentration leached over the first 8:1 liquid-to-solid ratio.

The MLC target values were determined from the average maximum concentration measured over the 79 days in the four replicated industrial waste leachates. Control concentrations for each replicate were subtracted at the MLC solid/liquid ratio. This rationale is consistent for all leaching curves, and it provides the desired measure—the maximum concentration of that constituent in the industrial waste leachate.

Target concentrations for AMC8, AMC20, AMC40, and IAC8 were determined as follows. First, the accumulative leaching curves for each of the 25 target constituents were fitted by using the best-fit curve from a selection of four basic single-variable models [4]:

$$Y = a + bX$$

$$Y = ae^{1/X}$$

$$Y = ae^{bX}$$

$$Y = a(\ln X)$$

where Y = total quantity of a constituent leached (in milligrams), X = liquid-to-solid ratio, and a and b are constants for a particular curve. Fitting the cumulative leaching curves to one of these four models provided equations that could be used to calculate quantities of the constituents leached over discrete liquid-to-solid ratios. Using these fitted leaching curves, the quantity of each target constituent was calculated for selected liquid-to-solid ratios. Target concentrations for the 25 inorganic and organic constituents in terms of MLC, AMC8, AMC20, AMC40, and IAC8 are listed in Table 3.

Summary and Conclusions

A target constituent has been defined as any inorganic element or organic compound that exhibited a distinct concentration maximum over the 79-day leaching period and whose total mass leached from the industrial waste was greater than that leached from the control MWL. A total of 25 target constituents (16 inorganic elements and nine organic compounds) were identified in the leachates of the industrial wastes:

- API sludge–incinerator ash—The inorganic elements calcium, chromium, potassium, molybdenum, sodium, and strontium and the organic compound naphthalene.

TABLE 3—*Target concentrations.*[a]

| | Target | | | | |
Waste and Chemical	MLC	AMC8	AMC20	AMC40	IAC8
API sludge-incinerator ash					
Calcium	1188	787	771	774	792
Chromium	63	22	11	6	21
Molybdenum	2.2	0.60	0.30	0.09	0.66
Potassium	562	164	76	40	164
Sodium	1904	617	258	105	617
Strontium	3.0	2.1	1.5	1.3	2.1
Dichloroethylene still bottoms					
Dichloroethane	49	26	26	25	30
Hexachlorobutadiene	2651	90	45	31	90
Trichloroethane	83	53	43	38	53
Trichloroethylene	97	26	21	26	22
Paint sludge					
Barium	2.3	0.35	0.14	0.07	0.35
Zinc	220	77	35	20	72
Ethoxyethanol	4729	1055	430	219	1055
Ethoxyethyl acetate	1892	405	165	84	405
Toluene	39	17	9.9	6.9	17
Xylenes	614	269	174	136	269
Electroplating waste					
Barium	0.47	0.28	0.28	0.30	0.25
Boron	148	51	22	11	51
Manganese	7.4	1.3	0.79	1.5	0
Nickel	147	107	93	98	80
Potassium	125	30	39	41	29
Sodium	7058	1209	472	209	1209
Strontium	1.1	0.32	0.34	0.36	0.31
Zinc	149	85	85	79	49

[a]Concentrations in milligrams per litre.

- Dichloroethylene still bottoms—No inorganic elements, but four organic compounds: dichloroethane, trichloroethane, trichloroethylene, and hexachlorobutadiene.
- Paint sludge—The inorganic elements barium and zinc and the organic compounds ethoxyethanol, ethoxyethyl acetate, toluene, and xylenes.
- Electroplating waste—The inorganic elements boron, barium, potassium, manganese, sodium, nickel, strontium, and zinc but no organic compounds.

Thirty-two laboratory extraction methods were ranked according to their ability to simulate the five sets of target concentrations. The difference between the laboratory concentrations and their respective target concentrations was defined as follows:

$$\text{difference} = [\text{abs}(TC - LC)/TC] \times 100$$

where TC is the target concentration, LC is the laboratory concentration, abs( ) is the absolute value, and the difference is expressed as a percentage.

An average difference between the target concentration and the laboratory concentration was determined for each of the 32 treatments and each of the 25 target constituents. An overall average difference was then determined for each of three categories—(1) inorganic, (2) organic, and (3) inorganic and organic target constituents—for each of the five sets of target concentrations—MLC, AMC8, AMC20, AMC40, and IAC8. These average differences were ranked from the lowest to the highest. For the 32 replicated extraction methods (all but the EP and bisequential extraction methods were duplicated), significant differences among extractions could be determined in an analysis of variance testing procedure [8].

The highest-ranked extraction methods for estimating MLC target concentrations were those that used ORNL/MWL as an extraction medium. There appeared to be no preference relative to the type of extraction procedure (upflow-column or rotary-batch) with ORNL/MWL; however, liquid-to-solid ratios at 10:1 and less ranked consistently better than those at 20:1. The only synthetic media among the ten top-ranked methods were sodium acetate and carbonic acid extraction media in the rotary-batch extraction procedure at liquid-to-solid ratios of 2.5 to 10:1. Sodium acetate at liquid-to-solid ratios of 5 and 10:1 ranked slightly better than carbonic acid at 2.5 and 5:1 liquid-to-solid ratios. Statistically, there were no significant differences between these sodium acetate and carbonic acid extractions, suggesting that any of the four extraction methods would be satisfactory. Carbonic acid was the extracting medium in the five top-ranked methods for estimating AMC8 target concentrations (Table 4). It is quite evident that to simulate AMC8 target concentrations for wastes containing both organic and inorganic constituents, the preferred extracting medium would be carbonic acid. Because of the operational constraints of the upflow-column procedure (inherently slow flow rates with wastes of low hydraulic conductivities) and the relatively small differences in ranking between the two procedures, the rotary-batch extraction procedure would be selected over the upflow-column procedure. Choice of liquid-to-solid ratio appeared to be less important: the first five ranked methods (those using carbonic acid) included 5, 10, and 20:1 liquid-to-solid ratios. For AMC target concentrations at higher liquid-to-solid ratios (20 and 40:1 as compared to 8:1), the preferred extraction methods generally used 20:1 liquid-to-solid ratios and less aggressive extracting media (de-ionized distilled water or carbonic acid rather than sodium acetate or ORNL/MWL; see Tables 4 and 5. The best-ranked extraction method for both AMC20 and AMC40 (inorganic and organic target constituents) was carbonic acid in an upflow-column at a 20:1 liquid-to-solid ratio. Over all AMC target sets (AMC8, AMC20, and AMC40), carbonic acid in a rotary-batch extraction procedure

TABLE 4—*Ranking of 34 laboratory extraction methods to simulate AMC8 inorganic and organic target concentrations from the field lysimeter test facility.*

Rank	Medium	Type	Liquid-to-Solid Ratio	Difference, % Average	Difference, % Minimum	Difference, % Maximum	Coefficient of Variation, %
1	carbonic acid	batch	20.0	54.8	0.7	101.9	53.2
2	carbonic acid	column	10.0	56.0	9.7	112.6	52.2
3	carbonic acid	column	5.0	59.9	9.1	214.1	74.8
4	carbonic acid	column	20.0	62.3	12.6	99.0	39.4
5	carbonic acid	batch	10.0	64.4	21.7	298.2	87.6
6	distilled water	column	10.0	67.1	13.7	119.2	45.2
7	sodium acetate	column	20.0	69.0	22.8	151.0	44.3
8	distilled water	column	20.0	71.5	23.8	99.7	31.6
9	sodium acetate	column	10.0	71.8	28.6	189.2	52.3
10	MWL	column	20.0	73.0	7.2	168.6	60.4
11	distilled water	batch	20.0	75.1	19.2	113.3	30.9
12	distilled water	column	5.0	75.8	6.8	182.0	58.1
13	distilled water	batch	10.0	76.4	21.9	241.3	67.0
14	sodium acetate	batch	20.0	78.7	17.9	468.9	111.5
15	sodium acetate	batch	10.0	79.4	1.4	397.8	106.2
16	sodium acetate	column	5.0	86.0	18.2	203.0	54.3
17	MWL	batch	20.0	88.7	4.6	604.3	132.6
18	EP	batch	20.0	90.3	7.6	406.9	84.0
19	carbonic acid	column	2.5	90.4	9.1	375.4	83.2
20	MWL	batch	10.0	90.8	24.6	321.4	84.3
21	MWL	column	10.0	92.3	10.6	311.3	82.0
22	distilled water	batch	5.0	92.9	7.4	412.0	96.9
23	carbonic acid	batch	5.0	95.9	3.9	511.5	110.7
24	sodium acetate	column	2.5	106.7	20.6	283.1	63.0
25	distilled water	column	2.5	117.4	7.5	338.5	73.4
26	MWL	column	5.0	128.3	11.6	556.1	111.6
27	carbonic acid	batch	2.5	156.3	21.6	639.6	97.8
28	MWL	batch	5.0	162.4	14.2	498.5	81.8
29	MWL	column	2.5	198.8	14.9	810.2	110.2
30	distilled water	batch	2.5	200.4	2.1	1546.9	155.0
31	sodium acetate	batch	5.0	200.9	6.8	2235.6	225.7
32	MWL	batch	2.5	222.4	12.0	1096.1	109.6
33	bisequential	batch	2.5	314.0	2.9	3625.5	229.5
34	sodium acetate	batch	2.5	330.0	10.4	3758.1	227.1

at a liquid-to-solid ratio of 20:1 ranked, respectively, first, fourth, and fifth for extracting inorganic and organic target constituents from the four industrial wastes.

The outstanding observation relative to the ranking of the extraction methods to simulate IAC8 target concentrations (inorganic and organic) was the total dominance of deionized distilled water as the extraction medium in the top five rankings. The major differences in the target concentrations for IAC8 and AMC8 were the lower IAC8 values for nickel and zinc in the electroplating waste (Table 3). In retrospect, the high rankings for deionized distilled water were not surprising because the pH of the leachates from the two alka-

TABLE 5—*Ranking of 34 laboratory extraction methods to simulate AMC20 inorganic and organic target concentrations from the field lysimeter test facility.*

Rank	Medium	Type	Liquid-to-Solid Ratio	Difference, % Average	Difference, % Minimum	Difference, % Maximum	Coefficient of Variation, %
1	carbonic acid	column	20.0	58.8	10.7	213.1	72.0
2	distilled water	column	20.0	66.1	12.7	133.8	54.1
3	sodium acetate	column	20.0	69.5	12.2	150.3	55.4
4	carbonic acid	batch	20.0	71.2	1.9	411.3	125.0
5	carbonic acid	column	10.0	73.0	9.5	331.4	94.7
6	distilled water	column	10.0	81.2	9.2	190.6	61.0
7	distilled water	batch	20.0	82.1	2.7	440.1	98.8
8	sodium acetate	column	10.0	92.0	20.9	290.9	77.3
9	MWL	column	20.0	98.7	11.0	473.8	96.6
10	distilled water	batch	10.0	112.6	0.9	764.1	133.5
11	EP	batch	20.0	114.8	1.1	405.5	89.2
12	carbonic acid	batch	10.0	115.6	8.8	908.1	159.9
13	MWL	batch	20.0	116.2	9.7	762.8	151.0
14	carbonic acid	column	5.0	116.9	7.1	537.4	102.7
15	sodium acetate	batch	20.0	132.3	4.8	1340.2	197.5
16	sodium acetate	column	5.0	134.8	24.5	509.4	87.7
17	distilled water	column	5.0	140.9	8.4	433.0	81.1
18	sodium acetate	batch	10.0	157.5	5.9	1160.2	148.3
19	MWL	column	10.0	164.2	4.5	941.3	123.4
20	MWL	batch	10.0	170.4	28.6	785.2	108.0
21	distilled water	batch	5.0	190.0	6.6	1196.2	126.4
22	sodium acetate	column	2.5	191.8	31.9	764.1	99.7
23	carbonic acid	column	2.5	191.8	7.2	864.7	108.7
24	carbonic acid	batch	5.0	214.2	10.4	1448.2	143.5
25	distilled water	column	2.5	248.8	6.5	918.9	106.1
26	MWL	column	5.0	271.7	11.8	1561.0	135.5
27	carbonic acid	batch	2.5	342.3	20.0	1772.3	124.2
28	MWL	batch	5.0	345.0	24.4	1366.7	111.2
29	distilled water	batch	2.5	420.2	44.2	3064.3	158.4
30	MWL	column	2.5	420.9	13.0	2204.4	133.0
31	sodium acetate	batch	5.0	436.6	14.5	4501.8	217.8
32	MWL	batch	2.5	473.6	5.8	2269.3	124.3
33	bisequential	batch	2.5	483.3	4.3	2714.1	147.4
34	sodium acetate	batch	2.5	692.3	8.2	7542.3	223.3

line wastes (API sludge–incinerator ash mixture and electroplating waste) was relatively high during the first 8:1 liquid-to-solid leaching interval (pH values ranged from 9.5 to 8.1 and 8.4 to 6.4, respectively). Under these conditions, the interactions of the MWL with the wastes were predominantly the same as the interactions of distilled water: the water-soluble constituents were leached, but the acid-soluble metals such as nickel and zinc remained until the leachate pH became lower later in the leaching.

The objective of the research was to develop a laboratory extraction method for solid wastes that simulates the concentrations of inorganic and organic constituents in leachates that result from codisposing of industrial wastes with municipal wastes in landfills. The intent of the research was to produce a scientific rationale and data base supporting the selection of such an extraction method. It is obvious from the present and previous research that no one extraction method will be optimal for all wastes, waste constituents, or landfill scenarios. This research has, however, demonstrated the relative effectiveness of a number of extraction methods for a variety of wastes and target constituents and has indicated that certain extraction methods may be able to indicate potential problem wastes with reasonable accuracy. The final selection of any one method or combination of methods will depend on what leachate target concentrations are to be simulated. The data presented here and those in the full report [4] suggest that the use of carbonic acid as an extracting medium in a rotary-batch procedure should be thoroughly considered; such an extraction method would fulfill many of the previously mentioned criteria. The expected compatibility of carbonic acid extraction with numerous biotesting protocols would also be a significant advantage in evaluating the toxicity of solid waste leachates.

Acknowledgments

The authors wish to thank Llewellyn Williams of the Environmental Monitoring Systems Laboratory, Las Vegas, Nevada, and Todd Kimmell and David Friedman of the Office of Solid Waste, U.S. Environmental Protection Agency, Washington, DC, for their guidance and suggestions throughout the project. The authors also wish to thank Marilyn S. Hendricks and Joe Gooch of the Environmental Sciences Division, ORNL, and R. Wallace Harvey of the Analytical Chemistry Division, ORNL, for the persevering technical assistance that made operation of the lysimeters and handling of the large number of analytical samples proceed so well.

The research described in this paper was conducted under interagency agreements between the U.S. Department of Energy (40-1087-80) and the U.S. Environmental Protection Agency (AD-89-f-1-058). The Oak Ridge National Laboratory is operated by Martin Marietta Energy Systems, Inc., under Contract DE-AC05-840R21400 with the U.S. Department of Energy.

References

[1] U.S. Environmental Protection Agency, "Identification and Listing of Hazardous Waste," Environmental Protection Agency Hazardous Waste Management System, 40 CFR 261.24, U.S. Government Printing Office, Washington, DC, 1980.
[2] Epler, J. L., et al, "Toxicity of Leachates," EPA-600/2-80-057, Office of Research and Development, U.S. Environmental Protection Agency, Washington, DC, 1980.
[3] Millemann, R. E., Parkhurst, B. R., and Edwards, N. T., "Toxicity to *Daphnia Magna* and Terrestrial Plants of Solid Waste Leachates from Coal Conversion Processes," *Proceedings*, 20th Hanford Life Sciences Symposium on Coal Conversion and the Environment, Battelle Pacific Northwest Laboratory, Richland, WA, 1981, pp. 237–247.
[4] Francis, C. W., Maskarinec, M. P., and Goyert, J. C., "Mobility of Toxic Compounds from Hazardous Waste," ORNL/6044, Oak Ridge National Laboratory, Oak Ridge, TN, 1984.

[5] Schuetzle, D., Prater, T. J., and Roddell, S. R., *Journal of the Air Pollution Control Association*, Vol. 24, Nov. 9, Sept. 1975, pp. 925–932.
[6] Fungaroli, A. A. and Steiner, R. L., "Investigation of Sanitary Landfill Behavior," Vol. 1, Final Report, EPA-600/2-79-053a, Municipal Environmental Research Laboratory, U.S. Environmental Protection Agency, Cincinnati, OH, July 1979.
[7] Kimmell, T. A. and Friedman, D., "Model Assumptions and Rationale Behind the Development of Extraction Procedure III," *Hazardous and Industrial Solid Waste Testing: Fourth Symposium, ASTM STP 886*, American Society for Testing and Materials, Philadelphia, 1985, pp. 36–53.
[8] "SAS Users Guide Statistics," SAS Institute, Inc., Cary, NC, 1982.

Todd A. Kimmell[1] and David Friedman[1]

Model Assumptions and Rationale Behind the Development of EP-III

REFERENCE: Kimmell, T. A. and Friedman, D. A., "**Model Assumptions and Rationale Behind the Development of EP-III**," *Hazardous and Industrial Solid Waste Testing: Fourth Symposium, ASTM STP 886*, J. K. Petros, Jr., W. J. Lacy, and R. A. Conway, Eds., American Society for Testing and Materials, Philadelphia, 1986, pp. 36–53.

ABSTRACT: As part of the efforts of the U.S. Environmental Protection Agency (EPA) to expand the Extraction Procedure (EP) Toxicity Characteristic (40 CFR 261.24) to include additional organic toxicants, the Oak Ridge National Laboratory (ORNL) is developing for EPA a leaching procedure, known as EP-III, for use in determining the leaching potential of organic constituents from solid wastes. The intent is to develop a test method that would accurately and reproducibly model leachate production from industrial waste's codisposed with municipal refuse.

In addition to accurately modeling the mobility of organic constituents, other objectives of the test are that it be relatively inexpensive to conduct in terms of time, equipment, and personnel; that it be compatible with biological toxicity tests; and that it also model the mobility of inorganic species.

This paper discusses in detail the model, assumptions, and rationale used in developing the EP-III. Emphasis is placed on a discussion of the possible shortcomings of the protocol outlined in the current EP and the approach taken to develop a second generation mobility procedure that eliminates the short-comings. The data generated by ORNL in support of this program are discussed to illustrate the rationale behind selection of a draft leaching procedure. Conclusions are drawn as to which laboratory extraction procedure appears to best approximate the lysimeter leachate values. In light of these conclusions, selection of a draft EP-III is described.

KEY WORDS: leaching, mobility, landfill, sanitary landfill, RCRA, leaching methods, hazardous wastes

Under Section 3001 of the Resource Conservation and Recovery Act of 1976 (RCRA), the U.S. Environmental Protection Agency (EPA) is charged with identifying those industrial wastes that pose a hazard to human health and the environment if improperly managed. In order to carry out this mandate, EPA identified a number of properties that would indicate that a waste

---

[1]Environmental scientist and manager, Methods Program, respectively, Office of Solid Waste, U.S. Environmental Protection Agency, Washington, DC 20460.

requires regulated management. One of these properties relates to the degree to which toxic species might leach out of the waste and contaminate groundwater if the waste was disposed of in a nonsecure sanitary landfill. The test procedure promulgated by EPA for use in determining if an unacceptably high level of groundwater contamination might result under such conditions is called the Extraction Procedure (EP) (Identification and Listing of Hazardous Waste, Characteristic of Extraction Procedure Toxicity, 40 CFR 261.24).

The EP results in an extract that is analyzed for the eight metals (arsenic, barium, cadmium, chromium, lead, mercury, selenium, and silver), four pesticides (endrin, lindane, methoxychlor, and toxaphene), and two herbicides (2,4-D and 2,4,5-TP) for which National Interim Primary Drinking Water Standards have been established. Hazardous waste definition thresholds have been established for each of these species, taking into account attenuative processes expected to occur during the movement of leachate through the underlying strata and groundwater aquifer [1].

The EP was intended to be a first-order approximation of the leaching action of the low-molecular-weight carboxylic acids generated in an actively decomposing sanitary landfill. Acetic acid, the carboxylic acid most prevalent in municipal waste leachate, is added to distilled water to make up the extracting medium used in the EP. The acetic acid primarily models the leaching of metals from an industrial waste. The impetus behind the present research program is the concern that the EP may not be aggressive enough to model the leaching of organic compounds adequately.

In addition, we believe that the EP protocol can be improved in certain areas. For example, the need for continual pH adjustment, which is time-consuming and can cause problems for certain waste types, can possibly be eliminated. Also, the initial liquid/solid separation technique, currently involving 0.45 μm pressure filtration, warrants simplification. Finally, any procedure developed to model the leaching of organic compounds must adequately prevent volatilization of volatile organic compounds.

The Oak Ridge National Laboratory (ORNL), working under an EPA Office of Research and Development interagency agreement with the Department of Energy, has been responsible for the experimental phase of the development program. The intent is to develop a test method using the model and assumptions upon which the original EP was based. In our experience, the current EP has been found to yield a reasonable differentiation between hazardous and nonhazardous wastes. The major objective is thus to model accurately the mobility of organic constituents, including volatiles, from a waste. Other objectives are that the test be relatively inexpensive to conduct in terms of time, equipment, and personnel; that it yield an extract amenable to evaluation with biological toxicity tests; and that it also model the mobility of inorganic species. This last objective would permit EPA to expand the EP toxicity characteristic to encompass additional toxic organic compounds, yet still require only one test for both organics and inorganics.

The specific environment modeled by both the current EP and this new mobility procedure, known as EP-III, is codisposal with refuse in a sanitary landfill. The intent was to model the early stages of landfill decomposition (generally the first few years of landfill life), as it is during this stage that the leachate would be most aggressive in its mobilizing ability, both in terms of leachate acidity and in complexation ability. Specific features of the model are that the landfill receives predominantly domestic refuse (only 5% of the landfill is industrial in nature) and that the character of the leaching fluid that the waste will be exposed to is a function of the decomposing refuse in the landfill [1].

Briefly, the overall approach employed in the development program was as follows:

1. Large-scale field lysimeters were filled with domestic and commercial refuse and used to generate municipal waste leachate (MWL).
2. The MWL was used to leach four industrial wastes in large columns.
3. The concentration of a number of organic and inorganic species that were present in each waste were measured over time in the leachate.
4. A variety of laboratory tests, including the current EP, were run on the four wastes. These tests included both column and batch procedures using four media (0.1 M sodium acetate pH 5 buffer, carbon dioxide-saturated deionized distilled water, deionized distilled water, and actual municipal waste leachate) and four extractant-to-waste ratios (2.5, 5, 10, and 20 to 1).
5. Target concentrations were established for each constituent based on the leaching curves, and these concentrations were compared to those observed in the laboratory tests.
6. The two laboratory tests that best replicated the lysimeter results were selected for further evaluation.
7. These tests are now in the process of extensive evaluation and verification by comparison to a second series of lysimeter studies. At the conclusion of this second series of experiments, a draft mobility procedure will be selected for single laboratory evaluation and multilaboratory collaborative study.

An ORNL report [2] explains in detail the experimental approach and describes the results obtained during the initial series of tests. In addition, details regarding the experimental design, research results, and statistical treatment of the data may be found in Francis et al [3] in this volume.

Approach to Selection of Leachate Target Concentrations

As the ORNL report [2] indicates, a number of sets of lysimeter leachate target concentrations for the waste constituents were established with various criteria. These criteria are based on the concept of leaching interval (that is, liquid-to-solid ratio) and relate to the exposure time that EP-III is to model. Leaching procedures developed in the past generally employed liquid-to-solid ratios ranging from 4:1 to as high as 40:1 [4]. These ratios were selected primarily on the basis of practical considerations rather than on the basis of estimated time in terms of real-world leaching.

Exposure time or leaching interval, however, is critical in the establishment of EP-III, since the method is intended to be used in evaluating the migratory potential of chronically toxic organic compounds. Chronic toxicity generally implies a long exposure period of years. Hence, short exposure periods, such as those represented by test methods employing relatively low liquid-to-solid ratios, would be inappropriate for evaluating the hazard posed by chronically toxic compounds.

It is important, therefore, to establish a relationship between liquid-to-solid ratio and leaching time. Wilson and Young [5], assuming an annual rainfall infiltration rate of 250 mm and a 1-m-thick layer of waste, calculated that it would take about four years for the amount of leachate represented by approximately a 1:1 liquid-to-solid ratio to be generated in the field. Wilson and Young further concluded that the more normal ratios of 5:1 and 10:1 usually quoted for laboratory leaching tests correspond to the average concentration of a contaminant in leachate over many years.

Wilson and Young's [5] calculations, however, assume a monodisposal situation, whereas the EP-III is modeled after the codisposal situation, where only 5% of the landfill is assumed to be industrial in nature. Adjusting for this assumption, a 1:1 liquid-to-solid ratio corresponds to roughly one fifth of a year of leaching. Hence, applying Wilson and Young's assumptions to the codisposal scenario, a liquid-to-solid ratio of 20:1, such as that employed in the current EP, corresponds to approximately four years of leaching.

The assumptions used to arrive at this conclusion, such as a 1-m layer of waste and 250 mm of annual infiltration, simplify an extremely complicated site-specific phenomenon. For example, infiltration may vary tremendously with climatic and hydrogeological factors. In light of the uncertainties involved, it is preferable to identify a range, rather than a discrete value, when one is attempting to describe leaching in terms of time. Thus, the 20:1 liquid-to-solid ratio employed by the EP might represent a period of leaching as low as three years or as high as ten years or more, depending on landfill depth, infiltration rate, and other factors. The relationship between liquid-to-solid ratio and leaching time should therefore be interpreted very broadly as indicating relative differences and not absolute values.

As Table 1 indicates, a number of sets of leachate target concentrations were established. These target concentrations are based on the liquid-to-solid ratio/leaching time relationship, and correspond to short-term, medium-term, and long-term exposure periods.

The highest concentration of a constituent that would be expected to occur in the leachate is represented by the maximum leachate concentration (MLC)

TABLE 1—*Leachate target concentrations.*[a]

Waste and Chemical	MLC[b]	Liquid-to-Solid Ratio			
		8:1[c]	20:1[c]	40:1[c]	8:1[d]
API sludge/incinerator ash					
Calcium	1188	787	771	774	792
Chromium	63	22	11	6	21
Potassium	562	164	76	40	164
Molybdenum	2.2	0.60	0.30	0.09	0.66
Sodium	1904	617	258	105	617
Strontium	3.9	2.1	1.5	1.3	2.1
Naphthalene	0.92	0.21	0.18	0.17	0.21
Dichloroethylene still bottoms					
Dichloroethane	49	26	26	25	30
Trichloroethane	83	53	43	38	53
Trichloroethylene	97	26	21	26	22
Hexachlorobutadiene	2651	90	45	31	90
Paint Sludge					
Barium	2.3	0.35	0.14	0.07	0.35
Zinc	220	77	35	20	72
Ethoxyethanol	4729	1055	430	219	1055
Ethoxyethyl acetate	1892	405	165	84	405
Toluene	39	17	9.9	6.9	17
Xylenes	614	269	174	136	269
Electroplating waste					
Boron	148	51	22	11	51
Barium	0.47	0.28	0.29	0.30	0.25
Potassium	125	30	39	41	29
Manganese	7.4	1.3	0.79	1.5	0
Sodium	1904	617	472	105	617
Nickel	147	107	93	98	80
Strontium	1.1	0.32	0.34	0.36	0.31
Zinc	149	85	85	79	49

[a]All concentrations are in milligrams per litre; the data are from Ref 2.
[b]Maximum leachate concentration.
[c]Liquid-to-solid ratio based on a leaching interval that encompasses the MLC as its midpoint.
[d]Liquid-to-solid ratio based on initial leaching interval.

set of target concentrations. These concentrations are the highest concentrations observed in the waste leachates. The ORNL data indicate that these concentrations occur over a relatively short period of time, generally spanning only a few liquid-to-solid ratios [2]. Figures 1 and 2, the leaching curves for ethoxyethanol and toluene from a paint sludge, respectively, illustrate this point.

The MLC target concentrations thus represent a relatively short period of exposure. As previously indicated, EP-III is intended to be used in evaluating chronic exposure to toxic organic compounds, and so medium- or long-term exposure periods are more appropriate.

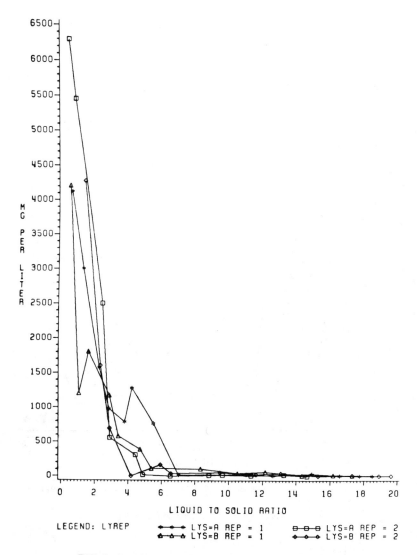

FIG. 1—*Leaching of ethoxyethanol from paint sludge [2].*

Consequently, leaching periods corresponding to longer exposure periods, represented by the 8:1, 20:1, and 40:1 liquid-to-solid ratios, were studied. As described earlier, the 20:1 ratio can be thought of as representing about three to ten years of leaching. Accordingly, the 8:1 ratio would represent approximately one to three years of leaching and the 40:1 ratio would represent a much longer period of leaching (six to 20 years). Target concentrations derived from these leaching periods represent an integrated average constituent concentration in the leachate over the particular leaching interval.

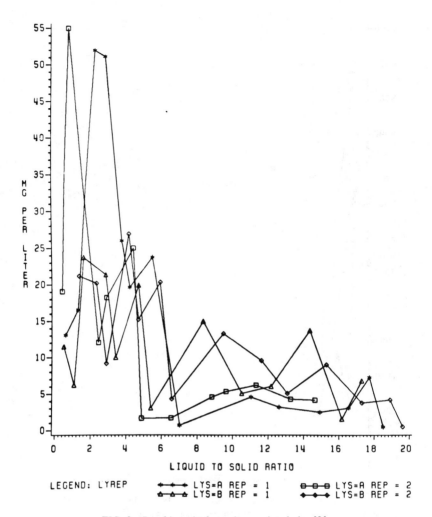

FIG. 2—*Leaching of toluene from paint sludge [2].*

The 8 : 1 liquid-to-solid ratio was used to represent a relatively short yet still environmentally significant exposure time. The target concentrations based on the 20 : 1 ratio constituted a moderate to long-term exposure period, and the target concentrations based on the 40 : 1 ratio were included to determine the consequences of longer time frames. Note that the target concentrations based on this long-term leaching period were included mainly for comparative purposes, since, as the report explains, they are predicted from substantial extrapolation of the leaching curves.

Selection of EP-III Target Concentrations

Since the EPA is most interested in applying the EP-III to toxic organic compounds, most of which present a chronic hazard, the medium- to long-term exposure periods represented by the 8:1 and 20:1 liquid-to-solid ratios are most appropriate. Additional sets of target concentrations based on other leaching intervals appeared to offer no advantages over those intervals already established.

In addition, as Table 1 indicates, longer exposure periods represented by the 40:1 liquid-to-solid ratio do not produce target concentrations substantially different from the target concentrations identified with the 20:1 ratio. Since extraction methods at liquid-to-solid ratios greater than 20:1 would require a relatively large extraction apparatus and offer no obvious advantages, they are not considered as viable options.

After considerable evaluation, the set of target concentrations based on the 20:1 liquid-to-solid ratio was chosen as the most appropriate leaching interval on which to model EP-III. There were basically three reasons for this. First, the longer exposure period associated with the 20:1 leaching interval was felt to be more appropriate to those compounds that are chronically toxic in nature. Second, extractions at the higher liquid-to-solid ratios were generally more precise.

The third reason is more operational in nature. The ORNL data indicate that some compounds, such as nickel and zinc in the electroplating waste, may experience an increase in leaching at liquid-to-solid ratios beyond 8:1 [2]. The leaching curve for zinc, for example, reaches its maximum point at 14:1 [2]. At the same time, other compounds, such as ethoxyethanol and toluene in a paint sludge (see Figs. 1 and 2, respectively), leach primarily over the first few liquid-to-solid ratios. The EP-III would need to identify target concentrations accurately in both situations. An extraction method based on the 8:1 ratio might not accurately identify both concentrations, whereas the target concentrations based on the 20:1 ratio appear to encompass both situations adequately.

Selection of Procedures for Further Testing

Tables 2 to 4 present the rankings of the various laboratory procedures according to their ability to reproduce the target organic, inorganic, and combined target concentrations at the 20:1 liquid-to-solid ratio. These tables utilize the concept of the absolute value of the percentage difference between laboratory test concentrations and lysimeter target concentrations. The absolute value of the percentage difference was determined for each target compound in each of the laboratory leaching tests. Next, these percentage differences were averaged and a coefficient of variation was determined. Average percentage differences were then ranked from lowest to highest, and signifi-

TABLE 2—*Ranking of 34 laboratory extraction methods to simulate organic target concentrations.*[a]

Rank	Medium	Type	LS Ratio[b]	Difference, %			Coefficient of Variation, %	Test for Significance[c]
				Average	Minimum	Maximum		
1	sodium acetate	batch	20.0	58.7	4.8	94.1	46.7	
2	carbonic acid	batch	20.0	61.0	20.9	98.8	48.3	
3	carbonic acid	column	10.0	61.0	9.5	98.8	56.5	
4	distilled water	batch	20.0	62.8	23.9	100.0	45.6	
5	distilled water	batch	10.0	65.2	24.0	96.2	36.1	
6	carbonic acid	column	20.0	67.8	37.5	98.8	40.5	
7	carbonic acid	batch	10.0	67.8	8.8	202.1	87.4	
8	EP	batch	20.0	71.0	24.4	100.0	36.7	
9	sodium acetate	column	20.0	71.1	12.2	98.8	46.0	
10	distilled water	column	20.0	72.8	27.6	133.8	51.8	
11	carbonic acid	column	5.0	75.6	42.8	121.8	37.9	
12	sodium acetate	column	10.0	76.6	37.7	98.8	27.2	
13	distilled water	column	10.0	78.6	27.9	117.8	41.5	
14	sodium acetate	batch	10.0	82.9	5.9	185.9	59.0	
15	ORNL/MWL	column	20.0	83.6	11.0	171.4	53.9	
16	sodium acetate	column	5.0	96.9	77.8	169.8	29.4	

17	carbonic acid	column	2.5	99.6	33.5	256.6	66.9
18	distilled water	batch	5.0	112.9	56.4	177.4	41.8
19	distilled water	column	5.0	114.2	22.0	325.2	81.2
20	sodium acetate	column	2.5	119.9	88.5	323.0	63.5
21	ORNL/MWL	batch	20.0	133.0	9.7	762.8	178.8
22	ORNL/MWL	column	10.0	139.3	4.5	486.0	103.4
23	carbonic acid	batch	5.0	142.5	36.0	490.3	112.7
24	ORNL/MWL	batch	10.0	156.7	35.5	416.3	88.9
25	bisequential	batch	2.5	183.1	4.3	636.5	130.5
26	distilled water	column	2.5	195.5	25.8	762.5	128.3
27	ORNL/MWL	column	5.0	224.4	15.1	1053.0	151.8
28	carbonic acid	batch	2.5	236.3	20.0	886.3	132.0
29	ORNL/MWL	column	2.5	358.1	33.9	1784.4	164.2
30	ORNL/MWL	batch	5.0	427.8	24.4	1366.7	115.3
31	distilled water	batch	2.5	483.5	55.9	3064.3	202.7
32	ORNL/MWL	batch	2.5	543.8	17.7	2269.3	149.2
33	sodium acetate	batch	5.0	626.3	50.8	4501.8	233.1
34	sodium acetate	batch	2.5	1020.0	53.6	7542.3	241.2

[a] Based on 20:1 liquid-to-solid ratio; the data are from Ref 2.
[b] Liquid-to-solid ratio.
[c] Any two laboratory extraction procedures connected together by a vertical line are not statistically different ($P < .05$) using Duncan's multiple range test. The EP and bisequential methods were excluded because they were not replicated.

TABLE 3—Ranking of 34 laboratory extraction methods to simulate inorganic target concentrations.[a]

Rank	Medium	Type	LS Ratio[b]	Difference, %			Coefficient of Variation, %	Test for Significance[c]
				Average	Minimum	Maximum		
1	carbonic acid	column	20.0	53.7	10.7	213.1	90.9	
2	distilled water	column	20.0	62.3	12.7	124.6	56.7	
3	sodium acetate	column	20.0	68.5	23.6	150.3	61.8	
4	carbonic acid	batch	20.0	77.0	1.9	411.3	143.0	
5	carbonic acid	column	10.0	79.8	11.4	331.4	104.0	
6	distilled water	column	10.0	82.6	9.2	190.6	70.0	
7	distilled water	batch	20.0	92.9	2.7	440.1	106.2	
8	sodium acetate	column	10.0	100.7	20.9	290.9	86.8	
9	ORNL/MWL	batch	20.0	106.8	12.0	567.2	128.6	
10	ORNL/MWL	column	20.0	107.1	12.9	473.8	107.5	
11	distilled water	batch	10.0	139.3	0.9	764.1	131.9	
12	EP	batch	20.0	139.5	1.1	405.5	86.7	
13	carbonic acid	column	5.0	140.1	7.1	537.4	103.5	
14	carbonic acid	batch	10.0	142.4	10.1	908.1	158.0	
15	distilled water	column	5.0	155.9	8.4	433.0	80.2	

16	sodium acetate	column	5.0	156.1	24.5	509.4	91.9
17	sodium acetate	batch	20.0	173.7	14.8	1340.2	185.5
18	ORNL/MWL	batch	10.0	178.1	28.6	785.2	117.3
19	ORNL/MWL	column	10.0	178.3	10.8	941.3	130.4
20	sodium acetate	batch	10.0	199.4	23.3	1160.2	142.5
21	sodium acetate	column	2.5	232.2	31.9	764.1	96.8
22	distilled water	batch	5.0	233.3	6.6	1196.2	125.4
23	carbonic acid	column	2.5	243.7	7.2	864.7	99.9
24	carbonic acid	batch	5.0	254.5	10.4	1448.2	143.0
25	distilled water	column	2.5	278.8	6.5	918.9	98.5
26	ORNL/MWL	column	5.0	298.2	11.8	1561.0	131.1
27	ORNL/MWL	batch	5.0	298.4	44.2	996.6	105.7
28	sodium acetate	batch	5.0	329.9	14.5	2132.4	159.4
29	distilled water	batch	2.5	384.6	44.2	1700.3	114.1
30	carbonic acid	batch	2.5	401.9	22.1	1772.3	118.5
31	ORNL/MWL	batch	2.5	434.1	5.8	1484.3	102.7
32	ORNL/MWL	column	2.5	456.2	13.0	2204.4	122.7
33	sodium acetate	batch	2.5	508.0	8.2	2780.5	138.5
34	bisequential	batch	2.5	652.1	14.6	2714.1	128.0

[a] Based on 20:1 liquid-to-solid ratio; the data are from Ref 2.
[b] Liquid-to-solid ratio.
[c] Any two laboratory extraction procedures connected together by a vertical line are not statistically different ($P < .05$) using Duncan's multiple range test. The EP and bisequential methods were excluded because they were not replicated.

TABLE 4—Ranking of 34 laboratory extraction methods to simulate inorganic and organic target concentrations.[a]

Rank	Medium	Type	LS Ratio[b]	Difference, % Average	Difference, % Minimum	Difference, % Maximum	Coefficient of Variation, %	Test for Significance[c]
1	carbonic acid	column	20.0	58.8	10.7	213.1	72.0	
2	distilled water	column	20.0	66.1	12.7	133.8	54.1	
3	sodium acetate	column	20.0	69.5	12.2	150.3	55.4	
4	carbonic acid	batch	20.0	71.2	1.9	411.3	125.0	
5	carbonic acid	column	10.0	73.0	9.5	331.4	94.7	
6	distilled water	column	10.0	81.2	9.2	190.6	61.0	
7	distilled water	batch	20.0	82.1	2.7	440.1	98.8	
8	sodium acetate	column	10.0	92.0	20.9	290.9	77.3	
9	ORNL/MWL	column	20.0	98.7	11.0	473.8	96.6	
10	distilled water	batch	10.0	112.6	0.9	764.1	133.5	
11	EP	batch	20.0	114.8	1.1	405.5	89.2	
12	carbonic acid	batch	10.0	115.6	8.8	908.1	159.9	
13	ORNL/MWL	batch	20.0	116.2	9.7	762.8	151.0	
14	carbonic acid	column	5.0	116.9	7.1	537.4	102.7	
15	sodium acetate	batch	20.0	132.3	4.8	1340.2	197.5	
16	sodium acetate	column	5.0	134.8	24.5	509.4	87.7	
17	distilled water	column	5.0	140.9	8.4	433.0	81.1	

18	sodium acetate	batch	10.0	157.5	5.9	1160.2	148.3
19	ORNL/MWL	column	10.0	164.2	4.5	941.3	123.4
20	ORNL/MWL	batch	10.0	170.4	28.6	785.2	108.0
21	distilled water	batch	5.0	190.0	6.6	1196.2	126.4
22	sodium acetate	column	2.5	191.8	31.9	764.1	99.7
23	carbonic acid	column	2.5	191.8	7.2	864.7	108.7
24	carbonic acid	batch	5.0	214.2	10.4	1448.2	143.5
25	distilled water	column	2.5	248.8	6.5	918.9	106.1
26	ORNL/MWL	column	5.0	271.7	11.8	1561.0	135.5
27	carbonic acid	batch	2.5	342.3	20.0	1772.3	124.2
28	ORNL/MWL	batch	5.0	345.0	24.4	1366.7	111.2
29	distilled water	batch	2.5	420.2	44.2	3064.3	158.4
30	ORNL/MWL	column	2.5	420.9	13.0	2204.4	133.0
31	sodium acetate	batch	5.0	436.6	14.5	4501.8	217.8
32	ORNL/MWL	batch	2.5	473.6	5.8	2269.3	124.3
33	bisequential	batch	2.5	483.3	4.3	2714.1	147.4
34	sodium acetate	batch	2.5	692.3	8.2	7542.3	223.3

[a] Based on 20:1 liquid-to-solid ratio; the data are from Ref 2.
[b] Liquid-to-solid ratio.
[c] Any two laboratory extraction procedures connected together by a vertical line are not statistically different ($P < 0.05$) using Duncan's multiple range test. The EP and bisequential methods were excluded because these methods were not replicated.

cant differences among laboratory tests were determined by using an analysis of variance [6].

As indicated in Tables 2 to 4, many of the procedures are identified as being equally predictive of the target concentrations. This is especially true for the organic compounds. This is an advantage in that it provides a choice of methods on which to base the new procedure, but it also presents a problem in that there is no clear-cut choice as to which procedure is best.

Therefore, in comparing these procedures, a number of factors (including absolute rank) were taken into account, such as ease and expense of operation, applicability to both organics and inorganics, and applicability to biological testing. The reader may recall that these factors were previously outlined as objectives for EP-III.

Considering that batch procedures are generally more precise, easier to conduct, less expensive, and applicable to a wider variety of waste types than column procedures [1], it was felt that if at all possible a batch procedure would be preferable to a column procedure. Column procedures were included in the study because researchers in the past have claimed that column procedures might provide the more accurate estimation of leaching in the real world [4]. Tables 2 to 4 indicate that there is probably no real difference between the two in terms of accuracy.

As far as biological applicability is concerned, either the water or the carbon dioxide–saturated water would be preferred as the extraction medium. Neither of these media are expected to interfere with biological systems. As earlier studies have indicated, the sodium acetate buffer extraction media may pose problems for some biological systems [2].

As indicated previously, the primary objective in establishing a new procedure is to determine the mobility of organic compounds. In Table 2, the top-ranked procedure for organics is the sodium acetate buffer batch extraction, and the second-ranked procedure is the carbon dioxide–saturated water batch extraction, both at the 20:1 liquid-to-solid ratio. Although the sodium acetate buffer medium may pose problems for some biological testing systems, both procedures offer the advantages of being relatively easy to conduct, inexpensive, and applicable to a wide variety of wastes.

In addition, Tables 3 and 4, which address inorganic compounds and organic and inorganic compounds combined, indicate that the carbon dioxide–saturated medium ranks relatively high (fourth in both cases) and might therefore be suitable for determining the mobility of inorganic constituents. The sodium acetate buffer medium also holds promise for the evaluation of inorganic compounds, in that it may more adequately address alkaline wastes. The problem is that an increase in leaching of the inorganics may be observed when the ability of the lime to resist acidity becomes exhausted. Since the carbon dioxide–saturated water has little acidity compared to actual sanitary landfill leachates, it may not adequately challenge the alkalinity of some industrial wastes. The sodium acetate buffer, as well as the current EP, will not suffer this problem. This phenomenon will be studied further by subjecting limed wastes to municipal waste leachate for longer periods of time and observing the leaching of inorganic constituents.

In view of these results, the sodium acetate buffer batch extraction and the carbon dioxide–saturated water batch extraction, both at the 20:1 liquid-to-solid ratio, have been selected for further work.

Briefly, a second series of lysimeter experiments has been completed, and laboratory extraction work comparing the two procedures to leachate target concentrations based on the 20:1 ratio is currently under way. In an effort to achieve greater statistical discrimination among the leaching procedures, and also to provide some indication of the direction of the statistical difference (positive or negative), different and more powerful tests of statistical significance will be applied. In addition, the data from the first series of lysimeter experiments will be reexamined by using these statistical tests. A comparison of the two procedures using the results obtained from all the statistical comparisons made on both series of lysimeter work will indicate which procedure should form the basis for a draft EP-III method.

Refinement of the Draft EP-III

After selection of a single draft method, the test will be subjected to a single-laboratory evaluation of ruggedness and precision. During this phase of evaluation, other aspects of the method will be studied in an attempt to improve on the current protocol, which is based on the existing EP (Fig. 3).

For example, the two most often mentioned trouble spots for the EP protocol are the initial liquid/solid separation and continual pH adjustment. The continual pH adjustment problem has been addressed through use of the carbon dioxide–saturated water system and the sodium acetate buffer solution. These media, while providing acidity as contained in municipal waste leachate, do not require continual pH adjustment as a waste is extracted.

Problems associated with the initial liquid/solid separation are due to some waste materials, such as oily wastes, clogging the 0.45 μm filter, even after considerable pressure (517 kPa [75 psi]) is applied. These problems are being addressed through use of a much simpler liquid/solid separation technique. Specifically, liquid/solid separation using RCRA Test 9095 (Paint Filter Liquids Test) is planned. A notice of this test has been published in the *Federal Register* for use in determining the free liquid content of wastes before landfill disposal [7]. It involves a simple 400 mesh conical paint filter for liquid/solid separation, and it appears to be of acceptable reproducibility [8].

Since the new method will be applicable to organic compounds, including volatile compounds, it is important that significant loss of volatiles during extraction be prevented. Hence, a closed system like those used in the rotary extractors will be important, and other precautions will be considered as well. For example, it may be necessary to narrow the temperature range allowed

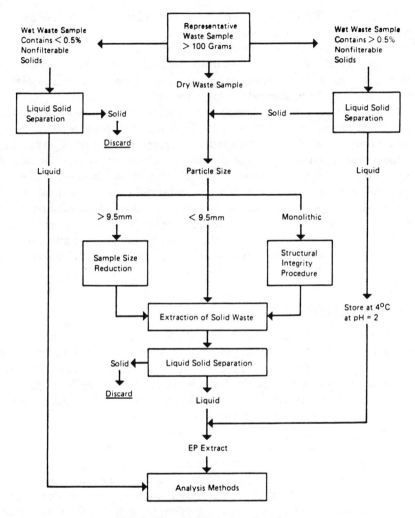

FIG. 3—*Extraction procedure flow chart.*

during extraction from the current range of 20 to 40°C. These and other aspects of the protocol, including extraction time and agitation method, will be studied extensively during single-laboratory evaluation. Finally, following successful single-laboratory evaluation, an extensive interlaboratory collaborative study is planned for early 1985.

Conclusions

Based on the data base being established by the research team at ORNL, a procedure for determining the potential leachability of organic compounds is being developed. The draft method will be applicable to the leaching of organic compounds and may be expanded to include inorganic compounds as well. Procedurally, it will be similar to the current EP; however, it should be more precise, less costly, and simpler to conduct and pose fewer problems when applied to biological test systems than the current EP.

Acknowledgments

We would like to express our appreciation to C. W. Francis, M. R. Maskarinec, J. C. Goyert, and the rest of the ORNL research team and to L. R. Williams, the EPA project officer. Through their expertise and dedication, an organic toxicant mobility procedure is becoming a reality.

References

[1] U.S. EPA, "Characteristic of Extraction Procedure Toxicity," Background Document, Section 261.24, PB 81/185-027, National Technical Information Service, Springfield, VA, May 1980.

[2] Francis, C. W., Maskarinec, M. P., and Goyert, J. C., "Mobility of Toxic Compounds from Hazardous Wastes," Report NTIS PB 85 117034, Oak Ridge National Laboratory/National Technical Information Service, NTIS, Springfield, VA, Jan. 1984.

[3] Francis, C. W., Maskarinec, M. P., and Goyert, J. C., "A Laboratory Extraction Method to Simulate Codisposal of Solid Wastes in Municipal Waste Landfills," *Hazardous and Industrial Solid Waste Testing: Fourth Symposium, ASTM STP 886*, J. K. Petros, Jr., W. J. Lacy, and R. A. Conway, Eds., American Society for Testing and Materials, Philadelphia, 1985, pp. 15–35.

[4] Lowenback, W., "Compilation and Evaluation of Leaching Test Methods," 600/2-78-095, U.S. EPA, Washington, DC, May 1978.

[5] Wilson, D. C. and Young, P. J., "Testing Methods for Hazardous Wastes Prior to Landfill Disposal," *Environment and Solid Wastes: Characterization, Treatment and Disposal*, C. W. Francis and S. I. Auerbach, Eds., Ann Arbor Science, Boston, 1983.

[6] *SAS User's Guide: Basics*, SAS Institute, Inc., Cary, NC, 1982.

[7] "Extension of 40 CFR 265.314(b)," 40 *Federal Register* 8304, 25 Feb. 1982.

[8] "Test Protocols for Determining the 'Free Liquid' Content of Hazardous Waste," final draft report, U.S. EPA, in press.

Leo R. Barsotti[1] and Thomas A. Palmer[1]

Comparison of Analytical Methods for the Determination of Total and Free Cyanide in Solid Waste

REFERENCE: Barsotti, L. R. and Palmer, T. A., **"Comparison of Analytical Methods for the Determination of Total and Free Cyanide in Solid Waste,"** *Hazardous and Industrial Solid Waste Testing: Fourth Symposium, ASTM STP 886,* J. K. Petros, Jr., W. J. Lacy, and R. A. Conway, Eds., American Society for Testing and Materials, Philadelphia, 1986, pp. 54–63.

ABSTRACT: Historically, standard reflux distillation methods have been used to determine the total cyanide in aqueous systems. This paper compares such a method with an automated total cyanide method and demonstrates good recovery and precision by both methods. The automated method has the advantages of speed and convenience, as well as being less labor-intensive.

Free cyanide, identified as the sum of hydrocyanic acid and cyanide ion in aqueous solution, is determined by a microdiffusion and a modified Roberts-Jackson method. A comparison of results obtained by both methods showed free cyanide was only about 2% of the total cyanide in an aqueous sample. Total cyanide methods are useful for screening the cyanide content of samples, but free cyanide methods increase the information available and improve the understanding about the kinds of cyanide compounds in a sample. Even greater selectivity of analytical methods for cyanide compounds in waste is desirable and a challenge to analytical chemists.

KEY WORDS: hazardous wastes, total cyanide, free cyanide, manual distillation, automated cyanide method, microdiffusion, modified Roberts-Jackson, complex cyanides, ferrocyanide, weak acid-dissociable cyanide

Historically, an acid reflux distillation has been used to dissociate complex cyanides and recover the evolved hydrogen cyanide free of the sample matrix. The analytical measurement of the captured cyanide is generally either a titrimetric, colorimetric, or a potentiometric procedure; see, for example, ASTM Tests for Cyanides in Water (D 2036) or the U.S. Environmental Protection

[1]Section head, Inorganic Analytical Section, and senior staff research chemist, respectively, Analytical Research Department, Center for Technology, Kaiser Aluminum and Chemical Corp., Pleasanton, CA 94566.

Agency's (EPA) Methods for Chemical Analyses of Water and Wastes (335.2).

More recently, an automated method has been developed which reduces the analytical time and cost of a total cyanide determination—Technicon Instruments Corp.'s Cyanides in Water and Waste Water (315-74W). This method is well suited for laboratories that require large numbers of cyanide analyses.

Free cyanide is generally recognized as the species of concern for the protection of freshwater aquatic life. The U.S. EPA has proposed 0.0035 mg/L free cyanide as a 24-h average value, with a maximum concentration not to exceed 0.052 mg/L at any time. For the protection of human health, the ambient water quality criterion is 0.200 mg/L cyanide, according to EPA's Ambient Water Quality Criteria for Cyanides (EPA 440/5-80-037). This requirement has been interpreted by many to mean free cyanide. Thus the determination of total cyanide, while easy and convenient with automated instrumentation, is inappropriate for free cyanide because it fails to distinguish between the several classes of cyanide compounds, each with varying degree of potential toxicity [1].

One problem has been the inability of analytical methods to distinguish selectively between the various classes of cyanide compounds. The introduction to ASTM D 2036 succinctly describes the difficulty facing analysts desirous of determining classes of cyanide species by stating, "These methods do not distinguish between cyanide ions and metallocyanide compounds or complexes."

Another problem has been the proliferation of terms used to describe various classes of cyanides. Some terms were and are necessary, but some seem inappropriate and analytically misleading. A listing of terms used to describe cyanide compounds is given below:

• *Total cyanide* refers to the sum of all the inorganic cyanides in a sample.
• *Free cyanide* refers to the sum of hydrocyanic acid (HCN) and cyanide ion (CN^-) in aqueous solution.
• *Simple cyanide* refers to a cyanide compound that dissociates in water to a cation and cyanide. Analytically, simple cyanide will behave as free cyanide.
• *Cyanides amenable to chlorination* are cyanides in solution that are oxidized by the addition of hypochlorite. These compounds include free and easily dissociable complex cyanides.
• *Weak acid dissociable cyanides* are frequently assumed to be similar to cyanides amenable to chlorination. Weak acid dissociable cyanides is a more technically correct and appropriate term, since methods such as ASTM D 2036 and the Roberts-Jackson method [2] use a pH 4.5 reaction medium to convert cyanides to HCN.
• *Complex cyanide* refers to a cyanide compound that dissociates in water to a cation and an anion comprised of two or more different species or atoms,

one of which is cyanide. The complex cyanide anion is subject to further dissociation.

This report compares the results obtained by four different analytical methods for free and total cyanide on an aqueous stream and an industrial waste. Problems with the various methods are also discussed.

Experimental Method

Total Cyanide

Two different methods were used for the determination of total cyanide: (1) the EPA-approved manual distillation method 335.2, followed by a silver nitrate titrimetric finish and (2) a modified version of the Technicon AutoAnalyzer® method 315-74W, which uses ultraviolet light and acid dissolution of complex cyanides, flash distillation of HCN gas, and photometric measurement of cyanide with pyridine-barbituric acid reagent. Photographs of the manual distillation and the AutoAnalyzer apparatus are shown in Figs. 1 and 2.

The titrimetric procedure was used to standardize stock cyanide solutions prepared from potassium cyanide (KCN) and stabilized with sodium hydroxide to pH > 12. Working standards were prepared by quantitative dilution of the stock solutions with appropriate pH control.

Total cyanide in industrial solid waste was determined by placing a 1-g sample, together with 500 mL deionized water, into the manual distillation flask and conducting the distillation as with an aqueous system.

Free Cyanide

Free cyanide was determined by a microdiffusion method [3] and by a modified Roberts-Jackson method [2].

The microdiffusion method is based on the volatility of HCN at pH 7 ± 0.1 and its absorption in caustic solution. A pH level of 7 was selected because hydrocyanic acid has a log acid dissociation constant pK_a of 9.3. Thus, greater than 99% of ionic cyanide exists at pH 7 as molecular HCN and is easily recoverable from the sample matrix. At pH values less than 7, partial dissociation of complex iron cyanides can be significant and should be avoided.

The microdiffusion method described by Kruse and Thibault [4] uses a Conway diffusion dish (see Fig. 3). Into the outer chamber is placed a sample solution buffered to pH 7 with 0.5 M potassium phosphate (KH_2PO_4) and into the inner chamber is placed dilute alkali absorbing solution. The dish is sealed with a glass plate and placed overnight in a dark cabinet. Six hours is adequate for the microdiffusion process, but overnight is convenient. The fol-

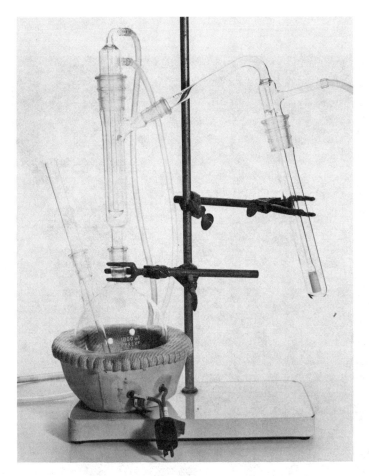

FIG. 1—*CN distillation apparatus.*

lowing day the absorbing solution is analyzed for cyanide by using either a manual pyridine-barbituric acid spectrophotometric procedure or the automated AutoAnalyzer procedure. Similar microdiffusion methods include the ASTM Test for Determination of Free Cyanide in Water and Wastewater by Microdiffusion (D 4282-83) and the American National Standards Institute's Method for Determining Free Cyanide in Photographic Effluents (pH 4.41).

For industrial solid waste, an appropriate slurry of sample, water, and buffer solution was prepared and placed into the outer chamber, and the microdiffusion process was carried out as with an aqueous sample.

The modified Roberts-Jackson method consists of a manual distillation conducted at pH 4.5 with acetic acid at atmospheric pressure with equal mo-

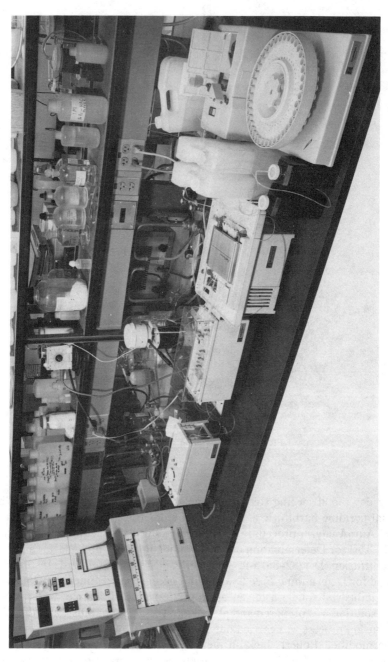

FIG. 2—*Technicon AutoAnalyzer.*

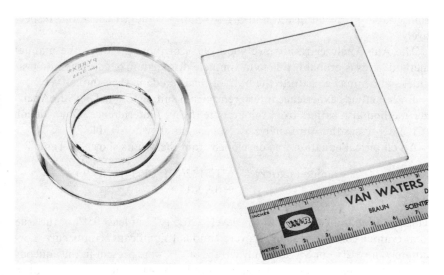

FIG. 3—*Conway microdiffusion dish.*

lar amounts of added zinc and lead acetates to minimize the dissociation of complex iron cyanides [5].

The distillate obtained by the above procedure was analyzed for cyanide by using either a manual pyridine-barbituric acid spectrophotometric procedure or the automated AutoAnalyzer procedure.

Free cyanide in industrial solid waste was determined by placing a 1-g sample and 500 mL deionized water into the distillation flask and conducting the analysis as with the aqueous sample.

Results and Discussion

Table 1 shows the results for total cyanide in an aqueuos sample by the manual distillation and AutoAnalyzer methods. Both methods had good and

TABLE 1—*Total cyanide in the aqueous stream by two methods.[a]*

Manual Distillation	Automated
$n = 14$	$n = 27$
$\bar{X} = 10.34$ mg/L	$\bar{X} = 9.87$ mg/L
$2S = 0.74$	$2S = 0.61$
RSD $= 3.6\%$	RSD $= 3.1\%$

[a]n = number of samples, $\bar{X}$ = sample mean, $2S$ = two standard deviations, and RSD = relative standard deviation.

comparable precision of 3.6 and 3.1% relative standard deviation, respectively. The AutoAnalyzer results are about 95% of those given for the manual method. This is probably due to incomplete dissociation of complex iron cyanides, which may constitute the bulk of the cyanide in the sample.

Both methods experience interference from sulfide (S^{2-}). The AutoAnalyzer method also suffers from thiocyanate (SCN^-) interference. Chloride ion (Cl^-) depresses the AutoAnalyzer response, as shown in Table 2.

An empirical equation was developed from the results shown in Table 2:

$$\% \text{ cyanide recovery} = -1.5122(\text{g/L } Cl^-) + 104.9$$
$$r = -0.9961$$

This equation indicates that at Cl^- levels <6.5 g/L, at least 95% of the true total cyanide concentration is recovered and not significant error occurs. Fortunately, no SCN^- or S^{2-} and only a trace of Cl^- were present in the aqueous sample, so no interferences were experienced. The AutoAnalyzer method has the advantages of speed (20 samples per hour) and convenience; it is also less labor-intensive than the manual distillation method.

The results obtained by the microdiffusion and modified Roberts-Jackson methods for free cyanide are shown in Table 3. The two methods gave essen-

TABLE 2—*Effect of sodium chloride on cyanide recovery.*

Chloride Ion Added, g/L	Cyanide[a] Recovery, %
0	101
1.8	102
5.5	99
9.1	91
14.6	83
18.2	79
21.8	74

[a]0.14 mg/L CN^-.

TABLE 3—*Free cyanide in the aqueous stream by two methods.*[a]

Microdiffusion	Modified Roberts-Jackson
$n = 10$	$n = 10$
$\bar{X} = 0.18$ mg/L	$\bar{X} = 0.20$ mg/L
$2S = 0.008$	$2S = 0.054$
RSD = 2%	RSD = 14%

[a]See Table 1 for meaning of symbols.

tially the same free cyanide concentration. This result was to be expected since both methods are largely unresponsive to complex iron cyanides, which are believed to constitute the bulk of the total cyanide in the sample. Ninety-eight percent of the total cyanide was unreactive to both methods and only about 2% was considered free cyanide.

The precision of the modified Roberts-Jackson method is poorer than that of the microdiffusion method. This phenomenon is ascribed to slight variations in pH, distillation medium, duration of distillation, or laboratory lighting, leading to dissociation of hexacyanoferrates.

Table 4 shows the extent to which both methods dissociate complex cyanides, leading them to behave analytically like cyanide ions. Weak metal cyanide complexes with a log dissociation constant of less than 20 dissociate completely at pH 7 and are determined as free cyanide by the microdiffusion method. The copper (I) cyanide complex is partially dissociated, while complexes with a log dissociation constant above 30 do not dissociate to any extent.

Dissociation of complex cyanides is even more pronounced when one is using the modified Roberts-Jackson method. The results reported by Ingersoll et al [5] and shown in Table 4 indicate that complexes other than iron and cobalt dissociate either to a large extent or completely. These results point out the need for an understanding of the scope of the analytical method used for the determination of free cyanide and a recognition that easily dissociable complexes behave analytically as cyanide ion.

Table 5 shows free cyanide recovery from potassium ferrocyanide [$K_4Fe(CN)_6$] solution by both methods. The free cyanide found in 10 mg/L $K_4Fe(CN)_6$ solution by the microdiffusion method is 0.018 mg/L, considerably less than the 0.058 mg/L found by the modified Roberts-Jackson method. These results indicate some dissociation of $Fe(CN)_6^{4-}$ occurs, espe-

TABLE 4—*Dissociation of metal-cyanide complexes.*

Metal Cyanide Comple	Log Dissociation Constant [1]	Cyanide Recovery, %	
		Microdiffusion [6] Method	Modified Roberts-Jackson [5] Method
$Cd(CN)_4^{2-}$	17	95	...
$Zn(CN)_4^{2-}$	17	103	...
$Pb(CN)_4^{2-}$	...	97	...
$Ag(CN)_2^{1-}$	21	<1	...
$Cu(CN)_3^{2-}$	28	48	100
$Ni(CN)_4^{2-}$	22	1	100
$Hg(CN)_4^{2-}$	42	...	75
$Fe(CN)_6^{4-}$	47	<0.5	1
$Co(CN)_6^{3-}$	64	0	0

TABLE 5—*Free cyanide recovery from $K_4Fe(CN)_6$ solution.*

Fe(CN)$_6^{4-}$, mg/L	CN$^-$ Added, mg/L	Microdiffusion Method	Modified Roberts-Jackson Method	Recovery, %
		CN$^-$ Found, mg/L[a]		
10	0	0.018 (3)[b]	0.058 (8)[c]	...
10	2.28	...	2.30 (4)	101
0	0.094	0.092 (4)	...	98
0	4.6	...	4.6 (4)	100
1.4	0.038	0.038 (2)	...	100
6.9	0.202	0.195 (2)	...	97
10	0.914	0.916 (3)	...	100

[a]The number in parentheses refers to number of replicate analyses.
[b]$S = 0.001$.
[c]$S = 0.021$.

cially with the modified Roberts-Jackson method. The modified Roberts-Jackson precision is relatively poor. Contrasting this precision with that obtained by distilling cyanide ion alone also suggests dissociation of complex iron cyanide is the probable cause of poor precision. Recovery of free cyanide from a $K_4Fe(CN)_6$ solution by the microdiffusion method ranged from 97 to 101%.

Both methods are free of interference from SCN$^-$ or S^{2-}. The microdiffusion method is relatively simple to perform and convenient, and it requires less direct analyst time than the modified Roberts-Jackson method.

Table 6 shows the total and free cyanide results for an industrial solid waste. The automated method was not used because the AutoAnalyzer cannot normally analyze solids. The precision of the three methods was excellent and no difficulties were experienced either in distilling solids as a slurry or in handling solids in the microdiffusion method. As with the aqueous samples, the free cyanide values are lower than the total cyanide value. There is also a

TABLE 6—*Total and free cyanide in industrial solid waste.*[a]

Total Cyanide	Microdiffusion Method	Modified Roberts-Jackson Method
	Free Cyanide	
$n = 4$	$n = 5$	$n = 4$
$\bar{X} = 2400 \ \mu g/g$	$\bar{X} = 280 \ \mu g/g$	$\bar{X} = 420 \ \mu g/g$
$2S = 150$	$2S = 22$	$2S = 13$
RSD = 3.2%	RSD = 3.9%	RSD = 1.5%

[a]See Table 1 for meaning of symbols.

difference between analytical results for the modified Roberts-Jackson and microdiffusion methods. This difference is under investigation, with the expectation that a better understanding of it will improve our knowledge of cyanide analytical chemistry.

Conclusions

Using a hierarchy of methods for total, acid-dissociable, and free cyanide provides analysts with a degree of analytical selectivity and allows a better understanding of the various cyanide species in a sample. In particular, the results show:

• Both total CN methods are satisfactory for most aqueous samples. The AutoAnalyzer is particularly useful when large numbers of samples must be analyzed.

• The microdiffusion method provides mild treatment of a sample under realistic conditions of temperature and pH. The combination of simplicity, precision, and accuracy make microdiffusion the method of choice for the determination of free CN.

• The modified Roberts-Jackson method does not completely prevent the dissociation of complex iron cyanides. This could lead to erroneously high free CN results, especially at trace levels when complex cyanides are also present. Its harsher operating parameters favor the dissociation of more acid-dissociable cyanides than the microdiffusion method; as a result, it should not be considered an equivalent method.

Acknowledgments

The authors acknowledge the technical assistance of Gisela Calkins and Judy Winsor for their analytical work.

References

[1] "Cyanide—An Overview and Analyses of the Literature on Chemistry, Fate, Toxicity, and Detection in Surface Waters," Ecological Analysts, Inc., Towson, MD, 1979.

[2] Roberts, R. F. and Jackson, B., "The Determination of Small Amounts of Cyanide in the Presence of Ferrocyanide by Distillation Under Reduced Pressure," *Analyst*, Vol. 96, No. 3, March, 1971, pp. 209-212.

[3] Palmer, T. A. and Skarset, J. R., "Determination of Free Cyanide Using a Microdiffusion-Photometric Technique," *Aluminium*, Vol. 59, No. 4, April 1983, pp. 292-299.

[4] Kruse, J. M. and Thibault, L. E., "Determination of Free Cyanide in Ferro- and Ferricyanides," *Analytical Chemistry*, Vol. 45, No. 13, November 1973, pp. 2260-2261.

[5] Ingersoll, D., Harris, W. R., Bomberger, D. C., and Coulson, D. M., "Development and Evaluation of Procedures for the Analysis of Simple Cyanides, Total Cyanides, and Thiocyanate in Water and Waste Water," EPA Contract No. 68-03-2714 (Draft), SRI International, Menlo Park CA, 1981.

[6] Palmer, T. A., "Determination of Total and Free Cyanide in Aqueous Systems," The Pittsburgh Conference, 5-9 March 1984, Spectroscopy Society of Pittsburgh, Pittsburgh, PA.

William J. Ziegler¹ and Richard Schlauch¹

A Rapid Method for Determination of Total Organic Halogens in Industrial Waste Samples

REFERENCE: Ziegler, W. J. and Schlauch, R., **"A Rapid Method for Determination of Total Organic Halogens in Industrial Waste Samples,"** *Hazardous and Industrial Solid Waste Testing: Fourth Symposium, ASTM STP 886,* J. K. Petros, Jr., W. J. Lacy, and R. A. Conway, Eds., American Society for Testing and Materials, Philadelphia, 1986, pp. 64-76.

ABSTRACT: A rapid method for determining total organic halogens (TOX) in industrial waste samples has been developed and validated. Total organic halogens is an indicator parameter specified by the U.S. Environmental Protection Agency (EPA) to be of environmental significance as representative of toxic organic halogenated compounds. The validation study included 36 compounds representing all halogenated organic classes on the EPA priority pollutant list. The recovery data indicate that the method is quantitative for all halogenated organic classes with a detection limit of 3.5 ppm (micrograms per gram) as inorganic chloride [approximately 6.5 ppm polychlorinated biphenyls (PCBs)]. The analysis can be performed within 20 min and can be used to screen hazardous waste samples for PCBs at the 50-ppm enforcement limit of the Toxic Substances Control Act.

KEY WORDS: total organic halogens (TOX), combustion bomb, ion chromatography, polychlorinated biphenyls, hazardous wastes

Existing federal regulations (40 CFR 264.13) require that all hazardous waste treatment, storage, and disposal facilities (TSDFs) maintain a Waste Analysis Plan. The purpose of this plan is to ensure that each waste shipment is identical to the waste specified on the manifest, and that each can be treated or disposed of safely at the facility.

Some organic halogenated compounds have been recognized to possess a high degree of toxicity [1]. It is significant that 71 out of 113 of the designated

---

¹Technical director and process development chemist, respectively, Stablex Corp., Radnor, PA 19087.

priority pollutants of the U.S. Environmental Protection Agency (EPA) and 126 out of 320 of the Appendix VIII listing of hazardous constituents of the Resource Conservation and Recovery Act (RCRA) are halogenated organics. Many TSDFs are regulated on the quantity of halogenated organics that may be accepted and must monitor the total organic halogens (TOX) as an indicator parameter in groundwater, as well as in treatment operations. In addition, the Toxic Substances Control Act provides for stringent regulatory control of polychlorinated biphenyls (PCBs) [2]. A rapid TOX screening technique would therefore be a beneficial component of waste analysis plans of TSDFs and waste oil recycling facilities regulated on quantities of halogenated organics that can be treated as a permit constraint.

Although commercial instrumentation exists for the analysis of TOX in water [3–5], these instruments are not suited to the analysis of waste samples. Existing methods require that the sample be passed through an activated carbon column to separate and concentrate organic halogens. There is therefore a strong dependency on the absorption efficiencies of different organic halogen compounds, and polar organic halogens have poor recoveries. These methods cannot be used for high-solid samples or samples containing inorganic halogens in excess of 0.2%.

Methods exist for determining the total chlorine in petroleum products by using an oxygen combustion bomb: the ASTM Test for Chlorine in New and Used Petroleum Products (Bomb Method) (D 808) and the ASTM Test for Chlorine in Coal (D 2361). Other combustion techniques have been applied to the determination of total organic halogenated compounds in organic materials using a Schoeniger flask [6–9], and a Dumas-Pregl apparatus [10]. These techniques are limited by small sample sizes and a high detection limit, 0.2% total halogens. Yamada [11] analyzed water samples by extraction with hexane followed by reduction of the extracted halogenated organic at 350°C in a palladium catalyst reduction tube. The hydrogen halides (HX) generated were absorbed in deionized water and determined coulometrically. This method could detect 120 ppm PCBs with 92 to 100% recovery, as opposed to 33 to 43% recovery for combustion techniques.

The method reported here is applicable to waste samples and makes use of selective extraction of organic halogens into a small volume of solvent, followed by combustion of the complete extract in a high-capacity Parr combustion bomb. Combustion converts organic halogenated compounds to hydrogen chloride, which is collected in the absorbing solution (standard anion eluent for ion chromatography) as inorganic salts. The inorganic halogens formed from the combustion of halogenated organics are then determined by using one of three methods of quantitation described below. The levels of total organic halogens down to 3.5 ppm can be detected (about 6 ppm PCBs) with recoveries of 72 to 120%. The limit of quantitation of the method is 12 ppm as chlorine.

Procedure

Equipment

A Parr oxygen combustion bomb apparatus (Model 1901) with high-capacity quartz-lined bomb is used for combustion of the sample. A recirculating deionized/distilled water bath at 4°C is used for rapid cooling following ignition. A centrifuge is used following extraction to separate the solvent phase.

Sample Preparation

A 3.00 ± 0.05-g aliquot of sample is weighed into a 10-mL screw-capped centrifuge tube. Five millilitres of nanograde *iso*octane is added, and a Teflon®-lined screw cap is placed on the tube. The tube is agitated vigorously for 2 min by using a vortex mixer, the sample is centrifuged for 3 min, and the *iso*octane layer is transferred to a separatory funnel. The *iso*octane layer is then washed three times with 25-mL aliquots of deionized/distilled water to remove inorganic chloride. Then 2.0 mL of the washed *iso*octane layer is transferred to a Parr oxygen combustion bomb sample boat. A 5.0-mL aliquot of a standard ion chromatography anion eluent (specified below) is added to the bottom of the quartz-lined bomb. The sample is sealed in the Parr bomb, pressurized with oxygen, and then ignited and burned. Following combustion, the contents of the bomb are quantitatively transferred to a 50-mL volumetric flask and diluted to volume.

Quantitation of Organic Halogens

After completion of sample preparation, any of three techniques can be used to determine the halogen concentration in the extract:

- *Ion Chromatography*—A Dionex Model 2010i ion chromatograph with an electrolytic conductivity detector is used to determine inorganic chloride or bromide. The following conditions are used: column—anion S-2 separator, 150 by 3 mm; precolumn—50 by 3 mm; anion fiber suppressor—regenerant, 0.025 N sulfuric acid (H_2SO_4); eluent—0.0030 M sodium bicarbonate—($NaHCO_3$)/0.0024 M sodium carbonate (Na_2CO_3); flow rate—3.0 mL/min; injection volume—50 μL; full-scale setting—10 s; and system pressure—4.69 MPa (680 psi). Quantitation is performed by using peak height measurements.

- *Argentometric-Turbidimetric Procedure*—To a 20-mL aliquot of sample is added 1 mL of concentrated nitric acid and five drops of a 5% solution of silver nitrate. The sample is vortex-mixed for 3 min and the turbidity of the silver chloride suspension is read on a Fisher Model DRT-150 turbidimeter, over a range of 0 to 300 nephelometric turbidity units (NTU).

• *ASTM Colorimetric Determination*—This method relies on the ASTM Tests for chloride ion in water (D 512). To a 10-mL aliquot of sample is added 2 mL of ferrous ammonium sulfate solution and 1 mL of mercuric thiocyanate in methanol solution. The sample is mixed and color allowed to develop for two minutes. The samples can then be visually compared with standards in Nessler tubes or read on a Perkin Elmer Model 550 spectrophotometer at a wavelength of 463 nm.

Calibration

Calibration is performed by using inorganic chloride standards over a range of 0 to 10 ppm chloride. The standards are prepared in the standard anion eluent.

Calculation

Concentration C of TOX in micrograms per gram (ppm by weight) as chlorine is given by:

$$C = \frac{A - B}{W} \times 125$$

where

A = ppm chloride measured in the extract,
B = ppm chloride measured in the preparation blank, and
W = weight of the sample, in grams.

For a quick spot test, the response of a 1.0-ppm chloride standard can be considered to be approximately equivalent to 50 ppm PCBs. After the ppm chloride in the preparation blank is subtracted, if the ppm chloride in a sample extract exceeds the response of a 1-ppm standard, then the waste in question may contain more than 50 ppm PCBs.

Safety Concerns

The Parr oxygen bomb is operated under pressure, and the instructions provided for operation of the bomb must be carefully followed [*12*]. A safety shield must be in place in front of the bomb before ignition. An ignition log must be kept, and after 200 firings the bomb must be returned to Parr for inspection and maintenance.

Experimental Work

Validation of Method

The method was validated by analyzing a large number of spiked samples of two standard wastes. One waste was an alkaline mix containing 13 different inorganic hydroxide sludges and cyanides and containing 0.54% organic carbon. The other waste was acidic, containing pickle liquor, hydrochloric acid, and about 10% organic carbon.

The following five spiking solutions were used in the validation:

- PCB spiking solution—19 880 mg/L Arochlor 1254, equivalent to 10 740 mg/L organic chlorine.
- Acid-extractable, base/neutral–extractable organic spiking solution—total equivalent organic halogen content, 2358 mg/L chlorine. The following compounds were used in the spiking solution:

bis(2-chloroethyl) ether	2-chlorophenol
1,4-dichlorobenzene	2,4-dichlorophenol
bis(2-chloroisopropyl) ether	4-chloro-3-methylphenol
2-chloronaphthalene	2,4,6-trichlorophenol
1,3-dichlorobenzene	pentachlorophenol
hexachlorobenzene	

- Volatile spiking solution—total equivalent organic halogen content 3066 mg/L chlorine. The following compounds were used in this spiking solution:

  trans-1,2-dichloroethylene
  carbon tetrachloride
  1,2-dichloropropane
  trichloroethylene
  1,1,2-trichloroethane
  2-chloroethyl vinyl ether
  tetrachloroethylene
  chlorobenzene

- Pesticide spiking solution—total equivalent organic halogen content 2226 mg/L chlorine. The following compounds were used in this spiking solution:

aldrin	endosulfan II	heptachlor epoxide
dieldrin	endosulfan sulfate	alpha BHC
p,p'-DDT	endrin	beta BHC
p,p'-DDE	endrin aldehyde	gamma BHC
p,p'-DDD	heptachlor	delta BHC
endosulfan I		

- Brominated spiking solution—total equivalent organic halogen content 2100 mg/L as chlorine. The following compounds were used in this spiking solution:

- bromoform
- 2-bromophenylphenyl ether

Known aliquots of these spiking solutions were added to portions of each standard waste. The samples prepared are summarized in Table 1.

Validation Results and Discussion

The results of the validation experiment are summarized in Tables 2, 3, 4, and 5, along with standard deviations and 95% confidence intervals. The confidence intervals were calculated using Student's t table for ($n - 1$) degrees of freedom where n is the number of data points [13].

Control charts for TOX results on PCB spikes, TOX results on the alkaline waste, and TOX results on the acidic waste are illustrated in Figs. 1, 2, and 3, respectively. The indicated control limits are based on two standard deviations for the upper control limit (UCL) and lower control limit (LCL) about

TABLE 1—*Samples analyzed in validation study.*

Organic Halogenated Compound Class	Range of Spike, ppm[a]	Acidic Waste	Alkaline Waste
PCBs	45–55	22	20
	90–110	10	5
	230–270	5	7
	450–500	5	12
Acid extractables	10–20	2	2
Base/neutral extractables	25–35	2	2
	40–60	2	2
	130–170	2	1
	250–350	1	1
Pesticides	7–12	2	2
	15–20	4	2
	70–80	2	2
	150–200	4	2
Volatiles	7–12	2	2
	20–30	2	2
	80–120	4	2
	180–220	2	2
Brominated organics	7–12	2	2
	40–60	1	1
	80–120	2	2
Total		78	73

[a]ppm = micrograms of chloride per gram of sample analyzed.

TABLE 2—*Summary of TOX results on alkaline waste.*

Organic Halogenated Compound Class	Range of Spike, ppm[a]	Determina- tions	Average Recovery, %	Standard Deviation	95% Confidence Interval
PCBs	45–55	20	94.9	17.9	23.7
PCBs	100–500	24	92.9	11.7	15.4
Acid extractables, base/neutral extractables	15–400	8	94.9	11.9	16.6
Volatiles	10–400	8	95.3	17.1	23.8
Pesticides	10–300	8	103	14.7	20.5
Brominated organics	10–100	5	101	14.6	21.5
Overall	…	73	95.6	14.6	18.9

[a]ppm = micrograms of chloride per gram of sample analyzed.

TABLE 3—*Summary of TOX results on acidic waste.*

Organic Halogenated Compound Class	Range of Spike, ppm[a]	Determina- tions	Average Recovery, %	Standard Deviation	95% Confidence Interval
PCBs	45–55	22	96.8	21.7	28.7
PCBs	100–500	20	101	9.9	13.1
Acid extractables, base/neutral extractables	15–400	9	94.2	14.0	19.4
Volatiles	10–400	10	99.5	17.4	23.9
Pesticides	10–300	12	97.8	16.5	22.4
Brominated organics	10–100	5	97.4	18.0	26.6
Overall	…	78	98.3	16.4	21.2

[a]ppm = micrograms of chloride per gram of sample analyzed.

TABLE 4—*Total organic halogen results—summary of all data.*

Organic Halogenated Compound Class	Range of Spike, ppm[a]	Determina- tions	Average Recovery, %	Standard Deviation	95% Confidence Interval
PCBs	45–55	42	95.9	19.8	25.8
PCBs	100–500	44	96.9	11.7	15.2
Acid extractables, base/neutral extractables	15–400	17	94.5	12.6	16.8
Volatiles	10–400	18	97.9	16.8	22.3
Pesticides	10–300	20	99.8	15.6	20.7
Brominated organics	10–100	10	99.4	15.6	21.4
Overall	…	151	97.0	15.6	20.1

[a]ppm = micrograms of chloride per gram of sample analyzed.

TABLE 5—*Comparison of three quantitation techniques.*

Parameter	Ion Chromatography	Argentometric- Turbidimetric	Colorimetric ASTM D 512
Average recovery, %	95.2	94.5	98
Standard deviation	15.7	19.6	11.5
95% confidence interval	20.3	25.3	14.8

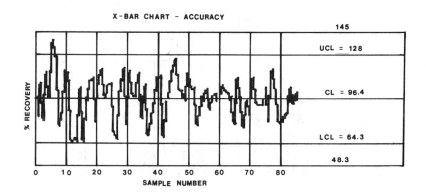

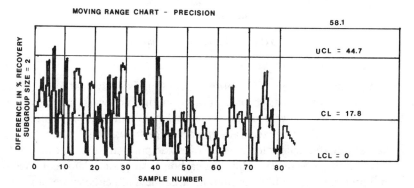

FIG. 1—*TOX control charts for PCB-spiked samples.*

the average or center line (CL). Precision and accuracy statements calculated from the moving range and X-bar control charts, respectively, are presented in Table 6. The moving range control charts are based on subgroup intervals of two [13].

The TOX results for PCB spikes presented in Tables 2, 3, 4, and 6 are calculated over two separate ranges. The 50 ppm level is of significance as the enforcement limit of the Toxic Substances Control Act (TSCA) [2]. The pre-

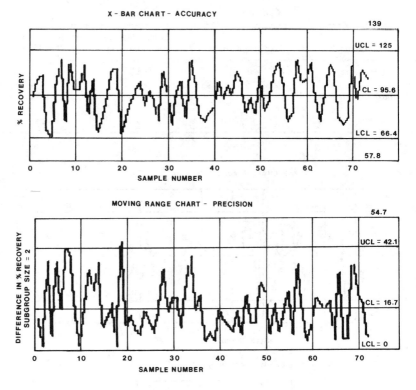

FIG. 2—*TOX control charts for alkaline waste.*

cision and accuracy at the TSCA enforcement limit is equivalent to the inter-laboratory variability of gas chromatography methods on waste samples [*14*].

Excellent recoveries were obtained, even for volatile priority pollutants. The method has equivalent accuracy and precision for essentially all classes of organic halogenated compounds on EPA's toxic pollutant list, in contrast to methods utilizing preconcentration via carbon absorption. Equivalent accuracy and precision were obtained on both alkaline and acidic waste samples. The acidic waste had a background of 10% organic carbon before spiking, which did not appear to interfere with the determination of TOX.

Table 5 presents a comparison of the three quantitation techniques used. The turbidimetric procedure is subject to greater variability and gave lower recoveries for brominated compounds, as well as for samples containing higher levels of TOX.

The colorimetric procedure gave better results for higher spikes and is as rapid as the turbidimetric procedure. In addition, the colorimetric method can be used as a spot test with the Nessler Tube, to screen out waste samples exceeding the 50-ppm TSCA limit for PCBs.

Ion chromatography requires a greater amount of time, but permits the

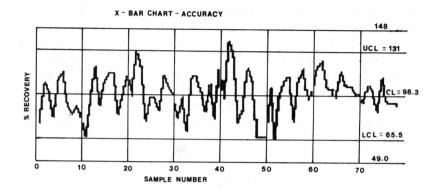

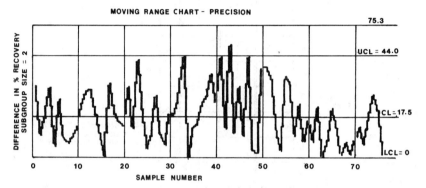

FIG. 3— *TOX control charts for acidic waste.*

TABLE 6—*Intralaboratory precision and accuracy statements
from control charts.*

Constituent	Range, ppm[a]	Determinations	X-Bar Chart 95% CI[b] Accuracy	Moving Range Chart 95% CI[b] Precision
PCBs	45–55	42	±25.4	±22.5
PCBs	100–500	44	±15.0	±12.1
Acid extractables, base/neutral extractables	15–400	17	±16.9	±16.3
Pesticides	10–300	20	±20.7	±16.4
Volatiles	10–400	18	±22.2	±21.3
Brominated organics	10–100	10	±21.5	±21.3
Alkaline waste	...	73	±19.0	±16.5
Acidic waste	...	78	±21.3	±17.2
Overall	...	151	±20.0	±17.0

[a]ppm = micrograms of chloride per gram of sample analyzed.
[b]Confidence interval.

separation of brominated and chlorinated response in a TOX analysis. A sample chromatogram is illustrated in Fig. 4.

A histograph of all TOX results is presented in Fig. 5. The interval widths were computed in accordance with Sturge's rule [13]. The distribution is almost bimodal, the cluster of data at lower recoveries being due to the turbidimetric procedure.

The detection limit of the method is about 3.5 ppm because of difficulties in achieving lower results on blanks. Great care is required as chloride con-

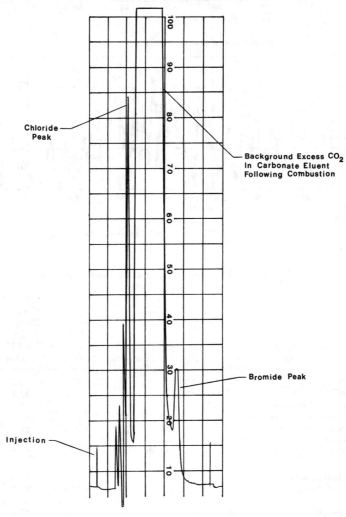

FIG. 4—*Ion chromatograph of TOX analysis on sample containing chlorinated and brominated organics (see text for chromatography conditions).*

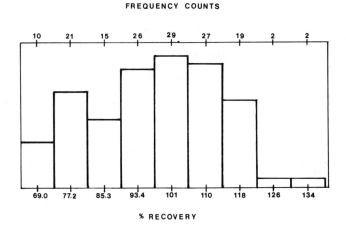

FREQUENCY COUNTS

FIG. 5—*Histograph of all TOX spike recoveries.*

tamination can easily occur during sample preparation. The limit of quantitation of the method is 12 ppm. These limits are variable, depending upon the sample size taken, which in turn depends upon the sample matrix and the partitioning volumes.

Conclusions

It is possible to analyze TOX in industrial waste samples to a limit of about 3.5 ppm as inorganic chloride. The precision of the method is ±17% and the accuracy is ±20%. The recovery appears to be equivalent for all halogenated priority pollutant classes. The analysis is rapid, requiring less than 30 min, and can be used to screen waste for PCBs at the 50 ppm Toxic Substance Control Act limit.

It is conceivable that this method can be used for water samples, particularly in RCRA groundwater monitoring for samples with high inorganic chloride, for which methods using activated carbon columns are difficult. A 1-L sample could be extracted with 10 mL of *iso*octane to give a limit of detection of approximately 0.035 ppm.

References

[*1*] Verschueren, K., *Handbook of Environmental Data on Organic Chemicals*, 2d ed., Van Nostrand, New York, 1983.
[*2*] *Federal Register*, Vol. 44, No. 106, 31 May 1979.
[*3*] Dohrman—Division of Envirotech, Application Literature, Santa Clara, CA.
[*4*] Takahashi, Y., "Determination of the Total Concentration of Organic Halides in Solid, Liquid, or Gaseous Samples," *Chemical Abstracts*, Vol. 92, June 1980, p. 226178a.
[*5*] OI Corp., Application Literature, College Station, TX.

[6] Calusaru, A., Cleper, M., and Domnisteanu, G., *Chemical Abstracts,* Vol. 92, April 1980, p. 121301g.

[7] Sienkows, E., *Chemical Abstracts,* Vol. 96, February 1982, p. 62349m.

[8] Burns, D. T., *Analyst,* Vol. 108, 1982, p. 452.

[9] Smith, F., McMurtrie, A., and Galbraith, H., *Microchemical Journal,* Vol. 22, 1977, pp. 45–49.

[10] Abramoyan, A., Tevosyan, A., and Megroyan, R., *Chemical Abstracts,* Vol. 87, July 1977, p. 15547d.

[11] Yamada, K. and Arimoto, H., *Chemical Abstracts,* Vol. 87, July 1977, p. 33299d.

[12] "Parr Oxygen Combustion Bomb Operation Manual," Parr Instrument Co., Moline, Illinois, 1982.

[13] Grant, E. L., *Statistical Quality Control,* McGraw Hill, New York, 1952.

[14] Ziegler, W. J., *Abstracts of the Pittsburgh Conference,* Paper No. 621, Pittsburgh Conference and Exposition on Analytical Chemistry and Applied Spectroscopy, Pittsburgh, PA, March 1982.

David Friedman[1]

An Overview of Selected EPA RCRA Test Method Development and Evaluation Activities

REFERENCE: Friedman, D., **"An Overview of Selected EPA RCRA Test Method Development and Evaluation Activities,"** *Hazardous and Industrial Solid Waste Testing: Fourth Symposium, ASTM STP 886,* J. K. Petros, Jr., W. J. Lacy, and R. A. Conway, Eds., American Society for Testing and Materials, Philadelphia, 1986, pp. 77–84.

ABSTRACT: This paper will present an overview of the U.S. Environmental Protection Agency's current and near-term future test method development and evaluation activities in light of the agency's regulatory activities under the Resource Conservation and Recovery Act. The paper will discuss the work in terms of the regulatory purpose to be served, the approach to be taken in conducting the studies, progress made to date, the responsible research organization, and the anticipated schedule.

Among the specific areas discussed are evaluation of current and proposed SW-846 (Test Methods for Evaluating Solid Waste) methods, development and evaluation of less costly methods for determining the presence and concentration of the Appendix VIII (40 CFR 261) hazardous constituents in groundwater, development and evaluation of sampling methods to determine the concentration of organic compounds in ambient air and stack emissions, and some aspects of the agency's activities in developing procedures to determine the leaching potential of waste materials placed in various environmental settings.

KEY WORDS: solid wastes, hazardous wastes, testing methodology, RCRA, sampling, hazard evaluation, leachates, mobility

Testing serves a very important function in the regulatory program mandated by the Resource Conservation and Recovery Act (RCRA). Accurate, cost-effective testing methods are important to the identification of hazardous wastes as well as to both the hazardous waste facility permitting and monitoring programs and the RCRA enforcement program. This paper will present an overview of selected current and near-term future RCRA test method development and evaluation activities of the U.S. Environmental

---

[1]Manager, Waste Analysis Program, Office of Solid Waste, U.S. Environmental Protection Agency, Washington, DC 20460.

Protection Agency (EPA). The paper will discuss the work in terms of the regulatory purpose to be served, the approach taken in conducting the studies, and progress made to date, concentrating on agency efforts to overcome the limitations of current methods and to develop additional ones.

Among the specific areas that will be discussed here are evaluation of current and proposed EPA Test Methods for Evaluating Solid Waste (SW-846), development and evaluation of less costly methods for determining the presence and concentration of hazardous constituents in groundwater [as specified in Appendix VIII (40 CFR 261)], and development and evaluation of sampling methods to determine the concentration of organic compounds in ambient air and stack emissions. In addition, I will briefly touch on some aspects of the agency's activities in developing procedures to determine the leaching potential of waste materials placed in various environmental settings.

SW-846 Methods Evaluation/Improvement

EPA Test Methods for Evaluating Solid Waste, otherwise known as SW-846, serves as a compilation of those testing methods that have been critically reviewed and approved by the agency for use in the RCRA program. Data collected using these methods are accepted by the agency without submitters having to demonstrate that the methods used are capable of yielding accurate data. Given the important role that they play in implementing the RCRA regulatory program, EPA has thus undertaken a major effort to determine the accuracy and precision of the methods.

Table 1 presents a list of some of the methods currently being evaluated, the anticipated date for the study to be completed, and the EPA organization that is responsible for conducting the study. These studies include challenging the methods with a variety of materials in order to uncover those matrices for which the method is not applicable. Once a method survives this single-laboratory evaluation, a multilaboratory collaborative study will be conducted to determine its precision and accuracy. If serious problems are uncovered during the single-laboratory evaluation, the problems will be solved and the method modified before a multilaboratory study is conducted.

In addition to the studies shown in Table 1, a number of other test methods are being evaluated as part of the process of developing new test methods.

Groundwater Monitoring

Current regulations promulgated under RCRA require owners and operators of hazardous waste facilities found to be contaminating the groundwater beneath the facility to determine the presence and concentration of the Appendix VIII constituents in the groundwater. Facility owners and operators

TABLE 1—*SW-846 methodology evaluation activities.*

Method No.	Method[a]	Completion Date	Responsible Organization
1.2.1.13	volatile organic sampling train (VOST)	1987	EMSL-RTP
1.2.1.9	semivolatile organic sampling train (MM5)	1987	EMSL-RTP
1010	flash point by Pensky-Martens closed cup	9/84	EMSL-Cinn
1020	flash point by Setaflash closed cup	9/84	EMSL-Cinn
1110	corrosivity by weight loss	12/84	EMSL-Cinn
1120	corrosivity by polarization	12/84	EMSL-Cinn
3010-60	metals dissolution techniques	3/85	EMSL-Cinn
XXXX	purgeable organics using tetraglyme	4/84	EMSL-Cinn
XXXX	methylene chloride extraction of organics	4/84	EMSL-Cinn
6010	inductively coupled plasma	1987	EMSL-Cinn
8030	GC/FID—acrolein, acrylonitrile, acetonitrile	10/84	EMSL-Cinn
8090	cleanup and GC for nitoaromatics and cyclic ketones	10/84	EMSL-Cinn
8150	GC for chlorinated herbicides	1987	EMSL-Cinn
8280	low/resolution GC/MS for "dioxins"	1985	EMSL-LV
8310	HPLC for polynuclear aromatic hydrocarbons	1986	EMSL-Cinn
8610	aromatics by ultraviolet absorption	9/84	EMSL-Cinn
8620	gas chromatographic method for N/P organics	9/84	EMSL-Cinn
9090	membrane liner compatability	9/85	OSW
	Ames Mutagenicity	9/86	EMSL-LV
	Daphnia magna chronic toxicity	1/85	EMSL-LV

[a]GC = gas chromatography, FID = flame ionization detection; MS = mass spectrometry; HPLC = high-performance liquid chromatography; and N/P = nitrogen/phosphorus.
[b]EMSL-RTP is the Environmental Monitoring and Support Laboratory in Research Triangle Park, NC; EMSL-Cinn is in Cinncinnati, OH; EMSL-LV is in Las Vegas, NV; and OSW is the EPA's Office of Solid Waste.

have complained that analyzing for each and every one of these compounds is exceedingly costly.

In order to improve the cost-effectiveness of the regulations, the EPA has undertaken a number of studies. Among these are the development and evaluation of the Hierarchical Analysis Protocol to screen groundwater samples and determine which types or groups of compounds are not present in order to reduce the number of analyses that have to be performed. This protocol is the subject of another paper in this volume [1] and will not be further discussed here.

A second effort is underway to evaluate a test scheme designed to identify and quantify the Appendix VIII compounds using the fewest number of procedures. The scheme consists of a number of SW-846 multicontaminant test methods selected for their broad coverage and applicability to groundwater analysis. Table 2 is a listing of the methods in the scheme.

The RCRA regulations do not require permittees to use a specific set of methods for conducting the required groundwater monitoring. However, the

TABLE 2—*Groundwater analysis scheme.*[a]

1. Volatile organics by GC/MS (purge and trap) SW-846 Method 8240

2. Waste-soluble volatile organics by GC/MS (direct aqueous injection)

3. Semivolatile organics by GC/MS SW-846 Method 8250 or 8270

4. Herbicides and pesticides by gas chromatography SW-846 Method 8140 or 8150

5. Selected organic compounds by HPLC SW-846 Method 8320 or 8330

6. Inorganics (metallic) by AA and ICAP SW-846 Method 6010 or 7000 series

7. Miscellaneous—sulfides, total cyanides, and TOX SW-846 Methods 9010, 9020, and 9030

[a]GC/MS = gas chromatography/mass spectrometry; HPLC = high-performance liquid chromatography; AA = atomic absorption spectroscopy; ICAP = inductively coupled argon plasma; and TOX = total organic halogens.

agency has received many requests for advice on what is the least costly approach to use in conducting the analysis. Since the scheme just described appears to offer an excellent means of accomplishing the required monitoring, the EPA has undertaken a program to determine the sensitivity and accuracy of the scheme using various types of groundwater.

As the first step in the evaluation, the detection limit for each Appendix VIII compound will be determined in distilled water. The single-laboratory precision will then be determined at the detection limit and at two higher concentrations in both distilled water and representative groundwaters. Based on the results of this study a decision will be made as to whether or not the scheme offers an acceptable means of conducting the required testing. If the decision is favorable, any shortcomings in the scheme that are uncovered will then be studied in order that they may be overcome. The initial single-laboratory evaluation was completed in August 1984. Notice of the availability of the data will be made in the *Federal Register*.

Ambient Air/Stack Monitoring

A key aspect of the RCRA hazardous waste management regulations is the requirement that the efficacy of the treatment process must be demonstrated before a permit can be issued to the facility operator.

In determining the efficacy of organic compound destruction in incinerators and boilers, it is critical that an accurate sample of the emissions be obtained. Toward this goal, the EPA has developed two techniques for conducting such sampling. These are the volatile organic sampling train (VOST) for the low molecular–weight purgeable compounds and the Modified Method 5 method for the nonpurgeable gas chromatographable compounds. Another paper in this volume describes the development of the VOST. The present work to evaluate the precision and accuracy of the VOST and MM5 is being performed under the direction of the Environmental Monitoring and Support Laboratory (EMSL) in Research Triangle Park, North Carolina. The performance of the methods will be evaluated through a combination of laboratory studies and field audits. As part of the effort a series of audit cylinders were prepared for use in assessing the accuracy of constituent measurements during trial burns. The cylinders contained a mixture of

carbon tetrachloride,
chloroform,
perchloroethylene,
vinyl chloride, and
benzene.

All the substances were present at parts-per-billion levels and the accuracy of the VOST was studied at four laboratories involved in incinerator efficacy evaluations.

Preliminary results indicate that within the limit of detection (100 ng/m^3) the VOST method can be used to sample for volatile hazardous constituents. The average percentage of accuracy found during the four-laboratory study was well within the range of +50 to −100% (Table 3). Given that it included errors due to both the sampling and analytical steps, the accuracy measurement was felt to be acceptable for the intended use. When the VOST cartridges were spiked and then analyzed, accuracy errors approximating that found for the total procedure were found. This indicates that analysis of the resin cartridges may be the limiting factor in some cases.

TABLE 3—*Summary of VOST audit results.*

	A	B	C	D	Average
	\multicolumn{5}{c}{Error, %}				
	\multicolumn{5}{c}{Laboratory}				
Both sampling and analysis					
Carbon tetrachloride	−50	27	10	−14	−7
Chloroform	0	28	20	−20	−7
Perchloroethylene	17	34	NA	12	21
Vinyl chloride	−50	−66	NA	−10	−42
Benzene	31	32	−9	6	15
Cartridge analysis only					
Carbon tetrachloride	−40	11	13	−7	−6
Chloroform	6	15	32	−20	8
Perchloroethylene	31	15	NA	19	22
Vinyl chloride	−47	−53	NA	15	−28
Benzene	46	25	17	0	22

Emissions of volatile organic toxicants from hazardous waste landfills, surface impoundments, and land treatment facilities is believed by Congress to represent a significant route of environmental contamination [2]. To bring this problem under control, the EPA is formulating emission control regulations and standards for such facilities. In order to properly gather the data necessary to develop such regulations and standards and to implement the standards once developed, ambient air monitoring methodology is under development by the agency.

Two methods currently under study for determining the presence and concentration of organic compounds are the Passive Air Samplers and the Tunable Atomic Line Molecular Spectrometer (TALMS). The Passive Air Samplers are badge-like devices that can either be placed on persons living or working near or at the facility or, more usually, attached to posts about the facility. After several hours or days of sampling the devices are collected and sent to a laboratory for analysis. This yields an average air concentration over the sampling period.

The TALMS is felt to offer a sensitive and selective method of conducting real-time monitoring of organic compounds in ambient air or stack emissions. TALMS relies on the splitting induced in an atomic vibrational spectrum line by a magnetic field and, after polarization of the spectral line, a measurement of the changes resulting from passage through a sample of the compound to be measured. This technique, which was invented by Dr. Hadeishi at the Lawrence Berkeley Laboratory (LBL), is being studied and developed into a usable tool by researchers at the EMSL Research Triangle Park (RTP) laboratory. A portable monitor, optimized for measurement of benzene, was constructed by LBL and delivered to the EPA's EMSL-RTP laboratory in December 1984. Tests with benzene at 184 to 189 nm gave approximate detection limits of 250 ppb. Many changes still appear necessary before this technique can be considered a useable analysis method.

Our problem that agency researchers have become more and more aware of in recent months is the limitations of the currently available Tenax resin collection medium, on which many of our sampling methods are based. In order to overcome these limitations, a program is underway at the EMSL-RTP laboratory to synthesize and evaluate new types of resins for use in organic compound sampling. We hope that this program will lead to the availability of improved resins within the next few years.

Leaching Methods

Present hazardous waste identification characteristics generally do not include a means of adequately identifying those wastes that pose a problem due to the presence of organic toxicants. The Extraction Procedure (EP) Toxicity characteristic primarily deals with inorganic materials. The characteristic is deficient both because the extraction procedure itself is not aggressive enough toward organic compounds and because the list of required analytes is limited to the compounds listed in the National Interim Preliminary Drinking Water Standards. To correct this deficiency, the agency is developing a new extraction procedure to assess better the mobility of organic chemicals codisposed with sanitary refuse and is also expanding the hazardous waste identification characteristic to include organic toxicants. Since two other papers [3,4] in this volume describe the work being done in this area in much greater detail than I could hope to cover in this overview I will not discuss this work any further.

Instead, I will discuss two other aspects of the mobility measurement problem: how to assess the long-term leaching potential of solidified or, as they are also known, stabilized wastes and how to assess the leaching potential of waste materials containing pyritic sulfur or other acid-forming species.

Congressional committees during their recent deliberations on the RCRA reauthorization have taken the position that reliance on land disposal, particularly landfills and surface impoundments, should be the least favored method for managing hazardous waste. Congress is therefore contemplating some restrictions on land disposal of hazardous waste. Inorganic toxicant-containing wastes remain hazardous forever. Thus, for such wastes not to present a hazard when disposed of on land, the toxicants would have to exhibit a low potential for migration. This might require banning the wastes from land disposal unless the mobilities of the toxic species are low enough that the potential environmental insult is not sufficient to justify banning the waste from land disposal.

The Extraction Procedure and its offspring (Multiple Extraction Procedure and Oily Waste Extraction Procedure) are designed to determine the leaching potential of wastes under specific situations. They are designed to model relatively worst-case conditions for the purpose of determining whether management of a hazardous waste is required. They may therefore not be suitable for evaluating the likely consequences of management under the conditions prevailing at a hazardous waste management facility. Test methods for assessing the site-specific consequences of disposal to determine whether a given method of land management is suitable for a given waste are thus needed. Alternatively, the methodology would serve as a means of determining when a hazardous waste deemed unsuitable for land management has been sufficiently stabilized or otherwise treated to permit its land disposal.

Among the factors that may not be adequately taken into account by the current tests are the physical and chemical stability of the waste, site-specific leaching medium, landfill design, and weather conditions that might prevail at the facility.

In order to understand better what factors must be modeled by such tests and to develop new or improved tests to use in evaluating solidified wastes, the EPA is involved in two studies. One of these is a study being conducted by the EPA's Municipal Environmental Research Laboratory (MERL) in Cincinnati at the Army Corps of Engineers Waterways Experiment Station in Vicksburg, Mississippi. The second is a study sponsored by the Canadian government aimed at evaluating available tests and, if necessary, developing new leaching methods.

The MERL study is designed to evaluate the effect of various materials from the standpoint of their effect on the solidification reactions and the chemical stability of the product of the solidification process.

The Canadian study, under the direction of Environment Canada and researchers at the Alberta Environmental Center and the Waste Water Technology Centre in Burlington, Ontario will, using wastes solidified by a number of commercially important processes, evaluate a variety of properties of the wastes and test procedures which may be suitable for their measurement.

It is the intention of both the EPA and our Canadian colleagues that these studies be conducted with the participation of both governments and with the active involvement and assistance of the affected industry.

Substantial quantities of toxic metal–bearing wastes are disposed of annually in the United States. Much of this waste, especially that derived from mining, contains materials such as pyritic sulfur, which under certain environmental conditions can oxidize to form acids. The degree of hazard posed by land management of such wastes is a function both of the type and amount of metal present in the wastes and of the waste's potential for forming acids that greatly increase the mobility of metals. Current test procedures do not offer a means of determining such acid formation.

References

[1] Poppiti, J. A. and Friedman, D., "Development of an Analytical Protocol for Groundwater Screening of Hazardous Waste Constituents," paper presented at the Fourth Symposium on Hazardous and Industrial Solid Waste Testing, Arlington, VA, 2-4 May 1984.
[2] Resource Conservation and Recovery Act, Section 3004(n), 1976, as amended 8 Nov. 1984.
[3] Kimmell, T. A. and Friedman, D., "Model Assumptions and Rationale Behind the Development of EP-III," *Hazardous and Industrial Solid Waste Testing: Fourth Symposium, ASTM STP 886*, J. K. Petros, Jr., W. J. Lacy, and R. A. Conway, Eds., American Society for Testing and Materials, Philadelphia, 1986, pp. 36-53.
[4] Francis, C. W., Maskarinec, M. P., and Goyert, J. C., "A Laboratory Extraction Method to Simulate Codisposal of Solid Wastes in Municipal Waste Landfills," *Hazardous and Industrial Solid Waste Testing: Fourth Symposium, ASTM STP 886*, J. K. Petros, Jr., W. J. Lacy, and R. A. Conway, Eds., American Society for Testing and Materials, Philadelphia, 1986, pp. 15-35.

Larry C. Michael, [1] *Rebecca L. Perritt,* [1] *Edo D. Pellizzari,* [1] *and Florence Richardson* [2]

Laboratory Evaluation of Test Procedures for Use in the RCRA Hazardous Waste Ignitability Characteristic

REFERENCE: Michael, L. C., Perritt, R. L., Pellizzari, E. D., and Richardson, F., **"Laboratory Evaluation of Test Procedures for Use in the RCRA Hazardous Waste Ignitability Characteristic,"** *Hazardous and Industrial Solid Waste Testing: Fourth Symposium, ASTM STP 886,* J. K. Petros, Jr., W. J. Lacy, and R. A. Conway, Eds., American Society for Testing and Materials, Philadelphia, 1986, pp. 85–105.

ABSTRACT: The purpose of this research project was to conduct a single-laboratory evaluation of three methods to determine and characterize the potential hazard posed by flammable materials disposed in landfills. The test procedures—(1) radiant heat ignition, (2) linear flame propagation and (3) extinguishability with water tests—were conducted in six replicates on 26 actual and simulated wastes, including three reference standards. Precision of the replicate analyses was approximately 30% and was consistent between actual samples and reference standards. Evaluation of the data from the three tests revealed a significant correlation between the radiant heat and flame propagation tests and hence a possible redundancy based on the samples analyzed. The data show that the ignitability of a waste is directly related to the rate of flame propagation and the volume of water necessary to extinguish the flame, and is inversely related to the time to ignition under a radiant heat source. These relationships were combined into an "ignitability factor" for assessing the overall ignition tendency for the samples tested.

KEY WORDS: ignitability, test procedures, radiant heat, flame propagation, extinguishability, hazardous wastes

The Resource Conservation and Recovery Act (RCRA), Subtitle C, was enacted in 1976 to assure that hazardous wastes are managed so that they pose

[1] Research analytical chemist, research statistician, and vice-president, respectively, Analytical and Chemical Sciences, Research Triangle Institute, Research Triangle Park, NC 27709.
[2] Chemist, Office of Solid Waste, U.S. Environmental Protection Agency, Washington, DC 20460.

no danger to human health and the environment. In this regard, ignitability was identified as a waste characteristic. The U.S. Environmental Protection Agency has defined an ignitable solid in the following terms:

> A solid waste is a hazardous waste if a representative sample of the waste: . . . is liable to cause fires through friction, absorption of moisture, spontaneous chemical changes, or retained heat from manufacturing or processing, or when ignited burns so vigorously and persistently as to create a hazard during its management.

The narrative definition was intended for temporary use until a quantitative ignitability test for solid waste could be developed and validated. In the absence of a quantitative test, waste management procedures are based on knowledge about the waste or its components or on experience with similar waste materials.

Although there are currently several methods for assessing the ignitability of solids, all but one of these procedures measures flame propagation exclusively [1-6]. Because of the complex nature of solid waste, no single test can provide an adequate and comprehensive evaluation of the ignitability of solid waste materials. The single-variant method [1-7] is actually a composite of three independently applied procedures and is the subject of this paper. The three components—(1) radiant heat ignition, (2) rate of flame propagation, and (3) difficulty in extinguishing with water—were intended to simulate the various ignition characteristics of solid waste.

The Radiant Heat Ignition Test was designed to measure the tendency of a solid waste to autoignite upon exposure to radiant heat. Once the waste is burning, regardless of the ignition source, the rate of advancement of the flame front is of concern not only because of the combustion of additional waste but of adjacent materials as well. The Flame Propagation Test was designed to measure the rate at which the flame from a burning waste would spread. The persistence of the flame from a burning waste and the ease of extinguishing the flame defines the threat it poses to igniting other materials. The Water Extinguishing Test was designed to measure the difficulty of extinguishing the flame with water.

The objective of the research reported here was to further refine and evaluate these three ignitability protocols—radiant heat ignition, flame propagation, and water extinguishing—using a variety of authentic hazardous waste samples. Considerable effort was devoted to minimizing errors in analyst judgement by carefully controlling test conditions and automating end-point detection.

Procedure

Sample Acquisition

The potentially ignitable solid waste samples to be used in this study comprised a wide range of physical and chemical properties so as to present a challenge in the administration of the three tests and in the interpretation of their results. An attempt was made to acquire samples that covered the range of the following textures and ignitability characteristics:

- Pastes—both resinous and of various viscosities.
- Gums—elastomeric and amorphous solids.
- Solids—particles of irregular size and hardness and particles of uniform size (free-flowing and non-free-flowing).
- Liquids—viscous and nonviscous.
- Ignitable by radiant heat.
- Difficult to extinguish.
- Possibly self-igniting.
- Slow-burning versus fast-burning.

Samples were therefore acquired on the basis of diversity of texture and ignition tendency, as well as availability. Additional reference materials were acquired to provide performance information on the tests and to provide test results on materials that are familiar to the fire safety community. Telephone contacts, followed by formal written requests, were made with numerous companies to acquire samples of solid waste.

Sample Descriptions

The following samples were utilized in the evaluations of the test procedures. Reference materials were included to provide test performance evaluation. Ambiguous sample names were deliberately assigned to ensure source confidentiality.

Wood Wool Excelsior—This material was selected as a reference material because of its use by Underwriter's Laboratories as a standard for Class A fire extinguisher evaluation. As its name would imply, it consisted of dried hardwood fibers, approximately 1 mm in diameter. The fibers were cut into 3/4-in. segments before testing.

Textile Lint—Floor sweepings were obtained from a textile manufacturer. It is unknown whether the fiber was natural or synthetic. The sample was light gray in color and was the texture and consistency of the contents of a vacuum cleaner bag. This sample was tested as received.

WOKSR-1—This paint manufacturing waste was gray-brown in color and had the texture of putty. A faint but undefined solvent smell was detectable although no separated solvent was observed. The sample was thoroughly mixed with a spatula before testing.

WOM-1—This waste was similar in source and characteristics to WOKSR-1, except for a more fibrous texture. The sample was mixed with a spatula before testing.

WOL-CM—This paint manufacturing waste was blue in color and had a sandy texture. Its mixed solvent smell was considerably stronger than that of the previous two samples. It was thoroughly mixed with a spatula before testing.

Clay/Oil—This refinery waste consisted of oil adsorbed onto a clay substrate. The sample had a pungent, oily smell and a fine, granular texture. No sample preparation was required.

Paint Waste J1—This composite of paint wastes had an overall amorphous consistency and a strong odor of ethyl acetate. In order to maintain sample homogeneity, vigorous mixing was required before withdrawal of each aliquot for testing.

Paint Waste J2—This composite paint waste had the texture of paste and a very strong odor of ethyl acetate. The sample was mixed with a spatula.

Paint Waste J3—This sample was nearly identical to Paint Waste J1 except that it was slightly darker in color and had a slightly greater amount of separated solvent. As with other paint wastes, this sample required repeated mixing during aliquoting.

Paint Waste J4—Of all the paint wastes, this one was by far the most atypical in consistency and odor. The general appearance of this sample suggests that it was composed of dried paint flakes in an amorphous milieu which was immiscible with its liquid phase. Attempts to homogenize this sample resulted in a heterogeneous mixture which retained its amorphous consistency.

Coconut/Toluene—This waste was comprised of slightly discolored, shredded coconut which had been extracted with toluene and had retained a significant amount of the extracting solvent. The sample was thoroughly mixed with a spatula before testing.

Styrene Polymer—In this sample, partially polymerized styrene was mixed with vermiculite. A strong odor of styrene was evident in this elastomeric solid. Chunks of sample were sliced to 10 mm with a spatula before they were placed in the sample test containers.

Still Bottom—This waste was comprised of a gray, pasty solid and a brown liquid that separated on standing for an extended time period. The odor suggested a complex mixture of solvents, including pyridine. The solid and liquid phases were mixed well with a spatula before testing.

Petroleum Waste—For the most part, this sample was comprised of vermiculite. Chunks of solidified petroleum waste were separated from the absorbent and tested without further preparation. A faint, unidentified solvent odor was associated with this sample.

Coal/Xylene—Lignite coal (North Dakota Beulah/Zap) was ground to approximately 2.0 mm (No. 10 mesh) and combined with reagent grade xylenes to yield a 25% (weight/weight [w/w]) mixture of xylenes and coal. This sample was thoroughly mixed with a spatula before testing.

Red Oak/Kerosene—Red oak sawdust, obtained from a local sawmill, was dried for 24 h in an ambient environment and then further dried at 60°C for 4 h. A mixture of 20% (w/w) reagent grade kerosene and dried red oak sawdust was prepared and exhaustively mixed with a spatula to ensure that the sawdust was evenly wetted by the kerosene.

Paint Filter—This sample consisted of a dry, red-brown dust (65% by weight) and a separate, three-dimensional, paper-like mesh (35% by weight). The mesh was cut with scissors to fit the sample container. Approximately 1 g of dust was sprinkled on top of the mesh immediately before testing.

Oil Pads—These oil pads were 72% oil (unknown type) and 28% absorbent pad. The pad was multilayer and appeared to be composed of a fabric, not paper, material. Sample aliquots were cut with scissors to the dimensions of the test containers.

Fuel Oil Pads—These pads had been used in the field to clean up a fuel oil spill. The pads had the appearance, texture, and flexibility of a wet chamois. Sample aliquots were cut to fit the sample test containers.

Sludge Barge Waste—This sample had the appearance and consistency of spent automobile motor oil. Sample test containers were filled to a depth of ~2 mm with sample.

Waste Oils—This sample was identical to the sludge barge waste, with the exception that it had a slightly lower viscosity.

Lighter Fluid—Ronsonol;® poured in the test container to a depth of approximately 1 mm.

Sterile Cotton—Red Cross;® cut to fit the test container.

Pipe Tobacco—Sir Walter Raleigh;® poured to a depth of 10 mm in the test container.

Polyurethane Foam—Density and source unknown; presence of fire retardant unknown; flexible; cut to fit the test container.

Polystyrene—Extruded, rigid polystyrene insulation (Dow Chemical, Styrofoam® Type TG, Lot AH810910); presence of fire retardant unknown; cut to fit the test container.

Application of the Tests

The three ignitability tests were conducted precisely as described in the analytical protocols provided in Appendix I. The apparatus used and its parts description is shown in Fig. 1. Six replicate measurements of each test were performed on each sample for statistical evaluation of the data. The thermocouple responses for the radiant heat ignition and the linear flame propagation tests were recorded on a strip chart recorder; data from the water extinguishing test were recorded manually in a laboratory notebook.

Radiant Heat Ignition

After positioning the heat source to provide an element to sample surface distance of 60 mm and positioning the thermocouples 5 mm above the top of the sample container, the radiant heat source was heated to 230°C by adjusting the voltage to the heater. The sample container (an aluminum weighing pan 60 mm in diameter by 20 mm deep) was filled with sample to a depth of 10 mm and centered under the preheated radiant heat source. Ignition time

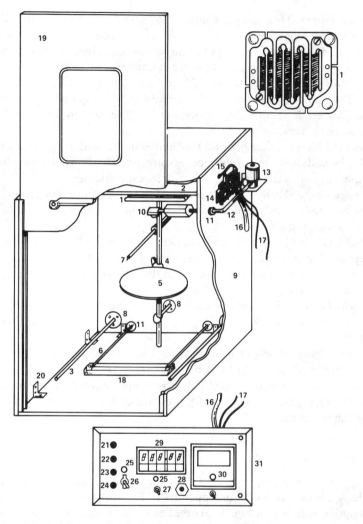

FIG. 1—*Ignitability test chamber and controller/sensor parts list (see Appendix II).*

was recorded by the thermocouple response on a strip chart recorder and calculated from the recorded distance and chart speed.

Rapidity of Flame Propagation

An aluminum foil trough, 32 by 25 by 250 mm or 32 by 6 by 250 mm, was filled with test material to a depth of 10 mm and the surface smoothed. The sample was ignited at one end with a methenamine tablet. This tablet is comprised of a combustible substance which provides a finite (~ 1 min) time for sample ignition to occur. The time required to burn between the two sensing thermocouples (190 mm) was recorded on a two-pen strip chart recorder. The linear flame propagation rate was computed in centimetres per minute.

Water Extinguishing

An aluminum pan, 114 mm in diameter and 10 mm deep, was filled with test material to a depth of 10 mm and the surface smoothed. The sample was ignited at the center with a methenamine tablet and allowed to burn until the entire surface was charred or until half the material present had burned. Water was applied from 60 mm above the sample by means of a stainless steel sprayer activated by a solenoid shut-off valve. The amount of water necessary to extinguish the fire was calculated from the flow rate and time the sprayer was activated.

Results and Discussion

Subjective Evaluation

Radiant Heat Ignition—A graphic display of the data in Fig. 2 illustrates the diversity of ignition times of the samples selected for this study. Ignition times ranged from 12 s to more than 300 s, with the responses of the reference standards occurring near midrange. In general, the ignition time tended to be significantly influenced by the presence of solvent in the sample. This observation, although subjective, was substantiated by the results of the application of this test to samples containing an obvious solvent component (such as paint wastes). Such samples tended to exhibit more rapid ignition than those lacking a solvent constituent. Heterogeneous samples whose liquid component was petroleum-based (moderate volatility fraction) ignited more slowly under the radiant heat source (for instance, fuel oil pad or sludge barge waste). Ultimately, with the exception of the wood wool excelsior, samples lacking a solvent or liquid petroleum component failed to ignite within 5 min.

Flame Propagation—Flame propagation rates, shown in Fig. 3, reveal that with only a few exceptions, samples tended to fall in one of three categories: (1) very fast propagation rates (> 10 m/min), (2) moderate propagation rates (< 500 mm/min) and (3) nonpropagating. Nonpropagating samples were designated NI (no ignition). Samples with fast propagation rates included, among others, the paint wastes and the toluene-containing coconut waste. Moderate propagation rates were observed for various matrix types and showed little regard for presence or absence of a flammable liquid component. Samples that failed to propagate included, among others, several oil-containing wastes and the textile lint.

Water Extinguishing—With only two exceptions, all of the samples that ignited with the methenamine tablet were extinguished with less than 6 mL of

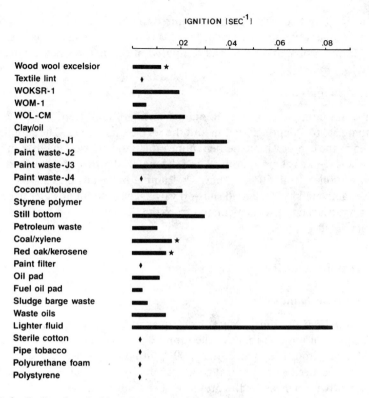

FIG. 2—*Radiant heat ignition of waste samples and reference materials. Star denotes reference standard; diamond indicates ignition time exceeded 300 s.*

water (Fig. 4). The coal/xylene reference material and the lighter fluid were considerably more difficult to extinguish because of the inherent flammability of the samples and a tendency for the samples to float on top of the water. Samples that failed to ignite (NI) with a methenamine tablet in the flame propagation test were excluded from the water extinguishing test.

Statistical Evaluation

Means, standard deviations, and 95% confidence intervals were calculated for six determinations by each test on each sample. The statistical formulas employed are presented below.

1. Mean X

$$X = \frac{1}{n} \left\{ \sum_{i=1}^{n} X_i \right\}$$

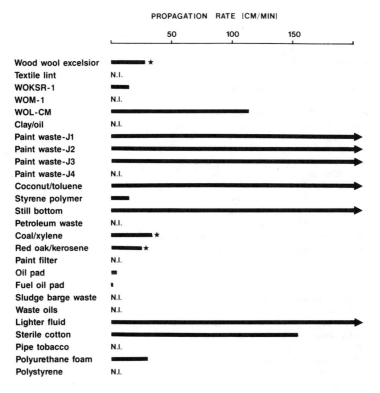

FIG. 3—*Flame propagation rate for waste samples and reference materials. Star denotes reference standard; NI indicates failure to ignite with a methenamine tablet.*

where

n = number of determinations
X_i = value of each determination

2. Coefficient of variation CV

$$CV = \frac{\frac{1}{n-1}\left\{\sum_{i=1}^{n}(X_i - \bar{X})^2\right\}^{1/2}}{\bar{X}}(100)$$

3. Confidence interval CI

$$CI = X \pm t_{N-1}(p)S_X$$

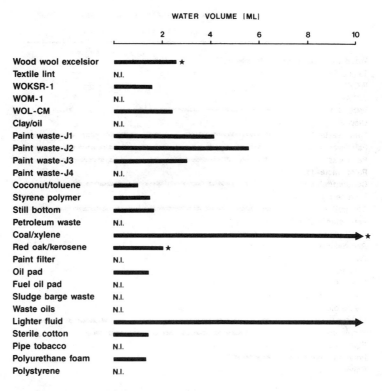

FIG. 4—*Effectiveness of water for extinguishing waste samples and reference materials. Star denotes reference standard; NI indicates failure to ignite with a methenamine tablet.*

where

t_{N-1} = tabulated t-statistic for $N - 1$ degrees of freedom
p = degree of confidence (90%, 95%, and so on)
S_X = standard deviation of the mean

Table 1 illustrates the results of the statistical evaluation of the data. Coefficients of variation CV and confidence intervals were not calculated where a discrete value was not reported (for example, <0.0033 s^{-1}, ~ 10 m/min, NI). Collective examination of the data reveals an overall CV of approximately 30%, although the precision results between samples ranged from 9 to 65%. The water extinguishing test tended to have slightly greater precision and a narrower range of CV values between samples; however, none of the tests demonstrated exceptional reproducibility. Conversely, all tests exhibited a level of precision which, in light of the nature of the measurement, was quite satisfactory. The mean CV for the reference standards (28%) was highly consistent with the precision observed for the actual waste samples. No relation-

TABLE 1—*Statistical data analysis.*[a]

Sample	Radiant Heat Ignition, s^{-1} @ 230°C			Flame Propagation, cm/min			Water Extinguishing, mL		
	$\bar{X}$	CI^b	CV, %	$\bar{X}$	CI^b	CV, %	$\bar{X}$	CI^b	CV, %
Wood wool excelsior[c]	0.012	0.002 4	16	28	6	21	2.6	0.8	31
Coal/xylene[c]	0.012	0.010 7	49	33	8	18	21	6	29
Red oak/kerosene[c]	0.016	0.004 7	29	26	10	38	2.1	0.4	19
Textile lint	<0.003 3	...	...	NI[d]	...	...	NI	...	...
WOKSR-1	0.020	0.004 2	20	15	4	27	1.6	0.4	25
WOM-1	0.006 3	0.003 4	35	NI	...	...	NI	...	...
WOL-CM	0.029	0.014	47	113	36	30	2.4	0.4	17
Clay/oil	0.008 0	0.000 78	9	NI	...	...	NI	...	...
Paint waste-J1	0.058	0.058	97	~1 000	...	...	4.2	0.7	17
Paint waste-J2	0.028	0.007 3	25	~1 000	...	...	5.6	0.9	14
Paint waste-J3	0.043	0.010	23	~1 000	...	...	3.1	0.6	19
Paint waste-J4	<0.003 3	...	...	NI	...	...	NI	...	...
Coconut/toluene	0.026	0.015	57	~1 000	...	...	1.0	0.2	20
Styrene polymer	0.017	0.008 7	48	15	2	13	1.5	0.5	33
Still bottom	0.031	0.006 4	19	~1 000	...	...	1.7	0.6	35
Petroleum waste	0.011	0.003 0	16	NI	...	...	NI	...	...
Paint filter	<0.003 3	...	...	NI	...	...	NI	...	...
Oil pad	0.013	0.003 5	26	4	2	50	1.4	0.6	43
Fuel oil pad	0.004 8	0.001 6	31	2	1	50	NI	...	...
Sludge barge waste	0.006 5	0.003 4	49	NI	...	...	NI	...	...
Waste oils	0.018	0.010	55	NI	...	...	NI	...	...
Lighter fluid	0.089	0.014	15	<2 000	...	...	<100	...	...
Sterile cotton	<0.003 3	...	...	154	105	65	1.4	0.6	36
Pipe tobacco	<0.003 3	...	...	NI	...	...	NI	...	...
Polyurethane foam	<0.003 3	...	...	30	3	10	1.3	0.2	15
Polystyrene	<0.003 3	...	...	NI	...	...	NI	...	...
Mean CV, (%)			30			32			25

[a] Six determinations.
[b] 95% confidence interval.
[c] Reference standard.
[d] Sample failed to ignite with a methenamine tablet.

ship among the magnitudes of the measured parameters (time to ignition, rate of flame propagation, and volume of water) was evident.

Correlation between the tests was assessed by two methods, Spearman and Pearson. Product-moment correlation coefficients were calculated using the following formulas:

$$r_s(\text{Spearman}) = \frac{\sum\limits_{i=1}^{n} \{(Q_i - \bar{Q})(R_i - \bar{R})\}}{\left[\sum\limits_{i=1}^{n} (Q_i - \bar{Q})^2 \sum\limits_{i=1}^{n} (R_i - \bar{R})^2\right]^{1/2}}$$

where

Q_i = rank of X_i
R_i = rank of Y_i

$$r_p(\text{Pearson}) = \frac{\sum\limits_{i=1}^{n} \{(X_i - \bar{X})(Y_i - \bar{Y})\}}{\left[\sum\limits_{i=1}^{n} (X_i - \bar{X})^2 \sum\limits_{i=1}^{n} (Y_i - \bar{Y})^2\right]^{1/2}}$$

The formula for calculating the Spearman product-moment correlation coefficient is used in its unreduced form since the data include ties in the assigned ranks. The results of these correlations, shown in Table 2, illustrate appreciable differences between the two treatments. The inclusion of additional observations in the Spearman correlation is facilitated by substituting a rank for the actual data in determining the relationship. Use of ranks permits incorporation of data points for samples lacking discrete values, either because their response was not measurable under the conditions of the test or because they failed to ignite with the methenamine tablet.

TABLE 2—*Correlation of ignitability tests [8-10].*

	Spearman		Pearson	
	Correlation[a]	Observations	Correlation	Observations
Radiant heat/flame propagation	0.732 7	26	0.798 7[a]	8
Radiant heat/water extinguishing	0.767 2	26	0.002 805	12
Flame propagation/ water extinguishing	0.876 0	26	0.091 7	9

[a] Significant at 95%.

Evaluation of data revealed significant correlations between the three tests when the Spearman correlation was used. Conversely, the Pearson correlation revealed that only the radiant heat and flame propagation results were significantly correlated. This apparent disparity between the two tests can be explained on the basis of the difference in the approach of the tests. The assignment of ranks enables inclusion of data in the Spearman test that cannot be used in the Pearson test, either because of nonignition or response outside the range of the test. This ranking process further enhances the strength of the correlation in the case of the flame propagation/water extinguishing tests since the relatively large number of nonignitable (NI) samples in both tests will be very strongly correlated.

The degree of correlation demonstrated by both the Spearman and Pearson methods between the radiant heat and the flame propagation ignitability evaluations suggests that these tests are providing very similar information. Furthermore, methods reveal a correlation that is significant at the 95% level. The conclusion drawn from this result is that for the samples analyzed in this study, measurement of the radiant heat ignition time permits prediction of the flame propagation rate and vice versa.

Conclusions

Application of three test procedures—(1) radiant heat ignition, (2) flame propagation and (3) water extinguishing—provides effective assessment of the ignitability of solid waste samples. Precision of replicate analyses was found to be approximately 30% and was consistent between actual samples and reference standards. Evaluation of the data from the three tests revealed a significant correlation between the radiant heat and flame propagation tests and hence a possible test redundancy based on the samples analyzed in this study.

The establishment of a decision level for classification of samples, in terms of burning properties, as hazardous or nonhazardous is not a direct result of this study, but can in part be provided on the basis of the analytical and statistical data generated. The data clearly illustrate that the ease of ignitability of a waste is directly related to the rate of flame propagation and the volume of water necessary to extinguish the flame, and is indirectly related to the time to ignition under a radiant heat source. These relationships were combined into an "ignitability factor" for assessing the overall ignition tendency for the samples tested. These factors (Table 3) were calculated as the combined product of the flame propagation rate (cm/min), the water volume necessary to extinguish the flame (mL) and the reciprocal of the radiant heat ignition time (s^{-1}). Table 4 shows the types of samples that fall under various ignitability factor categories. It should be noted that ignitability factors were not calculated for samples that did not ignite in the flame propagation and water extinguishing tests, and hence this result should be construed only as a means for

TABLE 3—*Calculated ignitability factors.*

Sample	Ignitability Factor[a]
Wood wool excelsior[b]	0.87
Textile lint	NI[c]
WOKSR-1	0.48
WOM-1	NI
WOL-CM	7.9
Clay/oil	NI
Paint waste-J1	240[d]
Paint waste-J2	150[d]
Paint waste-J3	130[d]
Paint waste-J4	NI
Coconut/toluene	26[d]
Styrene polymer	0.38
Still bottom	53[d]
Petroleum waste	NI
Coal/xylene[b]	15
Red oak/kerosene[b]	0.87
Paint filter	NI
Oil pad	0.073
Fuel oil pad	NI
Sludge barge waste	NI
Waste oils	NI
Lighter fluid	>17 000[e]
Sterile cotton	<0.71[f]
Pipe tobacco	NI
Polyurethane foam	<0.13[f]
Polystyrene	NI[c]

[a] Computed by

$$\frac{\text{(flame propagation rate, cm/min) (water volume, mL)}}{\text{(ignition time, s)}}$$

[b] Reference standard.
[c] Sample failed to ignite with a methenamine tablet.
[d] Flame propagation rate estimated at 100 cm/min.
[e] Flame propagation rate estimated at >2000 cm/min; water volume estimated at >100 mL.
[f] Radiant heat ignition time >300 s.

providing an estimate of the general ignition behavior of the sample. The results of this study do not support the assignment of a discrete hazardous/ nonhazardous decision level for solid wastes. However, the inclusion of representative "domestic" materials and reference standards with the waste samples ascribes significance to the data beyond a simple numerical value.

A multilaboratory study is currently in progress to further evaluate these three test procedures. The quantitative and narrative results of this study will not only provide performance data but also information that will facilitate improvements to the tests.

TABLE 4—*Categorization of waste samples and reference materials as a function of ignition tendency.*

Ignitability Range	Samples[a]
>0.01; <0.1	oil pad
>0.1; <1	wood wool excelsior[b]
	WOKSR-1
	styrene polymer
	red oak/kerosene[b]
	sterile cotton
	polyurethane foam
>1	WOL-CM
	paint waste-J1
	paint waste-J2
	paint waste-J3
	coconut/toluene
	still bottom
	coal/xylene[b]
	lighter fluid

[a] Samples that failed to ignite in either the flame propagation or water extinguishing test were excluded.
[b] Reference standards.

Acknowledgments

The authors would like to express their appreciation to R. W. Handy for his thoughtful contribution to the introduction. Although the research described in this article has been funded wholly or in part by the U.S. Environmental Protection Agency through Contract 68-03-3099, Work Assignments 1 and 5 to the Research Triangle Institute, it has not been subjected to the agency's required peer and administrative review and therefore does not necessarily reflect the views of the agency, and no official endorsement should be inferred.

APPENDIX I—Test Procedures

Radiant Heat Ignition Test Procedure

1. Scope

1.1 This method is applicable to the determination of the ignitability of solid wastes using a radiant heat source.

2. Summary of Method

2.1 An aluminum pan (60 mm in diameter by 17 mm deep) is filled with the test material to a depth of 10 mm. The sample is placed 60 mm beneath a preheated, temperature-controlled, radiant heat source and the time to ignition is measured on a strip chart recorder.

3. Apparatus and Reagents

3.1 Test chamber—a chamber with inside dimensions of 450 by 450 by 600 mm made of fireproof material not less than 6 mm thick, with hinged top and sliding door of the same material (Fig. 1).

3.2 Sample container—aluminum weighing pan, 60 mm in diameter by 20 mm deep (Fisher 08-732).

3.3 Radiant heat apparatus—Kjeldahl flask heater (A. H. Thomas, 5925-E60), heating element cemented in place with Sauereisen sealing cement (No. 33 powder) and mounted inverted in the test chamber.

3.4 Controller/sensor (Fig. 1).

3.5 Recorder—strip chart, variable speed and input impendance.

3.6 Fireproof boards, 200 by 200 mm.

3.7 Safety equipment

   3.7.1 Flameproof gloves—Zetex,® 360 mm (Lab Safety Supply, 1915M).

   3.7.2 Tongs—530 mm (Fisher, 15-207).

   3.7.3 Respirator (Fisher, 13-995-11).

4. Test Procedure

4.1 Locate the test chamber in a fireproof fume hood with the exhaust turned off. Put the test apparatus in the test chamber.

4.2 Adjust the position of the heat source to provide an element-to-sample surface distance of 60 mm. Adjust the position of the thermocouples to just above the sample surface.

4.3 Connect the thermocouples to the controller and recorder. Center the thermocouples 5 mm above the top of the sample container.

4.4 Set the controller to 232°C (450°F).

4.5 Position a dummy (blackened with carbon soot) sample container under the heat source. Preheat the radiant heat source to 232°C (450°F) by adjusting the voltage to the heater.

4.6 Fill a sample container with sample to a depth of 10 mm, making sure that the sample surface is level and even.

   CAUTION: Extreme care must be exercised in testing materials known to be or suspected of being extremely flammable. Preliminary tests using greatly diminished sample sizes should be conducted before performing the actual test, to ensure the safety of the analyst. A reduced test sample depth should be used in cases where sample ignition is extremely rapid or violent.

4.7 Start the recorder at a chart speed of 0.5 in./min and a full-scale sensitivity of 100 mV.

4.8 Remove the dummy sample container.

4.9 Center the filled sample container under the radiant heat source and immediately mark the recorder chart. Lower the doors on the test chamber and fume hood. Measure the time to ignition on the recorder.

4.10 Remove the ignited sample from beneath the radiant heater and carefully place it in the bottom of the chamber.

CAUTION: Raise the fume hood and test chamber doors just to a level that facilitates removal of the sample from beneath the radiant heat source. Flameproof gloves should be used in combination with tongs to protect the analyst from the burning sample.

4.11 Extinguish the fire with water or by covering with a fireproof board.

4.12 Close the fume hood door.

4.13 Clear the combustion products from the test chamber by turning on the fume hood exhaust for 5 min.

4.14 Prepare the chamber for the next sample by positioning the dummy sample container beneath the radiant heat source.

5. Calculations

5.1 Calculate the time to ignition (seconds) from the recorded distance and chart speed.

Flame Propagation Test Procedure

1. Scope

1.1 This method is applicable to the determination of the linear flame propagation rates for solid waste samples.

2. Summary of Method

2.1 An aluminum foil trough, 245 by 45 by 30 mm, is filled with test material to a depth of 10 mm and the surface is smoothed. The sample is ignited at one end with a methenamine tablet, and the time required to burn a premeasured distance between two sensing thermocouples is recorded on a two-pen strip chart recorder. The linear flame propagation rate is computed in centimetres per minute.

3. Apparatus and Reagents

3.1 Test chamber—a chamber with inside dimensions of 450 by 450 by 600 mm made of fireproof material not less than 6 mm thick with a hinged top and sliding door of the same material (Fig. 1).

3.2 Methenamine tablet—such as methenamine reagent tablet No. 1588 (Eli Lilly Inc., Indianapolis, IN) or hexamine tablets (available from most camping suppliers).

3.3 Testing troughs—prepared by molding aluminum foil in a 245 by 45 by 30 mm (inside dimensions) chromatography trough (Fisher, 05-729-54).

3.4 Recorder—strip chart, two-pen, variable speed and input impedance.

3.5 Safety equipment
3.5.1 Flameproof gloves—Zetex,® 360 mm (Lab Safety Supply, 1915M).
3.5.2 Tongs—530 mm (Fisher, 15-207).
3.5.3 Respirator (Fisher, 13-995-11).

4. Test Procedure

4.1 Locate the test chamber in a fireproof fume hood with the exhaust turned off. Connect the thermocouples to the recorder.

4.2 Fill the testing trough with sample to a depth of 10 mm, making sure that the sample surface is level and even.

CAUTION: Extreme care must be exercised in testing materials known to be or suspected of being extremely flammable. Preliminary tests using greatly diminished sample sizes should be conducted before performing the actual test, to ensure the safety of the analyst. A reduced test sample depth should be used in cases where sample ignition is extremely rapid or violent.

4.3 Place the test trough in the test chamber with the sensing thermocouples equidistant from the ends of the trough.

4.4 Start the recorder at a chart speed of 12.7 mm/min (0.5 in./min) and a full-scale sensitivity of 100 mV.

4.5 Ignite a methenamine tablet (held with tongs) with a lighted match and place the tablet on the sample at one end of the trough. Lower the doors on the test chamber and fume hood.

4.6 Measure the time required for the flame to propagate between the thermocouples (190 mm).

NOTE: If the flame does not propagate the full 100 mm distance, measure the time and distance propagated manually.

4.7 Raise the test chamber and fume hood doors slightly. Cover the sample with the fireproof board or extinguish the fire with water.

4.8 Lower the fume hood door.

4.9 Clear the fumes from the test chamber by turning on the fume hood exhaust for 5 min.

5. Calculations

5.1 Calculate the linear flame propagation rate of the waste by dividing the length of waste burned (in centimetres) by the burn time (in minutes) to yield a burn rate in centimetres per minute.

Water Extinguishing Test Procedure

1. Scope

1.1 This method is applicable to the determination of the effectiveness of water to extinguish flames on solid waste samples.

2. Summary of Method

2.1 An aluminum pan, 114 mm in diameter (top, inside) and 32 mm deep, is filled with test material to a depth of 10 mm and the surface is smoothed. The sample is ignited at the center with a methenamine tablet and allowed to burn until the entire surface is charred or until half the material present has burned. Water is applied from 60 mm above the sample by means of a stainless steel sprayer activated by a solenoid shutoff valve. The amount of water necessary to extinguish the fire is calculated from the flow rate and time the sprayer is activated.

3. Apparatus and Reagents

3.1 Test chamber—a box with inside dimensions of 450 by 600 by 600 mm made of fireproof material not less than 6 mm thick with a hinged top and sliding door of the same material (Fig. 1).

3.2 Methenamine tablet—such as methenamine reagent tablet No. 1588 (Eli Lilly Inc., Indianapolis, IN) or hexamine tablets (available from most camping suppliers).

3.3   Test container—aluminum tart pan, 114 mm diameter (top, inside) and 32 mm deep (available at most supermarkets).
3.4   Water, municipal supply.
3.5   Water pressure regulator (Fisher, 15-529).
3.6   Pressure gage, 0–103 kPa (0–15 psi) (Matheson, 63-3115).

4. Test Procedure

4.1   Set the flow rate of the sprayer to approximately 7 mL/s by adjusting the water pressure regulator. Measure the exact flow rate with a 250-mL graduated cylinder. Record the flow rate and the pressure.
4.2   Adjust the height of the sample platform to provide a sprayer-to-sample surface distance of 60 mm.
4.3   Fill the test container with waste to a depth of 10 mm, making sure that the sample surface is level and even.
   CAUTION: Extreme care must be exercised in testing materials known to be or suspected of being extremely flammable. Preliminary tests using greatly diminished sample sizes should be conducted before performing the actual test, to ensure the safety of the analyst. A reduced test sample depth should be used in cases where sample ignition is extremely rapid or violent.
4.6   Place an ignited methenamine tablet on the waste in the center of the container. Close the doors of the test chamber and fume hood.
4.7   Allow the fire to burn until the entire surface of the sample is charred or until half of the material present has burned.
4.8   Using the tongs, center the ignited sample directly beneath the sprayer.
4.9   Depress the switch that activates the sprayer and timer, in short bursts (~100 ms) approximately 1 s apart and observe for continued burning between bursts.
4.10 Record the time to extinguishment to the nearest 0.01 s.
4.11 If the water is not effective in extinguishing the fire, cover the sample with a fireproof board to smother the fire.

5. Calculations

5.1   Calculate the amount of water necessary to extinguish the fire from the elapsed time and measured flow rate. Report results as millilitres of water.

APPENDIX II—Parts List[3]

1. Heater, 750 W, 120 V (American Scientific Products, 61856); lower clay assembly for precision Kjeldahl flask heater; element cemented to ceramic base with Sauereisen sealing cement (No. 33 powder).
2. Heater support, Unistrut P-1000.
3. Aluminum rod, 1/2 in. diameter (Fisher, 14-666).
4. Right angle connector (Fisher, 14-666-20).
5. Plate support (Fisher, 14-666-24).
6. Thermocouple protection tube, Omegatite 450,® 3/16 by 1/4 by 12 in. (Omega Engineering, PTRA31614-12), with closed end sawed off.

[3]1 in. = 25.4 mm.

7. Thermocouple, subminiature, Type K, Inconel sheath, 0.020 by 12 in. (Omega Engineering, SCAIN-020U-12); thermocouple extension (Omega Engineering, EXTT-K-24); thermocouple connectors.

8. Foot plate (Fisher, 14-666-25) (Omega Engineering, SMP-K-MP).

9. Transite® box, 450 by 450 by 600 mm, hinged top.

10. Hollow spray nozzle, stainless steel, $1/4$ by $1^5/16$ in. male (McMaster-Carr Supply, 3405K76); female connector, $1/4$ by $1/4$ in. NPT (Swagelok®, SS-400-7-4).

11. Bulkhead union, stainless steel, $1/4$ in. (Swagelok,® SS-400-61).

12. Tubing, stainless steel, $1/4$ in. outside diameter (O.D.).

13. Solenoid valve, $1/4$ in., normally closed (McMaster-Carr Supply, 4639R58). Not shown: Copper tubing, $1/4$ in. O.D.; pressure gage, 0–103 kPa (0–15 psi) (Matheson, 63-3115); water pressure regulator (Fisher, 15-529); female branch tee, $1/4$ by $1/4$ in. NPT (Swagelok,® SS-400-3-4TTF).

14. Terminal board, double row (Kulka, 601-GP-2).

15. High-temperature cable, copper, 14 AWG (W. K. Hile, 9G-P14T-1); solderless connector, spade terminals.

16. Cable, 2 conductor, 18 AWG copper (power supply to heater).

17. Cable, single conductor, 18 AWG.

18. Chromatography trough (Fisher, 15-729, 54).

19. Door, Transite,® sliding; thermal-resistant glass window.

20. Angle bracket, four hole.

21. Fuse holder (Bussmann, Type HTA); fuse, 10 A, 32 V, slow-blowing (main power).

22. Fuse holder (Bussmann, Type HTA); fuse, 1 A, 250 V (timer).

23. Fuse holder·(Bussmann, Type HTA); fuse, 8 A, 32 V, slow-blowing (heater).

24. Fuse holder (Bussmann, Type HTA); fuse, 1 A, 250 V, slow-blowing (solenoid valve).

25. Power light, flush lens (Leecraft, 36EN2313).

26. Power switch, SPST, toggle.

27. Switch, momentary, DPDT, center off (C&K, 7205) (activates sprayer and timer simultaneously).

28. Reset button, SPDT (C&K, TP12HZQ).

29. Timer/Clock, 115 V AC, six-digit display (International Microtronics Model 402).

30. Digital thermometer, Type K (Omega Engineering Model 199).

31. Shadow cabinet, $13^1/4$ by $7^1/2$ by 9 in. (Budd, SB-2141). Not shown: Variable transformer, 115 V AC; recorder, strip chart, variable input impedance, and chart speed.

References

[1] Michael, L. C., Perritt, R. L., and Pellizzari, E. D., "Single-Laboratory Evaluation of Ignitability and Related Methods," final report, EPA Contract No. 68-03-3099, Work Assignment 5, Research Triangle Institute, Research Triangle Park, NC, May 1983.
[2] U.S. National Bureau of Standards, "Flammability Testing of Solids Under the Federal

Hazardous Substances Act," NTIS PB179047, National Technical Information Service, Springfield, VA, 1980.

[3] Hough, R., Lasseinge, A., and Pankow, J., "Hazards Classification of Flammable and Oxidizing Materials for Transportation—Evaluation of Test Methods (Phase II)," NTIS PB 277019, National Technical Information Service, Springfield, VA, 1973.

[4] Kucha, J. M. and Smith, A. F., "Classification Test Methods for Flammable Solids," NTIS PB 206463, National Technical Information Service, Springfield, VA, 1972.

[5] Kucha, J. M., Furno, A. L., and Imhof, A. C., "Classification Test Methods for Oxidizing Materials," NTIS PB 206889, National Technical Information Service, Springfield, VA, 1972.

[6] King, P. V. and Lasseinge, A. H., "Hazardous Classification of Oxidizing Materials and Flammable Solids for Transportation Evaluation of Test Methods," NTIS PB 220084, National Technical Information Service, Springfield, VA, 1972.

[7] Zweidinger, R. A., Michael, L. C., and Green, D. A., "Development and Evaluation of Test Procedures for Ignitability Criteria for Hazardous Waste," final report, EPA Contract No. 68-03-3099, Work Assignment 1, Research Triangle Institute, Research Triangle Park, NC, Oct. 1982.

[8] Mendenhall, W., Scheaffer, R. L., and Wackerly, D. D., *Mathematical Statistics with Application*, 2d ed., Duxbury Press, Boston, 1981.

[9] Snedecor, G. W. and Cochran, W. G., *Statistical Methods*, 7th ed., Iowa State University Press, Ames, 1980.

[10] Ray, A. A., Ed., *SAS User's Guide: Statistics*, SAS Institute, Cary, NC, 1982.

Robert W. Handy,[1] Edo D. Pellizzari,[1] and James A. Poppiti[2]

A Method for Determining the Reactivity of Hazardous Wastes That Generate Toxic Gases

REFERENCE: Handy, R. W., Pellizzari, E. D., and Poppiti, J. A., "**A Method for Determining the Reactivity of Hazardous Wastes That Generate Toxic Gases,**" *Hazardous and Industrial Solid Waste Testing: Fourth Symposium, ASTM STP 886,* J. K. Petros, Jr., W. J. Lacy, and R. A. Conway, Eds., American Society for Testing and Materials, Philadelphia, 1986, pp. 106–120.

ABSTRACT: An analytical system for measuring the evolution of hydrocyanic acid (HCN) and hydrogen sulfide (H_2S) from waste materials has been evaluated. The procedure involves acidifying the sample in an aqueous diluent and then drawing a current of air through the vapor space and onto a Draeger stain tube specific for each analyte. Estimates of method precision, analyte recovery, and interferences of one analyte on the other were carried out. A round-robin interlaboratory study was conducted on a variety of HCN- and H_2S-bearing wastes. The results of this study indicated that the presence of interferences and the poor method precision made this test method inappropriate for measuring HCN in wastes. On the other hand, study data suggested that, with some refinements, this procedure is a potentially useful technique for measuring H_2S in waste materials.

KEY WORDS: solid wastes, hydrogen cyanide, hydrogen sulfide, analysis, detector tube, hazardous wastes

The Solid Waste Disposal Act, as amended by the Resource Conservation and Recovery Act, Part 261, Subpart C, 261.23-a-5, states that a solid waste exhibits the characteristic of reactivity if "it is a cyanide or sulfide bearing waste which, when exposed to pH conditions between 2.0 and 12.5, can generate toxic gases, vapors, or fumes in a quantity sufficient to present a danger to human health or the environment." At the present time, there are no suitable validated procedures for determining the evolution of hydrocyanic acid (HCN) or hydrogen sulfide (H_2S) from a waste material.

[1]Principal scientist and vice president, respectively, Research Triangle Institute, Research Triangle Park, NC, 27709.
[2]Program manager, U.S. Environmental Protection Agency, Washington, DC 20460.

The EAL Corp. has developed a method originally proposed by the Office of Solid Wastes (OSW) of the U.S. Environmental Protection Agency [1]. The procedure involves acidifying the waste material to pH 2.0 and measuring the HCN or H_2S evolved by means of a stain tube (National Draeger, Inc., Pittsburgh, PA) specific for each gas. The present study was designed to further validate this method and to develop the protocol as a possible standard method for measuring reactive cyanide and sulfide. This paper reports on an in-house (intralaboratory) evaluation of the OSW method and an interlaboratory study of the method's precision.

Description of Method

Scope and Application

The OSW method is designed to measure only the HCN or H_2S evolved under the test conditions and not to reflect the total concentration of these gases or their precursors in the sample. Variations in the temperature, ionic strength, and total volume of the test solution will affect the amount of gas evolved. The total volume of the solution is kept constant from test to test. Temperature and ionic strength are not controlled, as they are inherent properties of each waste and standard solution.

Summary of Method

An aliquot of the waste is acidified to pH 2 in a closed system. The gas generated is swept from the reaction chamber by using a pump and passed through a calibrated gas detector tube. The detector tube reading is used to calculate the concentration of gas evolved.

Apparatus

The necessary apparatus includes the following:

- Three-neck round bottom flask with 24/40 ground glass joints, 250 mL.
- Magnetic stirrer bar with magnetic stirring apparatus.
- Addition funnel with pressure equalizing line and 24/40 ground glass joint, 124 mL.
- Straight glass adapter tubes with 24/40 ground glass joint and rubber adapter sleeve, two of each.
- Flexible tubing to make the connection from detector tube to pump.
- Detector tubes: HCN detector tube with a range of 10 to 120 μL (Draeger 67-28441 or equivalent); H_2S detector tube with a range of 5 to 60 μL (Draeger 67-28141 or equivalent). Detector tubes were supplied to each participating laboratory.

- A pH meter and pH electrode of sufficient length to reach the liquid level (approximately 180 mm).
- Pump capable of pulling 60 ± 3 mL/min (MSA Model C-210 or equivalent).
- Bubble meter for calibrating the pump.
- Stopwatch.

Reagents

- Sulfuric acid (H_2SO_4)—1 N.

Procedure

Place 10 g (or 10 mL) of the waste to be tested in a beaker. Add approximately 80 mL of deionized water and determine the amount of 1 N H_2SO_4 required to adjust the pH to 2.0.

Place a second 10 g (or 10 mL) aliquot of the waste in the 250-mL round bottom flask. Add enough deionized water so that the sum of the volume of water and the volume of 1 N H_2SO_4 required for pH adjustment will equal 100 mL. Use the same volume of 1 N H_2SO_4 for subsequent measurements of the same sample.

Assemble the apparatus as shown in Fig. 1, using the proper detection tube.

Calibrate the pump in line to a flow of 60 ± 3 mL/min using the bubble meter attached to the pump outlet. The flow will be checked before and after the measurement period with the Draeger tube attached to the inlet end of the pump.

Begin stirring and make sure all connections are tight.

Carefully adjust the test solution to pH 2.0 using the 1 N H_2SO_4.

Note the readings of the stain length at 5-min intervals. If the length of the stain exceeds the capacity of the detector tube, reanalyze it using a smaller aliquot. If very little or no stain develops, perform triplicate measurements and report the results.

After 30 min stop the pump and record the final reading on the detector tube.

Calculations

Total air volume = pump flow rate (final + initial)/2
ppm H_2S or HCN in gas evolved = tube reading/total volume
HCN evolved, μg = tube reading $\times$ 1.1
H_2S evolved, μg = tube reading $\times$ 1.4
HCN or H_2S, μg per gram of sample = μg (HCN or H_2S)/weight of sample, g

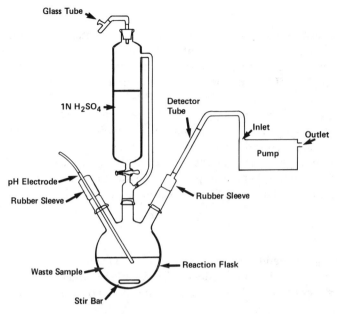

FIG. 1—*Test apparatus.*

Tube Characteristics

After acidification of the waste sample in the prescribed reaction flask, the evolved gas is purged through a graduated Draeger tube. The contents of the tube interact with the pollutant of interest to yield a colored stain front. The HCN Draeger tube (No. 67-28441) contains mercury chloride, which on reaction with HCN forms hydrochloric acid (HCl). In the presence of methyl red, the HCl gives a red stain front on a tube scale of 0.0 to 120.0 μL. Although a precleanse layer traps several acidic and basic interferring gases, high concentrations of ammonia (NH_3) and nitrogen dioxide (NO_2) can result in a yellow or dark orange coloration, respectively [1]. The H_2S Draeger tubes (No. 67-28141) contains a lead compound that reacts with H_2S to give a brown stain front on a tube scale of 0.0 to 60.0. The major interferent is reported to be sulfur dioxide [1].

Intralaboratory Study

The in-house evaluation of the method was carried out in two independent laboratories at the Research Triangle Institute (RTI). The investigation consisted of (1) analysis of standard solutions, (2) an interference study, (3) waste analysis, (4) a recovery study, and (5) attempts at increasing HCN yield. The standard solutions were diluted 1:10 in the measurement flask in each study.

Analysis of Standard Solutions

The following solutions were prepared: aqueous potassium cyanide (KCN) equivalent to 50 μg/mL HCN (Solution A) and 150 μg/mL HCN (Solution B); and aqueous sodium sulfide (Na_2S) equivalent to 2.5 μg/mL H_2S (Solution C) and 7.5 μg/mL H_2S (Solution D). Each solution was analyzed in triplicate using 10-mL aliquots. Check solutions of HCN and H_2S (100 μg/mL HCN and 5.0 μg/mL H_2S) were prepared and measured at the start of each day of analysis to monitor system performance. Ten millilitres of each check solution gave a stain approximately halfway up the tube.

Based on these determinations, 5 to 10% of the total HCN content in the added standard solution was trapped in the Draeger tubes. The similar recovery of H_2S on the Draeger tubes was estimated at 75 to 100%.

Interference Study

Solution A was added to an equal volume of (1) Solution C and (2) Solution D. Similarly, Solution B was added to (3) Solution C and (4) Solution D. The four mixtures (A + C, A + D, B + C, and B + D) were analyzed for both analytes.

The presence of H_2S resulted in a positive bias or enhancement of the HCN tube reading. The presence of 0.05 mol of H_2S per mole of HCN resulted in a tube reading twice that observed in the absence of H_2S. On the other hand, the presence of 20 mol HCN per mole of H_2S resulted in a negative bias or depression of the H_2S measurement (20 to 50% of the amount observed in the absence of HCN).

Waste Analysis

Two wastes containing sulfide (No. 1 and 2) and two with cyanide (No. 3 and 4) were analyzed in triplicate by both laboratories.

Precision estimates for the HCN measurements ranged between 5 and 35% relative standard deviation. There were not sufficient H_2S data to give precision estimates.

Recovery Study

The two HCN waste samples were each spiked with a low (Solution A) and high (check solution) level cyanide standard solution, and the resulting mixture was measured for evolvable HCN. Based on the measured HCN evolved from the standard spikes in the absence of waste material, the recovery obtained in these trials was 90 to 130%.

Similarly, the two H_2S waste samples were each spiked with the sulfide check solution and Solution C and measured for evolvable H_2S. No spike could be recovered from waste material No. 3. Based on the measured H_2S evolved from the standard spikes in the absence of waste material, the recovery obtained in these trials was 50 to 90%.

Attempts at Increasing HCN Yield

Attempts were made to increase the amount of HCN trapped on the Draeger tubes above the 5 to 10% of theory level. By extending the measurement time from 0.5 to 3.5 h, the amount of HCN evolved from Solution A and collected on the tube was increased by a factor of four. Performing the measurement at 50 ± 5°C instead of room temperature nearly tripled the amount of HCN collected. Continuing this experiment for an additional 30 min yielded an overall fourfold increase of HCN on the tube.

Interlaboratory Study

Overview

The reactivity intralaboratory study consisted of the analysis of eight sulfide-bearing and eight cyanide-bearing waste samples for evolvable H_2S or HCN, respectively. A total of twelve laboratories participated in this study. They performed triplicate measurements of the evolved HCN or H_2S, following a detailed set of instructions. In every case, two determinations were carried out on the same day and the third on a different day. In this way, estimates of within-day and across-day repeatability and reproducibility were obtained for sample measurements as described in the ASTM Practice for Evaluating Laboratories Engaged in Sampling and Analysis of Water and Waste Water (D 3856-80) and the Recommended Practices for Conducting Interlaboratory Studies of Methods for Chemical Analysis of Metals (E 173-80).

Methods and Procedures

The same general method described earlier for HCN and H_2S measurement was employed in the interlaboratory study. The analyses of all sulfide solutions and wastes were performed before any cyanide measurements were carried out. Table 1 presents a sample analysis schedule.

Sample Acquisition and Labeling

Waste materials were collected by ERCO (Cambridge, MA) and homogenized for use in the round-robin phase of this study. Sampling methods described in EPA Manual SW-846 [2] were used during the collection of the wastes.

TABLE 1—*Sample analysis schedule.*[a]

Day 1	H_2S-QC system check (1) (first ampoule)
	H_2S-PE1(2); H_2S-A(1); H_2S-B(2)
	H_2S-C(1); H_2S-D(2); H_2S-E(1)
Day 2	H_2S-QC system check (1) (first ampoule)
	H_2S-F(2); H_2S-G(1); H_2S-H(1)
	H_2S-B(1); H_2S-A(2); H_2S-PE1(1)
Day 3	H_2S-QC system check (1) (second ampoule)
	H_2S-C(2); H_2S-PE2(2)
	H_2S-D(1); H_2S-E(2)
Day 4	H_2S-QC system check (1) (second ampoule)
	H_2S-PE2(1); H_2S-F(1)
	H_2S-G(2); H_2S-H(2)
Day 5	HCN-QC system check (1) (first ampoule)
	HCN-PE1(2); HCN-A(1); HCN-B(2)
	HCN-C(1); HCN-D(2); HCN-E(1)
Day 6	HCN-QC system check (1) (first ampoule)
	HCN-F(2); HCN-G(1); HCN-H(1)
	HCN-B(1); HCN-A(2); HCN-PE1(1)
Day 7	HCN-QC system check (1) (second ampoule)
	HCN-C(2); HCN-PE2(2)
	HCN-D(1); HCN-E(2)
Day 8	HCN-QC system check (1) (second ampoule)
	HCN-PE2(1); HCN-F(1)
	HCN-G(2); HCN-H(2)

[a]The analysis schedule shown here was given for illustrative purposes only. It was not a required schedule of sample measurement. The number in parenthesis defines the number of measurements. There were eight sulfide wastes: H_2S-A to H_2S-H, and eight cyanide wastes: HCN-A to HCN-H.

The eight sulfide samples used in the round robin were prepared from five waste materials and included one pair of duplicates (H_2S-D and G) and one sample material (C) spiked at two different levels (H_2S-A and B). The eight cyanide samples were prepared from six waste materials and included two pairs of duplicates (HCN-D and G; HCN-E and H). The samples were sent to RTI where they were packaged for shipment (via Flying Tiger International) to the participating labs.

Each participating laboratory was assigned a number that was used to code waste samples sent to that laboratory. For example, Laboratory 02 received the following samples: eight sulfide samples (02-H_2S-A through 02-H_2S-H) and eight cyanide samples (02-HCN-A through 02-HCN-H).

Quality Control

At the start of each day of analysis, a quality control system check was carried out by measuring the evolvable HCN or H_2S from a reference solution.

The solution used contained the analyte being measured that day. These standards were supplied to each laboratory and were prepared and handled as follows:

Two ampoules containing Na_2S and two containing KCN concentrated stock solutions were shipped to each laboratory. The cyanide and sulfide solutions were labeled HCN-QC and H_2S-QC, respectively. After the first ampoule of each analyte was opened, the contents were transferred to one of the enclosed vials and placed in a refrigerator. One-half millilitre of the stock solution (H_2S-QC or HCN-QC) was diluted to 100 mL and 10 mL of the resulting solution was used as the system check. This diluted working standard was prepared daily. Depending on the volume of air pumped through the system, the stain front read between one third and two thirds of the full scale. The actual tube reading was recorded on the appropriate report form.

Two performance evaluation samples of each analyte were sent to each participating laboratory. The samples were labelled H_2S-PE1, H_2S-PE2 and HCN-PE1, HCN-PE2. Three ampoules of each test solution were provided.

Ten millilitres of H_2S-PE1 from two ampoules was analyzed for H_2S during the first day of H_2S measurements. In the same manner, two 10-mL aliquots of HCN-PE1 from two ampoules were analyzed for HCN during the initial HCN measurements. The third ampoule of each PE1 set was opened and analyzed on any other day when the analyte was being measured in waste samples.

Two ampoules of each PE2 sample set were opened and analyzed after 16 waste measurements had been performed (one sample analyzed in triplicate produces three measurements). The third ampoule from each PE2 set was opened and analyzed on any other day following the original measurements and during the time when the analyte was being measured in waste samples.

Results

The measurement data were recorded on a worksheet form (Fig. 2) and returned to RTI. The data were treated to give the standard deviation for repeatability and reproducibility and the estimate and coefficient of these parameters (see Tables 2 and 3). ASTM definitions of these terms (ASTM Recommended Practices E 173-80) are given below:

Repeatability: A quantitative measure of the variability of the results by a single analyst in a given laboratory using a given apparatus.

Reproducibility: A quantitative measure of the variability associated with different analysts and equipment in different laboratories.

The coefficients of repeatability and reproducibility for each sample material are summarized in Table 4. The precision for the sulfide measurements is clearly superior to that for the cyanide data.

REACTIVITY TASK WORKSHEET (31U-2258-5)

-Hydrogen Sulfide/Hydrogen Cyanide-

Date: _____

Analyst (denote by letter): _____

Laboratory No.: _____

Sample ID No. (delete Lab no.)*					
Sample Wt., g (or mL)					
Water added, mL					
Initial pH					
1N H_2SO_4 added to pH 2, mL					
Additional H_2SO_4 added, mL					
Pump No.					
Final Pump Flow Rate, mL/min					
Initial Pump Flow Rate, mL/min					
Total air volume, L					
Tube Reading, µL: 5 min					
10					
15					
20					
25					
30					
ppm H_2S/HCN in air pumped (Tube reading/Total volume)					
H_2S/HCN evolved, µg					
µg H_2S/HCN Per gm sample					

* Include all study data: daily QC system checks, PE samples, and waste
samples. For samples which require triplicate measurements, place an
ordinal number after the sample ID No.

FIG. 2—*Sample worksheet.*

TABLE 2—*Equations for repeatability calculations.*

Coefficient of repeatability $= 100 R_1 M_{ab}$

$R_1 =$ estimate of repeatability

$R_1 = S_1 K_2$

    $S_1 =$ standard deviation for repeatability

    $S_1 = X_1 K_1$

$R_1 = X_1 K_1 K_2$

    $X_1 = 1/n \sum\limits^{n} |(D_a - D_b) - (\bar{D}_a - \bar{D}_b)|$

        $D_a =$ first determination ($D1$ or $D2$ on Tables 5 and 6) reported by all laboratories

        $D_b =$ second determination ($D2$ or $D3$ on Tables 5 and 6) reported by all laboratories

        $\bar{D}_a =$ mean of first determinations

        $\bar{D}_a = 1/n \sum\limits^{n} D_a$

        $\bar{D}_b =$ mean of second determinations

        $\bar{D}_b = 1/n \sum\limits^{n} D_b$

        $n =$ number of laboratories or pairs of values

    $K_1 =$ constant; e.g., 0.934 for $n = 10$

    $K_2 =$ constant; e.g., 3.14 for $n = 10$

$M_{ab} =$ grand mean of average first and second determinations reported by all laboratories

$$M_{ab} = \frac{(\bar{D}_a + \bar{D}_b)}{2}$$

Coefficient of repeatability $= 50 K_1 K_2 (\bar{D}_a + \bar{D}_b) X_1$

The measurement of H_2S in sample H_2S-E (paper mill solid waste) possessed the best overall precision. The worst agreement was noted in the low-H_2S samples H_2S-A and H_2S-B (coke plant wastewater treatment waste).

For HCN, the measurement in samples HCN-E and H (quenching bath residues) possessed the best overall precision. The worst agreement was noted with HCN-A (lagoon sludge).

Except for single-laboratory H_2S results (repeatability), there was no appreciable difference between within-day and across-day precision. In nearly every case, precision between laboratories was poorer than that within laboratories. Sample HCN and H_2S report forms appear as Tables 5 and 6.

Conclusions

Based mainly on the likely presence of interferences and the poor analytical precision noted in this study, the use of this procedure for measuring HCN in waste materials is not recommended.

Based mainly on the absence of interferences and the reasonable analytical

TABLE 3—*Equations for reproducibility calculations.*

Coefficient of reproducibility $= 100 R_2 M_{ab}$

$R_2 = $ Estimate of Reproducibility

$R_2 = S_2 K_3$

$\quad S_2 = $ standard deviation for reproducibility
$\quad S_2 = [(X_1^2 + X_2^2)/2]^{1/2} K_1$

$R_2 = [(X_1^2 + X_2^2)/2]^{1/2} K_1 K_3$

$\quad X_1 = 1/n \sum\limits^{n} |(D_a - D_b) - (\bar{D}_a - \bar{D}_b)|$

$\quad X_2 = 1/n \sum\limits^{n} |(D_a - D_b) - (\bar{D}_a - \bar{D}_b)|$

$\qquad D_a = $ first determination ($D1$ or $D2$ on Tables 5 and 6) reported by all laboratories
$\qquad D_b = $ second determination ($D2$ or $D3$ on Tables 5 and 6) reported by all laboratories
$\qquad \bar{D}_a = $ mean of first determinations

$\qquad \bar{D}_a = 1/n \sum\limits^{n} D_a$
$\qquad \bar{D}_b = $ mean of second determinations

$\qquad \bar{D}_b = 1/n \sum\limits^{n} D_b$
$\qquad n = $ number of laboratories or pairs of values

$\quad K_1 = $ constant; e.g., 0.934 for duplicate data from 10 laboratories or 10 pairs of values
$\quad K_3 = $ constant; e.g., 3.12 for duplicate data from 10 laboratories and $X_1/X_2 = 0.2$

$M_{ab} = $ grand mean of average first and second determinations reported by all laboratories

$$M_{ab} = \frac{(\bar{D}_a + \bar{D}_b)}{2}$$

Coefficient of reproducibility $= 50 K_1 K_3 (\bar{D}_a + \bar{D}_b)[(X_1^2 + X_2^2)/2]^{1/2}$

precision noted in this study, this procedure represents a potentially useful technique for measuring H_2S in waste materials.

Certain test parameters (stirring, acidification, time) require further optimization before the H_2S measurement protocol can be employed routinely.

Acknowledgments

Although the research described in this article has been funded wholly or in part by the U.S. Environmental Protection Agency through Contract 68-03-3099 Work Assignment No. 3 to the Research Triangle Institute, it has not been subjected to the agency's required peer and administrative review and therefore does not necessarily reflect the views of the agency, and no official endorsement should be inferred.

The authors wish to acknowledge the contributions to this study made by Thomas A. Hinners, Environmental Monitoring Systems Laboratory, U.S. Environmental Protection Agency, Las Vegas, NV.

TABLE 4—*Precision summary.*

Waste Sample	Parameter	Coefficient, % Repeatability	Coefficient, % Reproducibility
		HYDROGEN CYANIDE	
A	within day	802	802
	across days	485;432	583;686
B	within day	166	296
	across days	105;176	280;255
D	within day	35.2	335
	across days	122;171	316;318
E	within day	73.4	216
	across days	81.2;139	204;196
F	within day	228	514
	across days	372;435	480;446
G	within day	350	412
	across days	213;255	332;367
H	within day	57.4	248
	across days	84.6;111	190;215
Average	within day	245	403
	across days	227	348
		HYDROGEN SULFIDE	
A	within day	57.5	463
	across days	227;338	427;447
B	within day	20.0	460
	across days	245;269	481;474
D	within day	26.6	106
	across days	64.6;81.5	70.7;80.7
E	within day	21.2	57.0
	across days	76.4;86.5	76.1;93.0
F	within day	95.6	180
	across days	142;146	231;196
G	within day	40.6	160
	across days	72.2;78.7	194;216
H	within day	59.7	119
	across days	380;434	336;374
Average	within day	45.9	221
	across days	189	264

TABLE 5—Analytical results for HCN waste sample D.

Laboratory	Concentration, µg HCN per gram of sample					
	$D1$	$D3$	$(D1 - D3)$	$\lvert (D1 - D3) - (\overline{D1} - \overline{D3}) \rvert$	$(D1 + D3)$	$\lvert (D1 + D3) - (\overline{D1} + \overline{D3}) \rvert$
01	8.2^a	8.9^a	−0.7	1.11	17.1	12.2
02	0.1	0.0	0.1	0.31	0.1	4.83
04	0.0^b	0.7^b	−0.7	1.11	0.7	4.23
05	2.5	2.0	0.5	0.09	4.5	0.43
07	3.3	3.0	0.3	0.11	6.3	1.37
08	0.0	0.0	0.0	0.41	0.0	4.93
09	0.4	0.6	−0.2	0.61	1.0	3.93
10	6.2^c	1.8^c	4.4	3.99	8.0	3.07
11	2.7	3.8	−1.1	1.51	6.5	1.57
14	3.3^d	1.8^d	1.5	1.09	5.1	0.17
Average	2.67	2.26	...	$X_1 = 1.03$	...	$X_2 = 3.67$

[a]Interference; tube initially red-orange, difficult to read.
[b]Diffuse front, difficult to read.
[c]Interference; tube difficult to read.
[d]Interference; tube initially red-orange.

Standard deviation for repeatability:

$$S_1 = X_1 K_1 = 1.03(0.934) = 0.962$$

Estimate of repeatability:

$$R_1 = S_1 K_2 = 0.962(3.14) = 3.02\% \text{ HCN}$$

Coefficient of repeatability = $(3.02/2.465) \times 100 = 122\%$
Standard deviation for reproducibility:

$$S_2 = \sqrt{(X_1^2 + X_2^2)/2}\, K_1 = 2.70(0.934) = 2.25$$

Estimate of reproducibility:

$$R_2 = S_2 K_3 = 2.52(3.09) = 7.79\% \text{ HCN}$$

Coefficient of reproducibility = $(7.79/2.465) \times 100 = 316\%$

TABLE 6—*Analytical results for H_2S waste sample D.*

Laboratory	D1	D2	(D1 − D2)	|(D1 − D2) − ($\overline{D1}$ − $\overline{D2}$)|	(D1 + D2)	|(D1 + D2) − ($\overline{D1}$ + $\overline{D2}$)|
				Concentration, μg H_2S per gram of sample		
01	531	507	24	26	1038	240
02	279	269	10	40	548	250
04	389	247	142	92	636	162
05	290	250	40	10	540	258
06	448	448	0	50	896	98
07	473	399	74	24	872	74
08	602	518	84	34	1120	322
09	266	182	84	34	448	350
11	474	438	36	14	912	114
12	491	479	12	38	970	172
Average	424	374	...	$X_1 = 36.2$	...	$X_2 = 204$

Standard deviation for repeatability:

$$S_1 = X_1 K_1 = 36.2(0.934) = 33.8$$

Estimate of repeatability:

$$R_1 = S_1 K_2 = 33.8(3.14) = 106\% \ H_2S$$

Coefficient of repeatability = $(106/399) \times 100 = 26.6\%$
Standard deviation for reproducibility:

$$S_2 = \sqrt{(X_1^2 + X_2^2)/2} \ K_1 = 146(0.934) = 136$$

Estimate of reproducibility:

$$R_2 = S_2 K_3 = 136(3.12) = 424\% \ H_2S$$

Coefficient of reproducibility = $(424/399) \times 100 = 106\%$

References

[1] "Development and Evaluation of a Test Procedure for Reactivity Criteria for Hazardous Wastes," final draft report, EPA Contract 68-03-2961, EAL Corp., Richmond, CA.
[2] Test Methods for Evaluating Solid Wastes, EPA Manual SW-846, 2nd ed., U.S. Environmental Protection Agency, Washington, DC, 1982.

Cary L. Perket[1] and Leo R. Barsotti[2]

Multilaboratory Analysis of Soil for Leads

REFERENCE: Perket, C. L. and Barsotti, L. R., **"Multilaboratory Analysis of Soil for Leads,"** *Hazardous and Industrial Solid Waste Testing: Fourth Symposium, ASTM STP 886*, J. K. Petros, Jr., W. J. Lacy, and R. A. Conway, Eds., American Society for Testing and Materials, Philadelphia, 1986, pp. 121–138.

ABSTRACT: A multilaboratory test program to determine lead content in soils produced aberrant results from two laboratories and indicated apparent problems with sample preparation. The program demonstrates the importance of quality control in remedial site investigation to determine soil contamination.

KEY WORDS: soil contamination, lead, hazardous wastes, sample preparation, sample digestion, sampling, quality control

It is routine practice to investigate the extent of soil contamination at a remedial site. The findings from such an investigation are used to develop a site plan for the removal of soils with unacceptable levels of contamination. The cost of completing the soil removal is directly proportional to the area and depth to which contamination is found. Reliable delineation of the extent of soil contamination is therefore an essential aspect of developing the most cost-effective remedial plan.

The results of this test program illustrate the importance of quality control in investigations of soil contamination. In particular, the program demonstrates the need for quality control in evaluating individual procedures used for sample preparation, laboratory procedures, and analysis. In so demonstrating, the paper also provides insight into what may or may not be reasonable goals of accuracy in delineating soil contamination.

[1]Senior engineer, Environmental Engineering and Management Ltd., Minneapolis, MN 55435.
[2]Section head, Analytical Research Department, Center for Technology, Kaiser Aluminum and Chemical Co., Pleasanton, CA 94566.

Scope

The multilaboratory program involved two types of soils: a medium-grain sand (sand) and organic surface soil (topsoil). For each soil type a set of 15 samples was prepared. Each set was composed of triplicate unspiked samples and triplicate samples spiked at one of four concentrations. Some laboratories analyzed both sets; others analyzed one set. Seven laboratories analyzed the sand set and six laboratories analyzed the topsoil set.

No instruction was provided to the laboratories regarding sample preparation, digestion of the sample, or analytical methods. Further, the laboratories were not given any specifications regarding internal quality control procedures to be followed. It was intended that the laboratory handle the samples as routine environmental samples submitted to determine total lead content. It is realistic, however, to expect laboratories to be particularly cognizant of quality control when participating in a multilaboratory test program.

Background

ASTM Subcommittee D34.02 on Physical and Chemical Characterization (a subcommittee of ASTM Committee D-34 on Waste Disposal) has a program under way to develop methods to dissolve solid samples of waste or soil for the ultimate purpose of determining the sample's total metal content. As part of the program, the subcommittee section undertook this test program to compare the methods. The objective was to determine whether significant differences in the results would occur if each laboratory used its own procedures.

The samples submitted to the participating laboratories were prepared from two soil types. One soil was a medium-grain sand that previous analysis had demonstrated to contain lead in concentrations less than 1 $\mu g/g$. The second was organic surface soil (topsoil) that was known to have lead and lead oxide contamination. Previous analysis of a topsoil sample obtained in the same area as those used in the test program revealed a lead concentration of about 300 $\mu g/g$.

Both soil types were submitted to laboratories in a set containing individual samples with and without lead oxide (Pb_3O_4) spikes. The laboratories analyzing the set of sand samples received 15 samples: three unspiked and three each spiked with 100, 600, 2000, and 5000 $\mu g/g$ of lead. The sets of topsoil sample were prepared similarly: an unspiked triplicate and triplicates spiked with 83, 493, 1640, and 4110 $\mu g/g$ of lead.

The sets of soil were sent to the laboratories without instruction regarding sample preparation or analysis. The samples were given identifying numbers, but the laboratories were not aware of the number of replicates or the concentrations of lead to expect.

Methods and Procedures

Sample Preparation

Topsoil—The topsoil used in this test program was homogenized both during sampling and at the laboratory. During sampling the topsoil was spread out on a clean surface, mixed, and then remixed. Over 30 repetitions of the remixing were done by dividing the sample into quarters, remixing the quarters into halves, and then combining the halves.

Ten kilograms of dried soil was delivered to the laboratory. At the laboratory it was crushed and passed through a 1.70-mm (No. 12) soil sieve. The sample was then divided into five groups. Four of the groups were spiked with Pb_3O_4. The samples were then further mixed by pouring the sample back and forth over 40 times between two metal pans. These samples were then divided into individual samples that were sent to the participating laboratories.

Sand—The sand used in this test program is sold commercially as Ottawa sand. It is a medium-grain sand. The sand was spiked with Pb_3O_4. The spike was homogenized in the sand by pouring the spiked sand back and forth between two metal pans over 40 times. Individual samples were then prepared for distribution to the participating laboratories.

Sample Digestion and Analysis

Each of the laboratories utilized its own choice of procedures for digestion and analysis of the soil samples. A summary of the digestion procedures and analytical instrumentation is presented in Table 1.

X-ray Analysis

The X-ray analysis was done on pulverized samples pressed into 32-mm (1¼-in.) pellets. The count per second was read for the lead alpha beta line. Measurement of the lead alpha beta line was done with a Phillips PW 1600 X-ray spectrometer.

Findings

The results reported by the laboratories are shown in Tables 2 and 3 for the sand and topsoil samples, respectively. The ASTM Practice for Determination of Precision and Bias of Methods of Committee D-19 on Water (D 2777) was followed in analyzing the results. ASTM D 2777 is intended to be used for the purpose of statistically evaluating a single method that has carefully prescribed procedures utilizing specified equipment. Its use in this study, which did not specify the procedure or equipment, therefore differs from the meth-

TABLE 1—Chemical methods for analysis of lead.

Parameter	Laboratory							
	A	B	C	D	E	F	G and I	H
Sample weight, g	25	0.5	0.5	0.05	0.5	0.5	0.5	0.5
Grind	no	no	yes; <65 mesh	yes; <60 mesh	no	no	no	no
Digestion	flask	Teflon® beaker	beaker^a	Parr bomb	beaker	beaker	EPA #3050	platinum crucible Na$_2$CO$_3$ fusion–dissolution in 20 mL HNO$_3$ + HF
Time	2 h	~2 h	2 h	overnight	overnight	1 h	...	...
Temperature	80–90°C	to HClO$_4$ fumes	89–90°C	125°C	89–90°C	100°C+ (air bath)	...	...
Acid	50 mL 1:1 HNO$_3$	2 mL HF-20 mL 85% HNO$_3$-15% HClO$_4$	5 mL HNO$_3$-2 mL HCl	3 mL HNO$_3$-5 mL HCL-0.5 mL HF; boric acid added after digestion	20 mL 1:1 HNO$_3$	2.4 mL HNO$_3$; 0.6 mL HCl	...	...
Analysis by^b	ICP	AAS	AAS	ICP	AAS	AAS	AAS (flame and furnace)	AAS

^{a}Also used method described by R. F. Farrell, S. A. Matthes, and A. J. Mackie, "A Simple, Low-Cost Method for the Dissolution of Metal and Mineral Samples in Plastic Pressure Vessels," Bureau of Mines, 1980.
^{b}After appropriate dilution of sample solution. ICP = inductively coupled plasma spectroscopy; AA = absorption spectroscopy.

TABLE 2—*Analysis of sand for total concentration of lead, in micrograms per gram.*

Set	Replicate	Lead Spike	Laboratory						
			F	B	G	D	E	H[a]	I[a]
Group 1	1	0	<5	BDL[c]	<1	BDL	BDL	447	BDL
	2	0	38[b]	BDL	2	BDL	BDL	1184	BDL
	3	0	37[b]	BDL	<1	BDL	BDL	842	BDL
	average		25	BDL	1	BDL	BDL	841	BDL
Group 2	1	100	73	90	60	70	72	342	4
	2	100	77	78	60	70	64	210	94
	3	100	101	114	80	350[b]	84	882	4
	average		84	94	67	163	73	478	34
Group 3	1	600	408	467	350	280	436	158	20
	2	600	425	429	400	270	436	290	20
	3	600	490	513	380	270	475	263	20
	average		441	470	377	273	449	237	20
Group 4	1	2000	1445	1533	1300	1400	1654	237	80
	2	2000	1495	1794	1560	1300	1793	237	76
	3	2000	1765	1611	1840	1400	1506	303	83
	average		1568	1646	1567	1367	1651	259	80
Group 5	1	5000	3735	3595	3230	3400	3686	368	242
	2	5000	4114	4387	3220	4000	3837	842	212
	3	5000	3905	4551	3380	4100	3699	540	243
	average		3918	4178	3277	3833	3741	583	232

[a]Laboratories with aberrant data.
[b]An outlier or aberrant data.
[c]Below detectable limit; used when laboratory did not report detection limit.

TABLE 3—Analysis of topsoil for total concentration of lead, in micrograms per gram.

Set	Replicate	Lead Spike	Laboratory					
			A	B	C	D	E	I[a]
Group 1	1	0	7100	6316	6500	5600	6436	201
	2	0	6000	4972	5500	5300	6717	260
	3	0	5400	5077	6100	6100	5775	250
	average		6167	5455	6033	5667	6309	237
Group 2	1	82	5500	5820	6000	5500	5677	244
	2	82	4600	4325	5900	5500	5417	293
	3	82	6000	5217	5400	5800	5134	248
	average		5367	5121	5733	5600	5409	262
Group 3	1	493	6600	5560	5900	6100	6398	322
	2	493	5700	6234	6000	6100	5887	308
	3	493	5400	6720	6300	5500	6976	249
	average		5900	6171	6067	5900	6420	293
Group 4	1	1640	5500	4757	5400	5200	5420	245
	2	1640	5100	6805	5300	7400	5293	303
	3	1640	6600	5711	5100	5400	6060	250
	average		5733	5758	5267	6000	5591	266
Group 5	1	4110	6900	6485	6900	6300	7035	428
	2	4110	7500	8705	8500	7700	8870	170
	3	4110	7100	7961	7800	7900	8018	260
	average		7167	7714	7733	7300	7974	286

[a]Laboratory with aberrant data.

ods intended purpose. However, D 2777 does provide a useful framework for evaluating data from this multilaboratory program.

The initial statistical evaluation in ASTM D 2777 determined whether any laboratory's data appeared to be aberrant when compared with the overall results reported by the participating laboratories. This evaluation utilized a t-distribution at the 5% significance level to identify such laboratories. The evaluation determined that the results for the sand sample provided by Laboratories H and I should be regarded as aberrant. Laboratory I was also found to have sufficiently different topsoil results from those generated by the other laboratories to be regarded as aberrant. The results from these laboratories for these soil specimens were not considered in subsequent evaluation of the data.

The ASTM evaluation subsequently compares each result for a sample to the overall results from all the laboratories for that sample. The purpose is to determine whether any individual result is different enough from the overall results to be regarded as an outlier. The evaluation identified three outliers with the sand samples, but no outliers with the topsoil. The outliers were two samples of unspiked sand reported by Laboratory F at 38 and 37 μg/g and one specimen of the sand spiked at 100 μg/g that was reported by Laboratory D to contain 350 μg/g. Statistical characteristics of the remaining results are in Tables 4 and 5.

Further evaluations compared the dispersion of the results among replicates (that is, samples of the same concentration). The dispersion is measured by the calculated standard deviation. Two different standard deviations are calculated. An overall standard deviation is calculated based on results from all the laboratories for a given group of replicates (for example, 15 unspiked topsoil samples). The overall standard deviation on this multilaboratory study includes dispersion caused by the use of different instruments and equipment. The second standard deviation is calculated to represent the average dispersion of results in a laboratory. It is the mathematical average of standard deviation exhibited by each individual laboratory for specific repli-

TABLE 4—*Statistical characteristics of lead analysis of sand after rejection of outliers.*

Factor	Group 1	Group 2	Group 3	Group 4	Group 5
Spike, μg/g	0	100	600	2000	5000
Number	13	14	15	15	15
Mean, μg/g	0.27	78.1	402	1560	3789
Standard deviation	0.58	15.3	78.5	170	398
Standard error	0.16	4.09	20.2	46.2	103
Minimum, μg/g	0	60	270	1300	3220
Maximum, μg/g	1.96	114	513	1840	4551
Range, μg/g	1.96	54	243	540	1331
Coefficient of variation, %	216	19.6	19.5	10.9	10.5

TABLE 5—*Statistical characteristics of lead analysis of topsoil after rejection of outliers.*

Factor	Group 1	Group 2	Group 3	Group 4	Group 5
Initial					
concentration, $\mu g/g$	unknown	5926	5926	5926	5926
Spike, $\mu g/g$	0	82	493	1640	4110
Sample					
concentration, $\mu g/g$	unknown	6008	6419	7566	10036
Number	15	15	15	15	15
Mean, $\mu g/g$	5926	5452	6091	5669	7578
Standard deviation	622	483	459	733	782
Standard error	161	125	118	189	202
Minimum, $\mu g/g$	4972	4325	5400	4757	6300
Maximum, $\mu g/g$	7100	6000	6976	7400	8870
Range, $\mu g/g$	2128	1675	1576	2643	2570
Coefficient					
of variation, %	10.5	8.86	7.52	12.9	10.3

cates (for example, three unspiked topsoil samples); this is a pooled standard deviation. A comparison of the two types of standard deviations in Tables 6 and 7 provides insight as to how the dispersion of results compares overall with the results of an average laboratory with a group of replicates.

A linear regression was performed independently on both the pooled and overall standard deviations in comparison to the mean concentration from the test results. The regression is used to develop an equation to describe the relationship between the pooled and overall standard deviations and the concentration. The regression results are presented for the sand and topsoil in Tables 8 and 9, respectively.

TABLE 6— *Tabulation of single laboratories' pooled and overall standard deviations for the sand.*

	Level of Spiking, $\mu g/g$				
Laboratory	0	100	600	2000	5000
B	0.0	18.3	42.1	134	511
D	0.0	0.0	5.8	57.7	379
E	0.0	10.0	22.5	144	83.7
F	0.0	15.1	43.3	172	190
G	0.68	11.6	25.2	270	89.6
Sum	0.68	55.0	139	777	1250
Pooled standard					
deviation	0.14	11.0	27.8	155	251
Overall					
standard deviation	0.58	15.3	78.5	179	398

TABLE 7—*Tabulation of single laboratories' pooled and overall standard deviations for the topsoil.*

	Level of Spiking, $\mu g/g$				
Laboratory	0	82	493	1640	4110
A	862	709	624	777	306
B	748	752	582	1025	1129
C	503	321	208	153	802
D	404	173	346	1217	872
E	483	272	545	411	918
Sum	3001	2228	2306	3583	4027
Pooled standard deviation	600	445	461	716	805
Overall standard deviation	622	483	459	733	782

TABLE 8—*Analysis of the relationship between precision and concentration for sand.*

Factor	Group One	Group Two	Group Three	Group Four	Group Five
Mean, $\mu g/g$	0.27	78.1	402	1560	3789
S_t	0.58	15.3	78.5	179	398
S_o	0.136	11.0	27.75	155.5	250.7

Linear Regression	$S_t = Ax + B$	S_t = overall precision
		x = concentration
$A = 0.102$	$B = 15.12$	$N = 5$
Standard error $= 0.0052$	Standard error $= 9.56$	Correlation $= 0.996$
		Significance level $= 0.001$
		Standard error estimate $= 16.54$

Linear Regression	$S_o = Ax + B$	S_o = single-operator precision
		x = concentration
$A = 0.067$	$B = 10.37$	$N = 5$
Standard error $= 0.0082$	Standard error $= 15.08$	Correlation $= 0.996$
		Significance level $= 0.001$
		Standard error estimate $= 26.11$

The final analysis used in ASTM D 2777 is a test for significant bias. Bias is evaluated by comparing the amount known or expected to be in the samples with the average amount reported. In the case of sand, the amount expected was the amount that was spiked. For the topsoil, the amount expected was the mean determined for the sample without spiking plus the spike. At the 5% significance level, all of the spiked samples for both sand (Table 10) and topsoil (Table 11) were found to have significant negative bias.

TABLE 9—*Analysis of the relationship between precision and concentration for topsoil.*

Factor	Group One	Group Two	Group Three	Group Four	Group Five
Mean, µg/g	5926	5452	6091	5669	7578
S_t	622.3	483.4	458.6	732.9	782.2
S_o	600.1	445.6	461.3	716.5	805.3

Linear Regression	$S_t = Ax + B$	S_t = overall precision
		x = concentration
$A = 0.10$	$B = 0.66$	$N = 5$
Standard error = 0.081	Standard error = 502	Correlation = 0.58
		Significance level = 0.0801
		Standard error estimate = 136

Linear Regression	$S_o = Ax + B$	S_o = single-operator precision
		x = concentration
$A = 0.125$	$B = -166.2$	$N = 5$
Standard error = 0.0801	Standard error = 496.3	Correlation = 0.671
		Significance level = 0.215
		Standard error estimate = 134.46

TABLE 10—*Summary of bias evaluation for sand.*

Factor	Group One	Group Two	Group Three	Group Four	Group Five
Amount of lead added, µg/g	0	100	600	2000	5000
Amount of lead detected, µg/g	0.27	78.1	402	1560	3789
± bias	+0.27	−21.9	−198	−440	−1211
± % bias	. . .	22%	33%	22%	24%
Number of observations	13	14	15	15	15
Standard deviation	0.58	15.3	78.5	179	398
Standard deviation of mean	0.16	4.09	20.2	46.2	103
Critical value from standard ± table	2.18	2.16	2.15	2.15	2.15
T-test value	1.69	5.35	9.80	9.52	11.75
Significance	no	yes	yes	yes	yes

Some of these findings suggest that sample preparation (that is, persistent heterogeneity) was a major influence in the test results:

1. The results from the analysis of spiked sand samples consistently showed a significant negative bias.
2. The dispersion of the unspiked topsoil samples, as measured by their overall and pooled standard deviation, was relatively large.
3. It was not statistically possible to demonstrate a difference between unspiked and spiked topsoil samples.

TABLE 11—*Summary of bias evaluation for topsoil.*

Factor	Group One	Group Two	Group Three	Group Four	Group Five
Initial concentration, $\mu g/g$	unknown	5926	5926	5926	5926
Amount of lead added, $\mu g/g$	0	82	493	1640	4110
Specimen concentration, $\mu g/g$	unknown	6008	6419	7566	10036
Amount of lead detected, $\mu g/g$	5926	5452	6091	5669	7578
$\pm$ bias	. . .	-556	-328	-1897	-2458
$\pm$ % bias	. . .	-9.3	-5.1	-25.1	-24.5
Number of observations	15	15	15	15	5
Standard deviation	622	483	459	733	782
Standard deviation of mean	161	125	118	189	202
Critical value from standard $\pm$ table	. . .	2.15	2.15	2.15	2.15
T-test value		4.45	2.77	10.03	12.2
Significance	. . .	yes	yes	yes	yes

4. The results from the analysis of spiked topsoil samples consistently exhibited a negative bias.

Hypothetically, the combined effect of heterogeneity of samples and loss of lead spike during mixing could be responsible for these findings. Heterogeneity in the topsoil sample would lead to dispersion in interlaboratory and intralaboratory results. Loss of spike would explain negative bias in both the sand and topsoil samples. And collectively, heterogeneity and loss of spike could be a reasonable explanation for the inability to differentiate among samples of the topsoil. There are, however, insufficient data to demonstrate these suggested cause-effect relationships.

Further illustration of the apparent lack of homogeneity in the topsoil is shown in Figs. 1 and 2. These figures illustrate the 95% confidence interval for typical results from a typical laboratory, E. Figure 1 illustrates that for the sand sample, Laboratory E can establish distinct differences between the sand samples based on the level of spike. However, Fig. 2 illustrates that for topsoil samples, Laboratory E could not differentiate among spiked samples or between unspiked and spiked samples.

Additional analytical work was performed on the samples by Laboratory E to compare the results of chemical and X-ray analyses. Linear regression performed on the results from both analyses of sand indicated a good correlation between results from X-ray and chemical analyses (Fig. 3) and between results from X-ray analysis and spike concentrations (Fig. 4). Corresponding linear regressions were performed for the topsoil samples (Figs. 5 and 6). Potential explanations for the difference in count per second per unit of lead for sand and topsoil include differences in mass absorption coefficients, the physical distribution and type of lead in the topsoil compared with surface coating

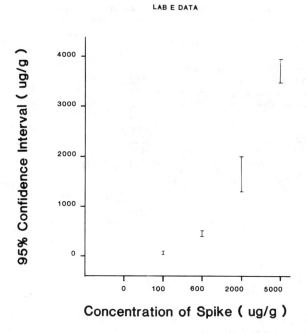

LAB E DATA

Concentration of Spike (ug/g)

FIG. 1—*95% confidence intervals at different spike concentrations in sand.*

of the sand with lead oxide, and different analytical time sequences for each set of samples. A comparison of Fig. 3 to Fig. 5, and of Fig. 4 to Fig. 6, illustrates the greater dispersion among topsoil data than among sand data.

Discussion

Over the last decade, environmental scientists have benefited from the availability of analytical instruments capable of more definitive detection of contaminants at lower concentrations. However, the ability to detect contamination at lower concentrations is shown in this study to be potentially offset by our ability to provide and prepare samples. The results indicate that use of sensitive analytical techniques may not be cost-effective in evaluating soil contamination.

Evaluating a soil sample for contamination involves the discrete steps of sampling, sample preservation, sample preparation (homogenizing), laboratory preparation (digestion), and analysis. Each of these steps must be performed correctly for the investigation to have a high degree of certainty in the results.

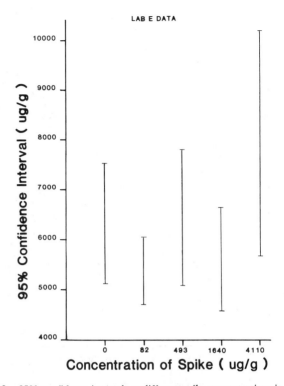

FIG. 2— *95% confidence intervals at different spike concentrations in topsoil.*

In this test program, both the sand and topsoil samples were subjected to some of these discrete steps. All the laboratories received samples prepared from the same sand and topsoil, thereby eliminating the variability from different sampling. The samples were all homogenized and spiked at one laboratory, making all the results subject to the same error. The digestion and analysis were performed by each laboratory according to their own procedures (see Table 1). An overview of sample handling is given in Table 12.

The statistical controls in this test program were not designed to separate the effects of the individual program steps. Consequently the assignment of confidence in each element is not possible. However, the results indicate there are problems in the performance of these steps in the test program. On an overall basis, two of seven laboratories analyzing the sand and one of six laboratories analyzing the topsoil were rejected for aberrant results. The specific reason why these laboratories produced aberrant results cannot be isolated by the study. However, the fact that it did occur testifies to the need for quality control programs in such analysis.

The results suggest that the sample preparation step has has a significant

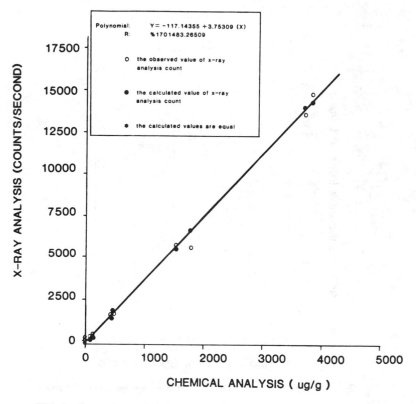

FIG. 3—*Regression analysis for sand X-ray analysis versus chemical analysis.*

effect on the results. In the particular case of the topsoil, the results suggest that the lead was still rather heterogeneously distributed even after the homogenization performed by the one laboratory. The significance of these findings extends beyond their impact on this test program, suggesting that the topsoil originally gathered for this sample was possibly even more heterogeneous. The combined effects of wide variations in the concentration of contaminant in the soil at a site being investigated and the lack of homogeneity in the prepared samples could lead to gross misrepresentation of the site conditions.

The bias reported with both sets in the test program could be from one or more causes: loss of Pb_3O_4 during blending, separation during shipment, inadequate digestion, or analytical error. The statistical controls in this program were not designed to isolate the effects of these steps. Individual laboratories did report observing the lead oxide separating from the sand samples. Some laboratories also provided results from duplicate analyses and spike re-

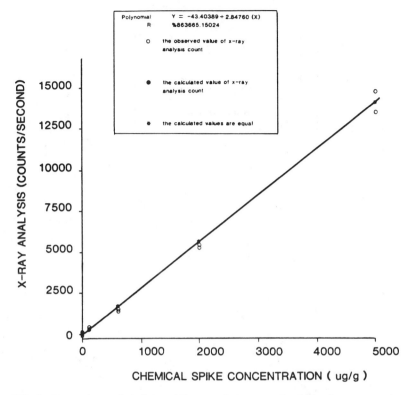

Polynomial	Y = −43.40389 + 2.84760 (X)
R	%863665.15024
O	the observed value of x−ray analysis count
●	the calculated value of x−ray analysis count
✷	the calculated values are equal

FIG. 4—*Regression analysis for sand X-ray analysis versus chemical spike concentration.*

TABLE 12—*Overview of sample handling.*

Step in Procedure	Sand	Topsoil
Sampling	sand was purchased from the supplier and had a uniform medium grain size	a single sample was taken from the surface; homogenization was done in the field
Sample preservation	none	none
Sample preparation	sand was mixed with Pb_3O_4 at various spiking concentrations by passing between containers 40 times—done by one central laboratory	topsoil was crushed and passed through a No. 12 sieve; five subsets were prepared and spiked at different concentrations mixed with Pb_3O_4 by passing between containers 40 times—done by one central laboratory
Laboratory preparation (digestion)	each laboratory utilized its own procedure	each laboratory utilized its own procedure
Analysis	each laboratory utilized its own procedure	each laboratory utilized its own procedure

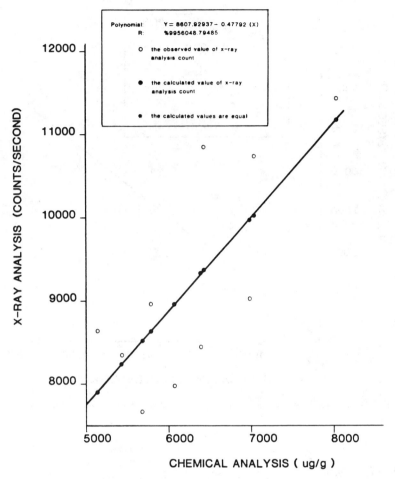

FIG. 5—*Regression analysis for topsoil X-ray analysis versus chemical analysis.*

coveries from the digestion solution submitted for analysis. The data from the duplicates and spikes substantiated the reliability of the analytical procedures used by these laboratories. However, overall there are insufficient data to specifically delineate the cause of the observed bias.

Conclusion

This test program demonstrates the potential for aberrant laboratory results, dispersion of individual results, and bias in the determination of lead content of soils. It further illustrates the need for ongoing quality control in similar investigations as well as the potential benefits of reliable standard procedures.

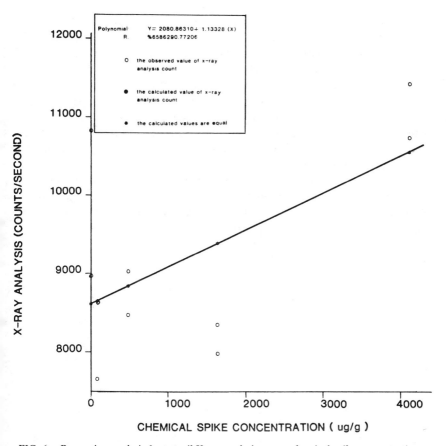

FIG. 6—*Regression analysis for topsoil X-ray analysis versus chemical spike concentration.*

It is not clear why entire sets of laboratory results were aberrant. Sample preparation is believed to be the primary reason for dispersion and bias reported. However, there is some evidence suggesting that laboratory preparation and analysis could have also significantly contributed to both dispersion and bias.

The test program did not include any provision for evaluating the influence of sampling or sample preservation. The results suggest, however, that the topsoil sample used to prepare replicates was still heterogeneous after efforts to homogenize it. Depending on the site the variation in the extent of contamination may be very significant in a relatively small area. Under such circumstances, the sampling procedures used would have a major influence on the confidence that one would have in the overall program's data.

Increasing the size and number of samples would be an obvious means of increasing confidence in a test program. Increasing the size of samples directly increases the importance of having a reliable procedure to homogenize the sample for laboratory use. Correspondingly, increasing the number of samples would probably result in compositing of individual samples and, therefore, also would require reliable procedures for homogenizing.

The growing use of sensitive analytical devices makes it possible to detect and quantitate concentrations well below those determined in this test program. These analytical devices create a challenge to develop procedures that can provide equivalent sensitivity and confidence to sampling, homogenizing, spiking, and digesting samples. In some instances it may not be possible to develop such procedures.

Realistically, an overall balance must be achieved in our programs for investigation of soil contamination. That balance will be achieved only through development and use of procedures that provide comparable levels of confidence in each step of the test program.

Donald F. Gurka,[1] Michael H. Hiatt,[1] and Richard L. Titus[2]

Nontarget Compound Analysis of Hazardous Waste and Environmental Extracts by Combined FSCC/GC/FT-IR and FSCC/GC/MS

REFERENCE: Gurka, D. F., Hiatt, M. H., and Titus, R. L., **"Nontarget Compound Analysis of Hazardous Waste and Environmental Extracts by Combined FSCC/GC/ FT-IR and FSCC/GC/MS,"** *Hazardous and Industrial Solid Waste Testing: Fourth Symposium, ASTM STP 886,* J. K. Petros, Jr., W. J. Lacy, and R. A. Conway, Eds., American Society for Testing and Materials, Philadelphia, 1986, pp. 139–161.

ABSTRACT: Complex environmental wastes have been analyzed by fused silica capillary column gas chromatography/mass spectrometry (FSCC/GC/MS) and fused silica capillary column gas chromatography/Fourier transform infrared spectrometry (FSCC/GC/ FT-IR). The two techniques used in conjunction are capable of determining or confirming the compound class of at least two thirds of all extract analytes. Detailed analytical results are presented that demonstrate that the relative strengths and shortcomings of each method are different and thus using both in conjunction leads to the identification of a far greater number of analytes than the use of either one alone. It is postulated that the present cost of using both techniques in conjunction can be substantially lowered by the direct interfacing of both spectrometers and the development of suitable computer software.

KEY WORDS: fused silica capillary column gas chromatography/mass spectrometer (FSCC/GC/MS), fused silica capillary column gas chromatography/Fourier transform infrared spectrometry (FSCC/GC/FT-IR), response factor, hit index, Gas Infrared Fourier Transform Software (GIFTS), Instrument Control Operating System (INCOS), hazardous wastes

The existence of uncontrolled and controlled hazardous waste sites [*1*] and the large quantities of industrial wastes generated within the United States [*2*]

[1]Research chemist, U.S. Environmental Protection Agency, Office of Research and Development, Environmental Monitoring Systems Laboratory, Las Vegas, NV 89114.
[2]Professor, Chemistry Department, University of Nevada, Las Vegas, NV 89109.

complicate the routine monitoring functions of the U.S. Environmental Protection Agency (EPA). Thus far, the burden of routine analysis for volatile organics has been borne by gas chromatography/mass spectrometry (GC/MS) techniques. However, the inadequacies of the GC/MS method (such as its inability to differentiate aromatic isomers without the use of authentic standards) and the desirability of increasing the reliability of compound identification by independent confirmation provide the impetus for the application of additional instrumental techniques. Committees of the American Chemical Society have advocated the use of independent confirmatory techniques, especially for regulatory agencies [3–5]. A recent editorial in *Environmental Science and Technology* has asked, "should public policies regarding the presence of individual chemicals be based on anything other than confirmed structural assignments?" [6]. The current EPA approach to independent confirmation usually involves the utilization of the gas chromatographic retention times of authentic standards of GC/MS identified analytes. This approach is of limited values since only target compounds can be confirmed and only if suitable standards are available and if those standards are chromatographically resolvable. Chromatographic resolution of standards is particularly difficult for isomers, a necessary prerequisite for isomeric differentiation by GC/MS.

The emergence of the gas chromatography/Fourier transform infrared (GC/FT-IR) method now provides a second spectral technique with the requisite sensitivity, rapidity, and selectivity for many types of environmental analysis [7]. The GC/FT-IR technique has been successfully demonstrated for the analysis of toxic organics [8], hazardous wastes [9], diesel particulates [10], and wastewater [11]. Using GC/FT-IR in conjunction with GC/MS can yield a greater number of identified analytes as well as providing confirmed identifications. Reports of the successful interfacing and operation of FT-IR and MS spectrometers suggest that complex mixture analysis by combined FT-IR and MS is evolving to a facile technique [12–14].

In the present paper, the capability of combined GC/FT-IR and GC/MS to characterize hazardous wastes is discussed. The complementary nature of these spectral techniques is demonstrated by a detailed analysis of several complicated hazardous waste extracts. The strengths, shortcomings, and future direction of the combined techniques is elaborated.

Procedure

Extract Preparation

Preparation of the extract of Sediment 1 has been previously described [9]. This sediment was gathered from the bank of a river located within a chemical dumpsite. Sediment 2 was gathered from an agricultural area that had been subjected to recent pesticide applications. It was prepared by adding

50 g of mixed soil to a 250-mL Erlenmeyer flask. Then 100 mL of pesticide grade acetonitrile were added and stirred for 1 min. The mixture was filtered into a 250-mL flask containing 5 g of sodium chloride and shaken vigorously. After separation, the filtrate was condensed to 10 mL in a Kuderna-Danish apparatus. The condensate was then solvent-exchanged into 10 mL of pesticide grade *iso*octane.

GC/FT-IR Instrumentation and Software

The previously described system [8] was updated to include a Digilab (Cambridge, MA) Model 099–0580 capillary interface and a narrow-band 1-mm^2 focal chip, mercury-cadmium-telluride detector (3800–700 cm^{-1}). Two scans per second were collected onto magnetic tape. Chromatography was performed with a Hewlett-Packard (Palo Alto, CA) Model 5880A gas chromatograph with a J&W Scientific (Rancho Cordova, CA) 1.0 μm DB-5 film fused silica capillary column (FSCC) (30 m by 0.32 mm) at a flow rate of 1 mL/min of helium. The end of the FSCC served as the transfer line to the light pipe which was maintained at 280°C for all measurements. A second FSCC line transferred the GC volatiles from the end of the light pipe to the 5880A flame ionization detector (FID). It was demonstrated that the FSCC transfer lines were capable of passing compounds boiling at up to 500°C. After an initial hold time of 30 s the GC oven was ramped from 40 to 280°C at 10°/min and held at 280°C for 20 minutes. All GC injections were 2 μL or less during a splitless injection time of 30 s. All spectra were referenced against the two scans of the chromatogram base line that immediately preceded the analyte peak file. The gas infrared Fourier transform software (GIFTS) has been previously described [15,16]. The EPA search library of 2300 vapor-phase FT-IR spectra was used.

GC/MS Instrumentation and Software

A Finnigan Model 4510 GC/MS spectrometer was used in the electron impact mode. The source was operated at 70 eV and maintained at 150°C. The spectrometer was tuned to EPA requirements using decafluorotriphenylphosphine [17]. Masses 38 to 550 were scanned in 1 s. Version 4.03 of the Finnigan INCOS software was used with a 33 133-mass spectrum library. The GC column and operating conditions were identical to those used for the FSCC/GC/FT-IR runs.

Results and Discussion

Nontarget Compound Analysis

To test the capability of combined FSCC/GC/FT-IR and FSCC/GC/MS to identify every GC volatile analyte in complex environmental sample extracts, two samples representing a spectrum of analytical difficulties were analyzed. The first sample (Sediment 1) was obtained from an uncontrolled dumpsite and contained many isomeric chlorinated aromatics. The second (Sediment 2) was collected from an area contaminated by agricultural pesticide applications. The complexity of these extracts would severely challenge the capability of any single spectral method. The ion and infrared reconstructed chromatograms (RIC and IRC) for these two sample extracts are shown in Figs. 1 and 2. Visual inspection of these chromatograms indicates that, if extracts for FSCC/GC/FT-IR analysis are concentrated about tenfold relative to the concentration required for mass spectral analysis, then it is possible to confirm spectrally about two thirds of the analytes. This is a twofold improvement over our previously reported packed-column GC/FT-IR system.

The analysis results for these sediments are listed in Tables 1 and 2. For identification or class confirmation it can be seen that the GIFTS hit index (HI) was generally smaller than 0.7 while the three Instrument Control Operating System (INCOS) parameters (PURITY, FIT and RFIT) were generally

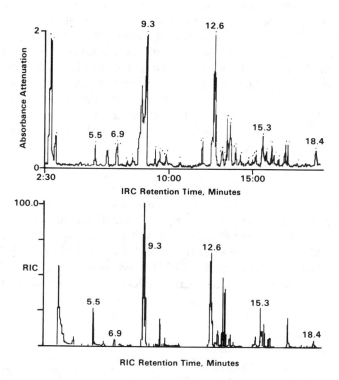

FIG. 1—*FSCC/GC/FT-IR infrared reconstructed chromatogram (IRC) and FSCC/GC/MS reconstructed chromatogram (RIC) of Sediment Extract 1. Retention time markers have been added to each chromatogram. Dots in the IRC denote FT-IR detectable analytes.*

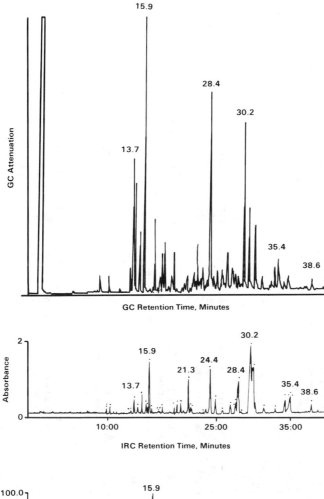

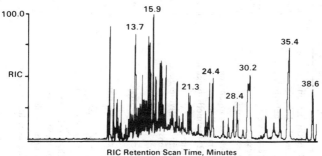

FIG. 2—(Top) *Flame ionization detector chromatogram:* (middle) *FSCC/GC/FT-IR infrared reconstructed chromatogram (IRC);* (bottom) *FSCC/GC/MS reconstructed chromatogram (RIC) of Sediment Extract 2. Retention time markers have been added to each chromatogram. Dots in the IRC denote FT-IR detectable analytes.*

TABLE 1—*Comparison of FSCC/GC/FT-IR and FSCC/GC/MS results for Sediment 1.*

Retention Time, min	GC/FT-IR Hit	GIFTS HI	GC/MS Hit	Finnigan INCOS			Class	Agreement[a]	
				PURITY	FIT	RFIT		On-Line	Visual
2.69	3-methylpentane	0.4066	in solvent front						
	2-methylbutane	0.4152							
	2,3,4-trimethylpentane	0.4229							
2.90	pentane	0.0748	in solvent front						
	hexane	0.0850							
	3-ethylpentane	0.1557							
3.24	methlycyclopentane	0.1766	methylenecyclopentane	845	866	941	alkane	+	+
	2-methylbutane	0.1945	bicyclo[3.1.0]hexane	877	903	939			
	2,2,4-trimethylhexane	0.2187	ethenylcyclobutane	878	942	929			
5.45*	toluene	0.2559	toluene	948	989	955	aromatic hydrocarbon	+	+
	2-chloroethylbenzene	0.4839	phenylmethylhydrazine	896	968	922			
	phenylacetonitrile	0.4868	2-propenylidenecyclo-butene	908	963	921			
6.16[b]	tetrachloroethylene	0.4824	tetrachloroethylene	951	985	960	tetrachloro-alkene	+	+
	hexachloroethane	0.7443	4,6-dichloro-1,3,5-triazine-2-amine	484	670	671			
	1,2-difluoro-1,1,2,2-tetrachloroethane	0.9869	1,2-dichloro-3,3,3-trifluoropropene	481	672	662			
6.87[b]	chlorobenzene	0.4891	chlorobenzene	930	980	940	chlorobenzene	+	+
	bromobenzene	0.5582	chlorofluorobenzene	810	853	898			
	iodobenzene	0.7249	1,2-dichloropropene-1	350	433	762			
6.92	2,2,2-trifluoroethanol	0.5533	no trifluoromethyl compound in this region of the RIC				trifluoromethyl alcohol or ether[c]		
	1,3-di-trifluoro-methylbenzene	0.5917							
	1,1,1-trifluoro-2-propanol	0.6369							

RT	Compound	Value	Identification				Class	trifluoromethyl aromatic[d]	no trifluoromethyl compound in this region of the RIC
6.97	4-fluorotrifluoromethylbenzene	0.7514							
	4-chlorotrifluoromethylbenzene	0.7621							
	trifluoromethylbenzene	0.7847							
7.28[b]	1,3-dimethylbenzene	0.7065	1,2-dimethylbenzene	908	972	929	dimethyl benzene	+	+
	3,3'-bitolyl	0.7295	1,4-dimethylbenzene	905	968	924			
	3-methylbenzonitrile	0.7730	ethylbenzene	900	963	917			
8.12	methylhydroquinone	1.0118	1,1,2,2-tetrachloroethane	931	986	939	chloroalkane	—	+
	1,1,1-trifluoro-2-propanol	1.0285	1,2-dichloro-2,2-difluoroethane	532	717	736			
	2-naphthol	1.0431	2,2-dichloro-1,1,1-trifluoroethane	580	809	713			
9.08[b]	2-chlorotoluene	0.3169	4-chlorotoluene	909	985	909	chlorotoluene	+	+
	2-chlorobenzyl chloride	0.5461	2-chlorotoluene	900	972	900			
	2-bromotoluene	0.5756	3-chlorotoluene	890	977	893			
9.29[b]	benzaldehyde	0.3709	benzaldehyde	952	983	967	benzaldehyde	+	+
	4-methylbenzaldehyde	0.4421	1,2,4-trioxolane-3-phenyl	765	799	939			
	2-bromobenzaldehyde	0.5759	1,2,4-trioxolane-3,5-diphenyl	766	801	935			
9.40	heptane	0.3669	1-methyl-4-(1-methylethyl)cyclohexane-cis	885	965	911	alkane	+	+
	octane	0.3751	1-methyl-4-(1-methylethyl)cyclohexane-trans	883	962	909			
	2,6,10,14-tetramethylpentadecane	0.3788	m-menthane	877	974	890			
9.65	1,3-dimethylbutylamine	0.8049	2-methyl-propene-2	867	915	919	alkane	—	—
	1,1,3,3-tetramethylbutylamine	0.8208	2,2-dimethyldecene-4(Z)	827	905	889			
	isobutylsulfide	0.8215	2,2-dimethyldecene-4(E)	825	899	888			

TABLE 1—Continued.

Retention Time, min	GC/FT-IR Hit	GIFTS HI	GC/MS Hit	Finnigan INCOS			Class	Agreement[a]	
				PURITY	FIT	RFIT		On-Line	Visual
9.87	2-chloro-5-trifluoromethylacetophenone	0.5137	no chlorotrifluoromethyl aromatic in this region of the RIC				chlorotrifluoromethyl aromatic[e]		
	3-trifluoromethyl-benzylchloride	0.6631							
	6-chloro-3-trifluoromethylphenol	0.6714							
10.11	4-chlorobiphenyl	0.7810	no chloroaromatic in this region of the RIC				dichlorobenzene and benzyl chloride coelute[f]		
	α-4-chlorophenyl benzylchloride	0.7875							
	benzylchloride	0.7933							
10.23[b]	6-isopropylquinoline	0.8537	3-cymene	900	977	913	isopropyltoluene	+	+
	4-isopropylbenzylamine	0.8572	4-cymene	896	979	910			
	4-cymene	0.8580	2-cymene	896	978	909			
10.35	benzylfluoride	1.0693	benzyl alcohol	890	978	903	benzyl alcohol	−	+
	amitrole	1.1061	2-cresol	850	927	871			
	pyrrole	1.1134	3-cresol	821	912	868			
10.55[b]	1,2-dichlorobenzene	0.5515	bis-dichlorobenzoyl peroxide	751	796	939	dichlorobenzene	+	+
	1-bromo-2-chlorobenzene	0.5598	1,2-dichlorobenzene	936	994	936			
	1,2-dibromobenzene	0.7663	1,4-dichlorobenzene	936	995	936			
11.68	3-bromo-4-chlorotrifluoromethylbenzene	0.9483	no halotrifluoromethylbenzene in this region of the RIC				halotrifluoromethylbenzene[g]		
	2,5-dichlorotrifluoromethylbenzene	0.9568							
	2-ethoxyethylmethyl acrylate	0.9568							

11.79[b]	propyl benzoate	0.3534	methyl benzoate	953	986	964	alkyl benzoate	+	+
	ethyl benzoate	0.3704	benzoylhydrazide	892	934	949			
	butyl benzoate	0.3901	N-aminocarbonyl-benzamide	616	669	889			
12.50[b]	1,3-dichloro-2-butene	0.5838	2-chlorobenzyl chloride	853	965	872	dichlorotoluene	+	+
	2,5-dichlorotoluene	0.5860	2,6-dichlorotoluene	773	989	780			
	2,4-dichlorotoluene	0.6513	3-chlorobenzyl chloride	761	992	761			
12.56[b]	4-chlorobenzaldehyde	0.3683	3-chlorobenzaldehyde	910	998	910	chlorobenzalde-hyde	+	+
	4-bromobenzaldehyde	0.4907	4-chlorobenzaldehyde	907	999	907			
	4-methylbenzaldehyde	0.5861	2-chlorobenzaldehyde	900	991	900			
12.68[b]	benzyl chloride	0.6166	3-chlorobenzyl chloride	887	962	917	benzal chloride	+	+
	benzal chloride	0.6498	benzal chloride	870	946	916			
	benzhydryl chloride	0.6584	2-chlorobenzyl chloride	796	869	910			
12.93	3,4-dichlorotoluene	0.7482	2,6-dichlorotoluene	874	972	892	dichlorotoluene	+	+
	4-chloro-1,3-dimethyl-benzene	0.7847	2-chlorobenzyl chloride	800	900	886			
	6-chloro-2-isobutyro-toluidide	0.7965	3-chlorobenzyl chloride	857	971	877			
13.28	2,2,3-trimethylpentane	0.2347	bicyclo[2.2.1]heptane-2-chloro-1,7,7-trimethyl	748	965	748	alkane	+	+
	2,3,3-trimethylpentane	0.2359	bicyclo[2.2.1]heptane-2-OL-1,7,7-trimethyl-(endo)	685	924	704			
	2,3,4-trimethylpentane	0.2479	bicyclo[2.2.1]heptane-2-OL-1,1,1-trimethyl (exo)	608	893	655			
13.43[b]	1,2,4-trichlorobenzene	0.3920	1,3,5-trichlorobenzene	943	995	943	trichlorobenzene	+	+
	2,4,6-tribromoaniline	0.7470	1,2,3-trichlorobenzene	929	999	929			
	2,5-dichlorobenzyl alcohol	0.8393	1,2,4-trichlorobenzene	918	996	918			

TABLE 1—Continued.

| Retention Time, min | GC/FT-IR Hit | GIFTS HI | GC/MS Hit | Finnigan INCOS | | | Class | Agreement[a] | |
				PURITY	FIT	RFIT		On-Line	Visual
13.65	2,5-dimethylhexene-3	0.7015	no alkane in this				alkane[h]		
	2,2,3-trimethylpentane	0.7132	region of the RIC						
	isobutyldisulfide	0.7152							
13.75[b]	4-chlorobenzyl chloride	0.7903	dichloromethyl benzene	858	969	884	chlorobenzyl	+	+
	4,4'-dichlorobiphenyl	0.8786	3-chlorobenzyl chloride	783	887	872	chloride		
	4-chlorotoluene	0.8843	2-chlorobenzyl chloride	855	979	862			
14.01[b]	1,2,3-trichlorobenzene	0.5690	1,2,3-trichlorobenzene	890	997	890	trichlorobenzene	+	+
	2,6-dichlorobenzonitrile	0.8322	1,3,5-trichlorobenzene	882	980	887			
	2,6-dichlorotoluene	0.8928	1,2,4-trichlorobenzene	869	991	869			
14.53	methyl (2-methyl)- benzoate	0.7688	no ester in this region of the RIC				aryl ester[i]		
	dibutyl phthalate	0.7845							
	dipropyl phthalate	0.7846							
14.74	2-iodothiophene	1.0075	no monohalo aromatic				monohalo aromatic[c]		
	2-nitroiodobenzene	1.0086							
	2-nitrobiphenyl	1.0088							
14.90	1,1,2-trichloroethane	0.5306	2,4-dichlorobenzyl chloride	898	981	905	trichlorotoluene	+	+
	bromochloromethane	0.5862	trichloromethylbenzene	877	987	884			
	2-chlorobenzyl chloride	0.5999	3,4-dichlorobenzyl chloride	862	982	871			
15.30[b]	2,4,5-trichlorotoluene	0.3327	2,4-dichlorobenzyl chloride	897	987	904	trichlorotoluene	+	+
	1,2,4,5,-tetrachloro- benzene	0.8037	3,4-dichlorobenzyl chloride	847	978	862			
	2,4,5-trichlorophenol	0.8623	phenylphosphonic dichloride	639	872	718			

15.39	cyclohexanethiol	0.3540	1,1'-bicyclohexyl	894	952	926	cycloalkane	−	+
	2-bromoethylcyclohexane	0.3634	1,1'-(1,2-ethanediyl)-bis-cyclohexane	746	879	798			
15.50[b]	2-cyclohexylcyclohexanol	0.3682		701	759	798			
	2,3,6-trichlorotoluene	0.4017	2,4-dichlorobenzyl chloride	877	978	886	trichlorotoluene	+	+
	1,2,3,4-tetrachlorobenzene	0.7663	3,4-dichlorobenzyl chloride	827	970	845			
	2,6-dichlorobenzonitrile	0.8128	trichloromethylbenzene	663	812	670			
15.83[b]	1,2,4,5-tetrachlorobenzene	0.5265	1,2,3,5-tetrachlorobenzene	903	982	916	tetrachlorobenzene	+	+
	2,3,4-trichloroanisole	0.9029	1,2,4,5-tetrachlorobenzene	871	975	883			
	2,3,6-trichlorotoluene	0.9386		852	988	861			
15.94	2,3,6-trichlorotoluene	0.8048	2,4-dichlorobenzyl chloride	868	982	877	trichlorotoluene	+	+
	adiponitrile	0.8616	3,4-dichlorobenzyl chloride	816	971	833			
	2-methyl-4-chlorotoluene	0.8676	trichloromethylbenzene	679	839	684			
16.19[b]	butyl benzoate	0.7533	butyl benzoate	688	928	733	alkyl benzoate	+	+
	isobutyl benzoate	0.7622	isobutyl benzoate	714	939	727			
	dibutyl terephthalate	0.7729	5-benzyloxypentanol	652	899	700			
16.61[b]	1,2,3,4-tetrachlorobenzene	0.3641	1,2,3,5-tetrac hloro-benzene	926	987	935	tetrachlorobenzene	+	+
	1,2,3-trichlorobenzene	0.8714	1,2,3,5-tetrachlorobenzene	896	978	909			
	N-methyl-4-toluene sulfonamide	0.8827	1,2,4,5-tetrachlorobenzene	877	989	885			

TABLE 1—Continued.

Retention Time, min	GC/FT-IR Hit	GIFTS HI	GC/MS Hit	Finnigan INCOS PURITY	FIT	RFIT	Class	Agreement[a] On-Line	Visual
16.76[b]	phenyl ether	0.4196	phenyl ether	855	955	886	aryl ether	+	+
	4-bromophenylphenyl ether	0.4205	diphenyl carbonate	523	581	836			
	2-phenoxybiphenyl	0.5516	3-ethoxybiphenyl	651	754	758			
17.36	1,10-dichlorodecane	0.6633	1-chlorodecane	717	943	740			
	1,9-dichlorononane	0.6656	1-chlorononane	726	976	727	fatty alkyl halide	–	+
	1-butylcyclohexanol	0.6670	1-chlorotetradecane	691	900	715			
18.35	cyclohexanethiol	0.2407	1,1'-bicyclohexyl-3-OL	717	467	730	cycloalkane	–	+
	ethynylcyclohexane	0.2839	1,1'-bicyclohexyl-4-OL	703	967	705			
	dicyclohexylketone	0.3205	1,1'-bicyclohexyl-2-OL	685	946	692			

[a] + denotes agreement; – denotes disagreement.
[b] Identity confirmed.
[c] Reconstructed ion chromatogram (RIC). FT-IR bands at 1340, 1287, and 1170 cm^{-1}.
[d] FT-IR bands at 1333, 1275, 1170, 1075, 925, and 840 cm^{-1}.
[e] FT-IR bands at 1335, 1275, 1230, and 1160 cm^{-1}.
[f] FT-IR bands at 1470, 1270, 1090, 1015, and 820 cm^{-1}.
[g] FT-IR bands at 1320, 1165, and 1145 cm^{-1}.
[h] FT-IR bands at 2975, 2930, and 2890 cm^{-1}.
[i] FT-IR bands at 1750, 1280, 1263, and 680 cm^{-1}.
[j] FT-IR bands at 1555, 1450, 850, and 785 cm^{-1}.

TABLE 2—Comparison of FSCC/GC/FT-IR and FSCC/GC/GC/MS results for Sediment 2.

Retention Time, min	GC/FT-IR Hit	GIFTS HI	GC/MS Hit	PURITY	FIT	RFIT	Class	On-Line	Visual
9.71	1,3-diethylbenzene	0.4559	1-ethyl-3-methylbenzene	905	992	905	dialkylbenzene	+	+
	1,2,3-trimethylbenzene	0.5179	1-ethyl-4-methylbenzene	904	990	904			
	1,2,3,4-tetramethyl-benzene	0.5216	1-ethyl-2-methylbenzene	899	988	899			
9.84[b]	1,3,5-trimethylbenzene	0.4121	1,3,5-trimethylbenzene	919	997	919	trimethylbenzene	+	+
	1,3-dimethylbenzene	0.6647	1,2,4-trimethylbenzene	918	996	918			
	1,2,4-trimethylbenzene	0.6983	1,2,3-trimethylbenzene	904	991	904			
10.36[b]	1,2,4-trimethylbenzene	0.2125	1,3,5-trimethylbenzene	914	994	914	trimethylbenzene	+	+
	1,2,3,4-tetramethyl-benzene	0.2962	1,2,4-trimethylbenzene	914	994	914			
	1,2,4,5-tetramethyl-benzene	0.3761	1,2,3-trimethylbenzene	901	989	901			
13.01	butylbenzene	0.5015	1-methyl-2-(2-propenyl)-benzene	728	967	740	alkylbenzene	–	+
	1-phenyldodecane	0.5069	1-ethenyl-3,5-dimethyl-benzene	690	923	699			
13.29	sec-butylbenzene	0.5105	1-ethenyl-3-ethylbenzene	689	865	699	inconclusive		
	octachlorocyclopentene	1.0043	tert-amylbenzene	895	983	895			
	hexachlorocyclohexane	1.0175	5-methyl-1(3H)-isobenzo-furanone	785	925	789			
	2,5-dichlorothiophene	1.0185	2-xylylethanol	613	780	779			
13.71[b]	naphthalene	0.4859	azulene	800	994	800	naphthalene	–	–
	1-naphthalene sulfonic acid, dihydrate	0.6161	naphthalene	800	994	800			
14.22	1-chloromethylnaphtha-lene	0.6566	2-H-thiete-2-methylene-4-phenyl-1,1-dioxide	407	541	741			
	2-ethylbutanal	0.6872	no carbonyl compound in this region of the RIC				aliphatic aldehyde[c]		
	2-ethylcaproaldehyde	0.6875							
	isovaleraldehyde	0.7166							

TABLE 2—Continued.

Retention Time, min	GC/FT-IR Hit	GIFTS HI	GC/MS Hit	Finnigan INCOS			Class	Agreement[a]	
				PURITY	FIT	RFIT		On-Line	Visual
14.79	2,4-dibromophenol	0.5930	no phenol in this region of the RIC				2,4-dihalophenol[a]		
	4-bromo-2-chlorophenol	0.6061							
	2,4-dichlorophenol	0.7125							
15.40	2,6-dimethylnaphthalene	0.6722	2-methylnaphthalene	939	990	945	methylnaphthalene	+	+
	phenanthrene	0.6728	1-methylnaphthalene	933	988	940			
	1-chloro-2-methyl-naphthalene	0.6859	1,4-methanonaphthalene-(1,4-H)	916	973	916			
15.68	1-methylnaphthalene	0.4240	2-methylnaphthalene	905	986	912	methylnaphthalene	+	+
	naphthalene	0.5077	1-methylnaphthalene	892	979	905			
	1-ethylnaphthalene	0.5243	1,4-methanonaphthalene-(1,4-H)	882	970	882			
15.89	N,N-dipropylnicotinamide	0.5225	dipropylcarbamothioc acid-S-ethylester	655	796	800	amide	+	+
	tetrabutyl urea	0.5532	N-butyl-N-nitro-1-butanamine	242	318	712			
	2-chloro-N,N-di-isopropyl acetamide	0.6252	4-octanone	263	390	636			
16.20	undecanenitrile	0.2768	heptadecane	868	956	905	alkane	−	+
	3-decyne	0.2794	tetradecane	873	957	902			
	1-chlorodecane	0.2868	heneicosane	839	931	896			
17.05	1,2,3,4-tetramethyl-benzene	0.7490	2-ethylnaphthalene	882	996	882	mono or dialkyl naphthalene	+	+
	hexamethylbenzene	0.7514	1-ethylnaphthalene	874	988	881			
	butylbenzene	0.7534	1,2-dimethylnaphthalene	850	953	850			
17.15	2,6-dimethyl-4-heptanol	0.9264	2,7-dimethylnaphthalene	934	996	934	dimethyl-naphthalene	−	−
	3,3-dimethyl-1-butanol	0.9379	1,7-dimethylnaphthalene	924	985	934			
	2-isopropyl-5-methyl-1-hexanol	0.9397	2,6-dimethylnaphthalene	925	985	934			

17.31	benzene	0.9057	1,8-dimethylnaphthalene	920	978	938	polynuclear aromatic	—	+
	benzyl mercaptan	0.9139	1,7-dimethylnaphthalene	921	982	936			
	bibenzyl	0.9206	1,3-dimethylnaphthalene	936	997	936			
17.35	1,2-diphenylbenzene	1.1146	1,7-dimethylnaphthalene	760.	979	776	polynuclear aromatic	—	+
	α-methylstyrene	1.1239	2,7-dimethylnaphthalene	764	984	774			
	chlorodiphenylmethane	1.1255	2,6-dimethylnaphthalene	756	972	773			
17.74	α-propylcyclohexane methanol	0.3469	pentadecane	883	967	910	alkane	—	+
	undecanenitrile	0.3523	heptadecane	862	954	901			
	3-decyne	0.3532	heneicosane	818	904	840			
19.41	3-decyne	0.4121	2,3,6-trimethylnaphthalene	714	961	738	inconclusive		
	undecanenitrile	0.4199	1,6,7-trimethylnaphthalene	720	965	736			
	1-chlorodecane	0.4218	1,3,6-trimethylnaphthalene	585	982	585			
19.58	1-phenylhexane	0.4224	heptadecane	904	965	927	alkane	—	+
	pentylbenzene	0.4422	heneicosane	869	931	920			
	3-decyne	0.4691	hexadecane	885	980	897			
20.17	1-methyl-2-pyrrolidinone	0.7265	4-chloro-2-methyl formanilide	840	995	840	amide	+	+
	N-methylformamide	0.7821	2-chloro-4-methyl benzenamine	605	976	605			
	phenyl-2-propanone	0.8360	2-chloro-1,4-dimethylbenzene	375	646	436			
20.51	2-ethylhexylamine	0.7139	heptadecane	819	915	877	alkane	+	+
	heptane	0.7172	nonadecane	787	887	872			
	hexanethiol	0.7207	6-methyloctadecane	778	877	862			
20.91	2-methyl-2-butene	0.8428	cannot correlate RIC and IRC peaks				inconclusive		
	1-isopropyl-4-methylbenzene	0.8441							
	2-methyl-2-pentene	0.8506							

TABLE 2—Continued.

Retention Time, min	GC/FT-IR Hit	GIFTS HI	GC/MS Hit	Finnigan INCOS			Class	Agreement[a]	
				PURITY	FIT	RFIT		On-Line	Visual
21.25	2-chloro-5-nitro-trifluoromethylbenzene	0.7158	2,6-dinitro-4-(trifluoro-methyl)-N,N-dipropylani-line	785	944	811	nitrotrifluoro methyl aromatic	+	+
	2-nitro-trifluoro-methylbenzene	0.7608	2,6-dinitro-N-propyl, N-(2-chloroethyl)-4-(trifluoro-methyl)aniline	468	603	666			
	5-bromo-2-nitro-trifluoromethylbenzene	0.7958	2-(acetyloxy)-5-iodo-benzoic acid	203	346	370			
21.52	3-decyne	0.2630	heneicosane	864	929	905	alkane	−	+
	myristonitrile	0.2693	pentacosane	803	878	892			
	undecane	0.2720	nonadecane	833	930	883			
21.66	2,4-dimethylhexane	0.5036	2,6,10,14-tetramethyl-pentadecane	852	942	898	alkane	+	+
	3,3-dimethylhexane	0.5102	heneicosane	790	856	886			
	2-ethyl-4-methyl-1-pentanol	0.5125	nonadecane	792	875	883			
23.33	2-chloro-4-ethylamino-S-triazine	1.0581	cannot correlate RIC and IRC peaks				nitro compound[e]		
	1,3-dichloro-2-nitro-benzene	1.0754							
	2,4-bis-ethylamino-S-triazine	1.0755							
23.73	myristonitrile	0.4490	heneicosane	759	921	814	alkane	−	+
	tetradecane	0.4578	eicosane	735	892	814			
	1-dodecene	0.4579	pentacosane	710	869	804			

24.39	6-methylpicoline aldehyde	0.6751	0,0-diethyl-6-methyl-2-(1-methylethyl)-4-pyrimidyl-phosphorothioc acid	856	997	856	alkyl phosphoro-thioate	+	+
	phosphoramidic acid, dibutyl ester	0.6999	3,5,7-trihydroxy-4-methoxyflavanone	178	465	325			
	ethyl phosphorothioate	0.7004	6-methyl-2-(1-methylethyl)-4-pyrimidone	277	904	284			
25.08	phosphonic acid, ethyl, diethyl ester	0.5799	disulfoton 0,0-diethyl-S-[2-(ethylthio)]	901	996	901	alkyl phosphorus compound	+	+
	α-cyclopropylbenzyl alcohol	0.5876	ester of phosphorothioc acid	822	936	870			
	phosphoric acid, vinyl	0.6141	bis-[2-(ethylthio)-ethyl]	454	873	481			
26.14	1-phenylhexane	0.5793	nonadecane	885	969	909	alkane	–	+
	myristonitrile	0.5863	pentacosane	824	894	905			
	3-decyne	0.5875	heptadecane	886	974	904			
27.24	0,0-dimethylphosphoro-thioc acid	0.9332	methyl parathion	897	998	897	alkyl phosphoro-thioate	+	+
	3,4-dihydro-2-methoxy-2-H-pyran	0.9430	0,0-dimethyl-0-(3-methyl-4-nitrophenyl)phosphoric acid	668	888	749			
	methoxymethylphosphoric acid	0.9811	0,0-dimethyl-0-(trichloro-phenyl)phosphoric acid	240	446	520			
27.74	2-aminoethanol	1.0388	cannot correlate RIC and IRC peaks				inconclusive		
	hexyl phosphite	1.0586							
	2-amino-1-butanol	1.0663							
28.00	1-tert-butyl-4-dodecylbenzene	0.5849	4-tert-butylphenol	739	947	756	alkyl phenol	+	+
	2-(-4-tert-butyl-phenoxy) ethanol	0.6120	3-tert-butylphenol	733	950	741			
	2,4-di-tert-butyl-phenol	0.6396	4-(1,1,3,3-tetramethyl-butyl)phenol	677	933	714			

TABLE 2—Continued.

Retention Time, min	GC/FT-IR Hit	GIFTS HI	GC/MS Hit	Finnigan INCOS			Class	Agreement[a]	
				PURITY	FIT	RFIT		On-Line	Visual
28.39	2-methyl-2-nitro-propane	0.4694	no nitro compound in this region of the RIC				nitro compound[f]	—	+
	2,6-dinitrotoluene	0.5432							
	2-(furfurylamino)-ethanethiol	0.6080							
28.72	1-phenylhexane	0.8548	heptadecane	810	934	861	alkane		+
	pentylbenzene	0.8621	eicosane	737	844	857			
	undecanenitrile	0.8704	heneicosane	767	889	856			
30.20	ethylphosphorothioate	0.6929	ethyl parathion	530	662	798	alkyl phosphoro-thioate	+	+
	phosphonic acid cyano-methyl ester	0.7439	bis-(M-nitrophenyl)-phosphinic acid methyl ester	261	478	418			
	phosphonic acid vinyl, diethyl ester	0.7507	5-iodo-3-nitro-salicylic acid	132	354	365			
30.39	allyl phenyl ether	0.9306	ethyl parathion	854	932	905	aryl phosphorus ester	—	+
	1,2-diphenoxyethane	0.9369	silicic acid tetrabutyl ester	215	410	419			
	1,2-diepoxy-3-phenoxy-ethane	0.9396	bis-(M-nitrophenyl)-phosphinic acid	163	358	415			
30.70[b]	4,4'-dichlorobenzo-phenone	0.6941	4,4'-dichlorobenzo-phenone	685	981	691	bis-chloroaryl ketone	+	+
	4-chlorobenzophenone	0.8059	3,3'-dichlorobenzo-phenone	664	968	668			
			2,4'-dichlorobenzo-phenone	643	948	665			
31.81[b]	methyl oleate	0.5361	methyl oleate	663	989	669	fatty ester	+	+
	butyl palmitate	0.5368	2,3-dihydroxypropyl oleate	630	923	668			
	pentyl myristate	0.5384	methyl 10-octadecenoate	660	987	667			

| | | | | | | | − | + |
RT	Library hit	ratio	identification				ester[g]	
	1,2-dichloroethane	1.1333	methyl ester, 10-octadecenoic acid	702	954	732		
	1,1,1-trichloroethane	1.1838	methyl ester, 9-octadecenoic acid	696	948	731		
	1,1,2,2-tetrachloroethane	1.2104	methyl ester, 8-octadecenoic acid	691	947	727		
34.63	benzyl mercaptan	0.9591	β-endosulfan	614	988	619	inconclusive	
	1,2-dichloroethane	0.9546	α-endosulfan	599	989	602		
	(2-bromoethyl)benzene	0.9694	chlorotripropyl stannane	125	462	260		
34.91	α-chlorotoluene	0.8668	cannot correlate RIC and IRC peaks				inconclusive	
	1,1-dichloroethane	0.9006						
	benzyl mercaptan	0.9113						
34.93	isobutyl chloride	0.9789	cannot correlate RIC and IRC peaks				inconclusive	
	1-chloropentane	1.0032						
	1-chloro-3-methyl-butane	1.0099						
35.34	2-methyl-2-propyl-1,3-propanediol	0.7823	cannot correlate RIC and IRC peaks				inconclusive	
	isobutyl alcohol	0.7836						
	2,2-dimethyl-1,3-butanediol	0.8047						
35.38	4-methyl-2-phenyl-valeronitrile	0.9982	cannot correlate RIC and IRC peaks				inconclusive	
	pentylbenzene	0.9997						
	butylbenzene	1.0031						
38.55	ethyl phosphite	0.9693	methyl azinphos	942	999	942	alkyl phosphorothioate	
	3-bromo-1-propene	1.0133	ethyl azinphos	748	873	821		
	ethyl-4-toluene sulfonate	1.0265	chloromethyl-1,2,3-benzotriazin-4(3H)one	726	877	799		

^{a} + denotes agreement; − denotes disagreement.
^{b} Identity confirmed.
^{c} FT-IR band at 1750 cm^{-1}.
^{d} FT-IR bands at 3585, 1275, and 1190 cm^{-1}.
^{e} FT-IR band at 1575 cm^{-1}. Three of first five library hits are nitro compounds.
^{f} FT-IR band at 1560 cm^{-1}. Four of first five library hits are nitro compounds.
^{g} FT-IR band at 1745 cm^{-1}.

greater than 850 (optimum GIFTS indexes are associated with small HIs while optimum INCOS indexes approach 1000). Mass spectral agreement with FT-IR was more closely correlated with FIT than with PURITY or RFIT. The FT-IR library usually did not contain spectra of those analytes of Sediment 2 corresponding to the first three GC/MS hits; however, the converse was true of Sediment 1. Although the primary FSCC/GC/MS problem was an inability to differentiate aromatic isomers, a total of five oxygen and four trifluoromethyl-containing compounds were not identified. This may be a result of weak molecular ions generated under electron impact conditions for oxygen compounds [18] or misidentification of coeluting analytes. The inability of mass spectrometry to differentiate azulene from naphthalene has been noted earlier [19], but the problem with trifluoromethyl-containing molecules is not understood. The primary FSCC/GC/FT-IR identification problems were an inability to differentiate members of homologous series and misidentifying alkanes as substituted alkanes. The latter error may be a result of high relative spectral intensities for the carbon-hydrogen (C-H) and functional group IR bands. For example, the strong nitrogen-hydrogen (N-H) IR band observed in the condensed-state spectra of aliphatic amines is almost absent in gas-phase spectra, leading to a gas-phase spectral similarity for aliphatic amines and their corresponding alkanes [20]. Thus the salient feature to note is the few instances when both spectral techniques experience identification problems with the same analyte. This feature clearly demonstrates the capability of the combined techniques to identify a greater number of analytes then either technique employed alone.

Comparison of FSCC/GC/MS and FSCC/GC/FT-IR Search Techniques

The INCOS and GIFTS search software index results were then determined by comparing the visual spectral check results with the on-line library search hits for each analyte. After visually determining the most likely compound class for each analyte the mathematical hit index difference between the corresponding search hit and the next most likely hit was tabulated. It was seen that this difference was generally greatest for those cases where the two spectral techniques agreed. It was also noted that the GIFTS hit index differences (ΔHI) were the largest for Sediment 1, which contains many analytes whose spectra are contained in the FT-IR search library. Sediment 2 apparently contains many analytes whose spectra are not in this library, which is small (2300 spectra); this gives credence to the suggestion of Isenhour et al [21] that the FT-IR search library be used as a functional group "prefilter" for the much larger mass spectral library (67 128 discrete mass spectra available [22]). It thus appears that spectral hit index mathematical differences may be used to determine the validity of on-line search hits and a substantial data base of such indexes should be tabulated to establish reliable on-line search acceptance/rejection criteria. The reliability of using such a data base of FT-IR and MS search indexes validated by visually-checked results can be ascertained from Table 3. It can be seen that the two techniques, when checked visually, agree on the compound class for 95% of all FSCC/GC/FT-IR and FSCC/GC/MS correlatable analytes (63 of 66) and 71% of all FSCC/GC/FT-IR detectables (63 of 89). Identification agreement was highest for Sample 1, which was a steam-distilled sample containing mostly chlorinated compounds, many of which are in the GC/FT-IR search library. Sediment 2 received a modest amount of cleanup but contained many organophosphorus pesticides, a compound class poorly represented in the GC/FT-IR search library. (Weiboldt [23] has reported 27 phosphorus compounds in this EPA vapor-phase library.) It may be further noted that these extracts contain very few of the EPA priority pollutants; thus a strict priority pollutant target compound approach would have missed most of the sample contaminants. This is particularly disturbing since the costs of sampling, shipping, workup, and storage have already been expended.

Conclusions

Combined capillary column GC/FT-IR and GC/MS is a powerful spectral tool with the capability to identify all or most of the GC-volatile analytes in environmental extracts. Using the much smaller vapor-phase infrared search library in conjunction with large mass spectral libraries can eliminate a principal drawback to the "stand-alone" GC/FT-IR method. To maximize the utility of this approach, future expansion of the EPA vapor-phase library should focus on the spectra of compounds not amenable to GC/MS (for example, aromatic isomers) and on the establishment of firm vapor-phase group frequencies for all toxic chemical classes. Another obstacle to implementation of the combined spectral techniques is their disparate sensitivities. Griffiths and Yang have proposed several straightforward ways to provide a tenfold increase in the sensitivity of GC/FT-IR, thus lowering its identification limit for the weakest infrared absorbers to the low nanogram range [24]. Among the suggestions are better FT-IR lightpipes, smaller focal area detec-

TABLE 3—*Summary of confirmation results for FSCC/GC/FT-IR and FSCC/GC/MS visually checked identifications.*

Sample	Identification	Compound Class[a]	Inconclusive	U.S. EPA Priority Pollutants[b]
Sediment 1	21/43	32/43	11/43	9
Sediment 2	5/46	31/46	15/46	3

[a]Includes positively identified compounds.
[b]U.S. EPA target compounds cited in Ref 28 identified by FSCC/GC/FT-IR or FSCC/GC/MS.

tors, and a double-beam approach [25]. The remaining obstacle to widespread application of the combined techniques is the development of a cost-effective approach that is amenable to the high sample throughputs of the EPA. We believe this can be achieved through a direct interfacing of both spectrometers. This approach has already been achieved in other laboratories [12-14]. Suitable computer software will be required to generate a single reconstructed chromatogram and to interactively search both spectral libraries and produce a single hard-copy hit list [21]. The combination of powerful software and a reduction in spectral visual checking (which substantially increases the cost of current GC/MS analysis) may reduce the cost of the combined techniques to that of current stand-alone GC/MS analysis. Increased sales of FT-IR spectrometers (expected to exceed the sales of dispersive infrared instruments for the first time in 1984 [26]), coupled with the availability of low-cost mass spectral detectors [27], should also provide a stimulus to lower the capital cost of a combined FT-IR/MS system below that of current individual floor model FT-IR or MS spectrometers.

Disclaimer

This paper has been reviewed in accordance with the U.S. Environmental Protection Agency's peer and administrative review policies and approved for presentation and publication. Mention of trade names or commercial products does not constitute endorsement or recommendation for use.

References

[1] Maugh, T. H., II, *Science*, Vol. 215, No. 4532, 29 Jan. 1982, pp. 490-493.
[2] *Federal Register*, Vol. 43, No. 243, 18 Dec. 1978, pp. 58946-59028.
[3] *Chemical and Engineering News*, Vol. 60, No. 23, 7 June, 1982, pp. 44-48.
[4] ACS Committee on Environmental Improvement, *Analytical Chemistry*, Vol. 52, No. 14, Dec. 1980, pp. 2242-2249.
[5] ACS Committee on Environmental Improvement, *Analytical Chemistry*, Vol. 55, No. 14, Dec. 1983, pp. 2210-2218.
[6] Christman, R. F., *Environmental Science and Technology*, Vol 16, No. 11, Nov. 1982, p. 594A.
[7] Griffiths, P. R., de Haseth, J. A., and Azarraga, L. V. *Analytical Chemistry*, Vol. 55, No. 13, Nov. 1983, pp. 1361A-1387A.
[8] Gurka, D. F., Laska, P. R., and Titus, R., *Journal of Chromatographic Science*, Vol. 20, No. 4, April 1982, pp. 145-153.
[9] Gurka, D. F. and Betowski, L. D., *Analytical Chemistry*, Vol. 54, No. 11, Sept. 1982, pp. 1819-1824.
[10] Newton, D. L., Erickson, M. D., Tomer, K. B., Pellizzari, E., and Gentry, P., *Environmental Science and Technology*, Vol. 16, No. 4, April 1982, pp. 206-213.
[11] Shafer, K. H., Bjorseth, A., Tabor, J., and Jakobsen, R. J., *Journal of High Resolution Chromatography and Chromatography Communications*, Vol. 3, No. 3, March 1980, pp. 87-90.
[12] Wilkins, C. L., Giss, G. N., White, R. L., Brissey, G. M., and Onyiriuka, E. C., *Analytical Chemistry*, Vol. 54, No. 13, Nov. 1982, pp. 2260-2264.
[13] Jakobsen, R. J. and Shafer, K. H., Paper 91, presented at the 8th Annual FACSS Meeting, Federation of Analytical Chemistry and Spectroscopy Societies, Philadelphia, PA, Sept. 1981.

[14] Crawford, R. W., Hirschfeld, T., Sanborn, R. H., and Wong, C. M., *Analytical Chemistry*, Vol. 54, No. 4, April 1982, pp. 817–820.

[15] de Haseth, J. A. and Isenhour, T. L., *Analytical Chemistry*, Vol. 49, No. 13, Nov. 1977, pp. 1977–1981.

[16] Hanna, A., Marshall, J. C., and Isenhour, T. L., *Journal of Chromatographic Science*, Vol. 17, No. 8, Aug. 1979, pp. 434–440.

[17] Budde, W. L. and Eichelberger, J. W., "Performance Tests for the Evaluation of Computerized Gas/Chromatography/Mass Spectrometry Equipment and Laboratories," EPA-600/4-80-025, U.S. EPA, Cinicinnati, OH, 1980.

[18] McLafferty, F. W., *Interpretation of Mass Spectra*, 2nd ed., W. A. Benjamin, Reading, PA, 1973, p. 31.

[19] Hites, R. A., *Environmental Science and Technology*, Vol. 11, Nov. 1982, p. 595A.

[20] Welti, D., *Infrared Vapour Spectra*, Heyden and Son, London, 1970, pp. 24–28.

[21] Williams, S. S., Lam, R. B., Sparks, D. T., Isenhour, T. L., and Hass, J. R., *Analytica Chimica Acta*, Vol. 138, June 1982, pp. 1–10.

[22] *Chemical and Engineering News*, Vol. 62, No. 4, 23 Jan. 1984, p. 14.

[23] Wieboldt, R. C., "Correlation and Interferometric Data Analysis Applied to Gas Chromatography Fourier Transform Infrared Spectroscopy," Ph.D. thesis, University of North Carolina, Chapel Hill, 1981, p. 46.

[24] Griffiths, P. R. and Yang, P. W. J., *Applied Spectroscopy*, Vol. 38, No. 6, 1984, pp. 816–821.

[25] Kuehl, D., Kemeny, G. J., and Griffiths, P. R., *Applied Spectroscopy*, Vol. 34, No. 2, March 1980, pp. 222–224.

[26] Borman, S. A., *Analytical Chemistry*, Vol. 55, No. 12, Oct. 1983, p. 1160A.

[27] Stafford, G. C., Kelley, P. E., and Bradford, D. C., *American Laboratory*, Vol. 15, No. 6, June 1983, pp. 51–58.

[28] Budde, W. L. and Eichelberger, J. W., *Analytical Chemistry*, Vol. 51, No. 4, April 1978, p. 567A.

Risk Assessment/Biological Test Methods

James Dragun[1] *and Theodore G. Erler*[2]

Human Exposure Assessment: Basic Elements and Applications in Remedial Engineering Actions at Sites Containing Contaminated Soil and Groundwater

REFERENCE: Dragun, J. and Erler, T. G., **"Human Exposure Assessment: Basic Elements and Applications in Remedial Engineering Actions at Sites Containing Contaminated Soil and Groundwater,"** *Hazardous and Industrial Solid Waste Testing: Fourth Symposium, ASTM STP 886,* J. K. Petros, Jr., W. J. Lacy, and R. A. Conway, Eds., American Society for Testing and Materials, 1986, pp. 165–175.

ABSTRACT: When a soil or groundwater contamination problem is discovered, the attention of the public and of regulatory agencies focuses on the potential effects on human health. However, human exposure to a chemical at some minimal concentration must occur before an adverse effect on human health is realized. The exposure assessment quantifies this exposure; an exposure assessment is the determination or estimation of the magnitude, frequency, duration, and routes of exposure.

This paper addresses the basic elements and applications of the exposure assessment. First, this paper will define and discuss the six basic elements of an exposure assessment. Second, this paper will discuss the usefulness of the exposure assessment in evaluating remedial actions to be implemented at hazardous waste sites. Third, this paper presents an actual case study, involving an existing Superfund site, in which an exposure assessment was performed to select a cost-effective remedial action plan that provided the greatest benefit at an acceptable risk.

KEY WORDS: human exposure, exposure pathways, exposure analysis, risk-benefit analysis, hazardous wastes, Superfund, remedial investigations, feasibility studies

Before undertaking remedial action at sites containing contaminated soil and groundwater, engineers and scientists must address the potential adverse

[1]Senior scientist, E. C. Jordan Co., Southfield, MI 48075; formerly with Kennedy/Jenks Engineers.
[2]Civil engineer, Kennedy/Jenks Engineers, San Francisco, CA 94105.

health effects that may result from the implementation of remedial actions. There can be no adverse health effects without chemical exposure to the population of concern; therefore, the extent of exposure is an important factor that determines which remedial actions are effective in abating adverse effects. In other words, remedial action should be a systematic engineering scheme, using various treatment processes and operations, for the abatement—or if possible the elimination—of exposure pathways [1].

The Exposure Assessment

Exposure is defined as the contact between a subject and a chemical. Exposure must occur before an adverse health effect is realized [2]. An exposure assessment is the determination or estimation—qualitative or quantitative—of the magnitude, frequency, duration, and routes of exposure. Although the structure of a specific exposure assessment will depend on its purpose and the sources of concern, an exposure assessment should address, in general, six key elements:

1. Sources
2. Environmental releases
3. Exposure pathways
4. Resultant environmental concentrations
5. Exposed population(s)
6. Integrated exposure analysis

The first section of this paper will briefly review their definitions and relate them to sites containing contaminated soil and groundwater. The next section will then discuss an actual case study, involving a Superfund site, in which an exposure assessment was performed to select a cost-effective remedial action plan that provided the greatest benefit at an acceptable risk.

Sources of Release

The first element—sources—identifies the points at which a chemical may enter the soil or groundwater at the site of concern. Possible sources of the chemical's release include areas of production or processing, storage or stockpiles, transportation, accidental/incidental production as a side reaction, temporary or permanent disposal, and the like.

Environmental Release

The element of environmental release characterizes and quantifies the mass release rates of the chemical to the environment. The release may be airborne, water-borne, or by spillage or leakage onto soil.

In many cases, environmental release of chemicals from industrial activity has been going on at the site for years or even decades before discovery; as a result, little is known about mass release rates. However, mass release rates should be estimated so that the magnitude of the overall problem can be ascertained.

Exposure Pathways

The element of exposure pathways addresses the question of how a chemical travels from a source to its ultimate sink, beginning with an identification and evaluation of all transport and transformation pathways of the chemical at the site. Dragun et al [3] have recently reviewed the chemical, physical, and biological reactions that organic chemicals undergo in soil-groundwater systems. This information is then integrated with information on sources and releases, and the environment fate of the chemical is predicted. An exposure pathway is any point on the environmental pathway that has (1) significant concentrations of the chemical and (2) a significant potential for human exposure.

The three primary routes of human exposure to chemicals are dermal contact (topical exposure), inhalation, and ingestion [4,5]. Topical exposure occurs when a chemical comes in contact with the skin. Topical exposure usually does not represent a serious problem with contaminated groundwater.

Exposure to airborne chemicals by inhalation is ubiquitous in most industrialized regions of the world [5]. Doull et al [4] believe exposure by inhalation to be the most significant route for eliciting a toxic response. Emissions of volatile chemicals and atmospheric dispersion of contaminated water droplets may originate from agricultural, domestic, and industrial applications where contaminated groundwater is sprayed or otherwise aerated. Volatilization of organics during storage, treatment of contaminated groundwater, or excavation of a site may represent a significant route of inhalation exposure of humans and animals to toxicants in localized areas. Also, inhalation of contaminant dust during remedial action or of smog resulting from release of volatile organics to the atmosphere may represent another source of respiratory exposure.

In the case of groundwater contamination, exposure by ingestion is a result of (1) drinking contaminated water, (2) ingesting food products that absorbed the contaminant during irrigation or processing with contaminated water, or (3) ingesting fish/shellfish that may have bioaccumulated the contaminant after its discharge to surface waters. Ingestion of contaminated drinking water is usually the route of exposure of greatest concern.

Resultant Environmental Concentration

The magnitude of the dose received by any member of the subject population is a function of the frequency and amount of a chemical available at the exchange boundaries (gastrointestinal tract, skin, lungs). It is therefore necessary to measure or estimate the fourth element, the exposure concentration of the chemical to the receptor population.

The resultant concentration is a function of the release rate at the source and the interaction of transport and transformation processes between the source and the subject. Resultant concentrations can be derived by two mechanisms: environmental pathway models and empirical methods.

The function of modeling is to predict concentrations of chemicals at the point where a receptor subject or population is potentially exposed. Mathematical modeling is a highly sophisticated discipline, and a detailed discussion of its importance and utility is beyond the scope and intention of this paper. However, Bonazountas [6] and Burns [7] have recently reviewed the usefulness of several soil-groundwater and aquatic mathematical models.

Empirical methods involve the measurement of a chemical in a real or simulated environmental medium (air, surface water, soil, groundwater, or sediment). These methods are generally accurate and can cut through the perplexing maze of variables and interactions that control the fate of a chemical at a specific site. However, they may be resource-intensive and require careful design and execution. Simulated environmental media include, for example, soil leaching columns and microcosms. The measurement of real media usually involves the collection of monitoring data.

When available, monitoring data are the preferred information resource for determining ambient concentrations at the point of human exposure.

Exposed Population

In the fifth element, populations that will be subject to exposure are analyzed primarily by the geographic distribution of the pollutant and the human activities associated with the primary exposure pathways. Some information regarding the characteristics of the population is usually needed and may include population size, location, and characteristics (for example, domestic residents, workers, schoolchildren, etc.). In order to completely evaluate the significance of exposure, one should determine not only who is exposed but how many individuals are exposed at various environmental concentrations and what demographic characteristics affect exposure. In this paper, effects of exposure to biotic populations other than humans are not addressed; however, they may be significant at some sites containing contaminated soil and groundwater.

Integrated Exposure Analysis

The sixth element, the integrated exposure analysis, compiles the total exposure "picture"; it integrates the information provided by the five elements listed above into a series of estimated exposure levels that may occur during one or more exposure scenarios.

Exposure occurs, in general, under six scenarios: (1) occupational exposure, (2) consumer exposure, (3) exposure resulting from disposal and cleanup operations, (4) exposure through food, (5) exposure through drinking water, and (6) exposure from the ambient environment. For example, an exposure assessment involving groundwater contamination may consider a scenario involving domestic water usage, which would cover the exposure to populations residing in the vicinity of the site and relying on wells for drinking water. It may also consider a scenario involving construction workers who would implement a plan of remedial action at the site. It may consider a scenario involving exposure of a subdivision's population to emissions from a hazardous waste treatment, storage, or disposal facility. In some assessments, only one scenario may be cause for concern; in others, multiple scenarios may be involved.

For each scenario, the elements of an exposure assessment—sources, releases, pathways, resultant environmental concentrations, and exposed populations—are identified and tabulated. Next, the cumulative population exposure for each scenario is calculated by multiplying the number exposed by the duration, the exposure level, and the fraction absorbed for each exposure route. Then, the cumulative population exposure for each scenario is summed to give an integrated exposure. An analysis of this integrated exposure along with action levels derived from human health effects data should result in a better understanding of the exposure problem as well as a better understanding of potential adverse effects on human health.

Structure and Level of Detail of Exposure Assessments

The six elements discussed above are appropriate for exposure assessments addressing global, national, regional, county, site-specific, and workplace environments. They are applicable to both single media and multimedia assessments. They can be utilized to consider past, present, or future exposure scenarios.

The structure of a specific exposure assessment will depend on its purpose, the sources of concern, and the exposure media evaluated. The order in which topics appear is arbitrary, and all the elements listed above may not necessarily be applicable to all exposure assessments. However, the assessment should generally proceed in a logical order from sources to exposure estimates.

The level of detail of the exposure assessment will depend on the scope and purpose of the assessment. Some exposure assessments may require the development of numerical values for sources, environmental releases, chemical transport and transformation pathways, resultant environmental concentrations, exposed populations, and so on. Other assessments may only require a qualitative description of these elements. The level of detail is determined by the analyst, who must take into account the end use of the assessment.

The case study presented below involved a very simple level of detail for an exposure assessment. It utilized some empirical methods to determine resultant concentrations and very basic mathematical modeling to calculate the migration of a contaminant groundwater plume. It involved a very qualitative identification of the exposed population, the population size, and the level of exposure. Yet, this level of detail was adequate for choosing the most cost-effective remedial action plan.

Case Study: Selecting a Remedial Action Plan for a PCB-Contaminated Superfund Site

Sources and Environmental Releases

The case study project site was an aluminum casting plant located on approximately five acres of land near the Pacific coast in northern California [8]. During the ordinary operation of the casting machines, leakage of cooling water and hydraulic fluid occurred; collector drains, which surrounded each machine, allowed these fluids to drain into a concrete sump. The fluids were occasionally removed by an oil collection service; however, they were usually pumped onto the field south of the plant.

Between 1966 and 1972, hydraulic fluid used by most casting machines contained polychlorinated biphenyls (PCBs; Aroclor 1242); therefore, PCBs were discharged onto the surface of the soil at the southern part of the site. The exact chemical composition of the hydraulic fluid in use from 1966 to 1981 is not known. However, in 1981, ethylene glycol was used as the hydraulic fluid for the machines.

In November 1981, a site investigation of the Superfund site [8] began. Groundwater and seismic refraction studies were performed to determine the groundwater flow characteristics and the depth to bedrock. Soil samples from various depths and locations, as well as samples of groundwater and surface water, were collected and analyzed for PCBs in order to determine the extent of contamination.

Exposure Pathways and Resultant Environment Concentrations

Mobility of PCBs in Soil—In general, PCBs are classified as immobile in soil-groundwater systems [9] because they are quickly and extensively adsorbed onto soils [10-14]. During the site investigation, however, PCB contamination was found at depths of 4.57 to 7.62 m [8]. Surface concentrations reached approximately 4000 ppm in small areas and subsurface soil concentrations reached from one to several hundred parts per million. In order to determine how PCBs migrated 7.62 m below grade, laboratory investigations were performed; these investigations determined that (1) solvents discharged onto the surface of soils at the plant site significantly increased the soil mobility of PCBs and caused PCBs to migrate to depths of 4.57 to 7.62 m, and (2) PCBs were immobile in the soil-groundwater system under present site conditions [8].

The laboratory investigations also revealed that (1) the velocity of the PCB front in both the horizontal and the vertical directions was negligible and that (2) the solvents were no longer present in the soil-groundwater system at the site and were no longer influencing the mobility of PCBs [8]. All vertical velocities were less than 9.67 pm/s (0.001 ft/yr). The horizontal velocities for subsoils ranged from 14.6 to 197 nm/s; for topsoil, from 253 to 206 nm/s. Therefore, groundwater transport should not be a significant exposure pathway. However, since PCBs were extensively adsorbed to soil at this site, atmospheric suspension of soil and soil erosion during rainfall events are potential pathways for PCB transport and subsequent human exposure.

PCB Biodegradation in Soil—Several environmental transformation pathways—biodegradation, acid and base hydrolysis, atmospheric oxidation, photolysis, and soil colloid–catalyzed reactions—are responsible for limiting the mobility and persistence of organic chemical in the environment. For PCBs, biodegradation is the dominant transformation pathway in soil systems. Laboratory investigations on PCB biodegradation were not conducted because many researchers have already reported that PCBs are biodegradable in soil [15-18]; based on these studies it was assumed that the PCBs would eventually degrade in site soil.

Exposed Populations

Since PCBs were shown to be immobile in the soil-groundwater system under present site conditions, exposure through drinking water was not expected to occur. However, limited exposure to workers and residents was possible via atmospheric suspension of PCBs adsorbed to soil particles at the surface because the site was covered with vegetation. Also, exposure to residents could have occurred through sediment transport (that is, soil erosion) of PCBs adsorbed to soil particles. The size of the potential exposed population was not estimated, but was assumed to be small because the population of the surrounding community was relatively small.

Selection of a Remedial Action Plan

Remedial action is a systematic engineering scheme, utilizing various treatment processes and operations for the abatement or, if possible, the elimination of exposure pathways. In this case study, two potential pathways were identified: atmospheric suspension of soil particles and sediment transport during rainfall. In addition, disposal and cleanup operations at the site could create potential pathways for PCB transport and exposure. The remaining sections of this paper will discuss (1) two potential plans that were developed and (2) the assumed effectiveness of these two plans in abating the potential exposure pathways listed above and in minimizing potential exposed populations.

The On-Site Remedial Action Plan—Because the PCBs in the subsurface soil were immobilized by their adsorption onto soil particles, and because PCBs should eventually degrade, an on-site remedial action plan that primarily addressed the atmospheric suspension and sediment transport of soil was developed. The on-site plan consists of four elements [8]:

1. The contaminated soil will be covered by a soil cap and seeded with native grasses to prevent contaminated soil erosion, to prevent atmospheric suspension of soil particles, and to provide a soil buffer if accidental solvent spills should occur.
2. A concrete-lined surface drainage ditch will be constructed to divert surface runoff.
3. Operational procedures will be instituted at the plant to minimize the potential for solvent spills.
4. The owners will voluntarily impose certain covenants, conditions, and restrictions on the property regarding excavations, financial guarantees, self-monitoring, inspection and repair of the surface cover, and the lease or sale or changes of use of the property.

The cost of the On-Site Plan—$230 000—places a financial burden on the company owning the case study site, but the burden could be carried by the company without adversely affecting plant operations and the jobs of its 105 employees.

The Off-Site Remedial Action Plan—The state of California has developed and abides by a set of policy guidelines and proposed regulations for the disposal of PCBs (Draft Regulatory Criteria for Identification of Hazardous and Extremely Hazardous Wastes, September 1983). In accordance with policy and proposed regulations, soil materials containing PCB concentrations below 50 ppm are considered nonhazardous. Materials containing PCB concentrations above 50 ppm are considered hazardous. Because the site described here contains PCB concentrations in soil as high as 4000 ppm, two California regulatory agencies requested that a plan to remove all contaminated soil containing over 50 ppm PCB be developed and considered for implementation at the site. The off-site plan, which was developed in response to this request, entails the removal of all contaminated soil for off-site disposal at a permitted PCB disposal facility. Since the soil underneath the plant itself contained PCBs, the plan includes provisions for dismantling the existing plant in order to allow removal of this soil. The direct cost of implementing the off-site plan—approximately $4 million—would place a financial burden on the company owning the site; the company would not be able to rebuild the casting plant. As a result, approximately 105 jobs would be lost, the community would lose approximately $1.3 to $2 million of household income per year, and the unemployment rate of the community would increase approximately 1.4%. In order to determine if the adoption of the off-site plan would lead to a significant decrease in potential human exposure, the exposed population for the two plans was compared. This comparison is discussed below.

Environmental Pathways and Exposed Populations—Table 1 lists the potential human exposure routes, potential environmental pathways, and potential exposed populations before and after implementation of the on-site and off-site plans. An analysis of the information presented in the table reveals that after the implementation of either plan, the level of human exposure to PCBs at the site would be the same. However, an analysis of the qualitative information, presented in Table 2, reveals that the implementation of the off-site plan, which would cost approximately 17 times as much as the on-site plan, would involve (1) additional sources of release, (2) additional human exposure, and (3) a larger exposed population of undetermined size that is exposed to an undetermined resultant environmental concentration of PCBs. These additional types of exposure should not occur during the imple-

TABLE 1—*Potential human exposure routes, environmental pathways, and exposed populations before and after implementation of both plans.*

		Potential Exposed Population		
Potential Human Exposure Route	Potential Environmental Pathway	Before Imple- mentation	After On-Site Plan Imple- mentation	After Off-Site Plan Imple- mentation
Dermal absorption and inhalation	atmospheric suspension of PCBs adsorbed to soil particles[a]	plant employees and residents	none	none
Ingestion	sediment transport (i.e., soil erosion) of PCBs, adsorbed to soil particles, to drinking waters during rainfall events[b]	residents	none	none
Ingestion	groundwater transport of PCBs to aquifers	none	none	none
Ingestion	uptake by agricultural and horticultural crops	none	none	none

[a]Because the site is presently covered with a vegetative cover, there is minimal potential for suspension of soil particles at the present time.
[b]Very minute amounts of PCB are leaving the site via sediment transport.[8].

TABLE 2—*Effects of the off-site plan.*[a]

Potential Source of Release	Potential Environmental Pathway	Potential Human Exposure Route	Exposed Population
Soil excavation	atmospheric suspension of PCBs adsorbed onto soil particles	inhalation, ingestion, dermal absorption upon physical contact	Plant employees, excavation crew, residents in vicinity
Soil excavation	sediment transport of PCBs, adsorbed to soil particles and to drinking water during rainfall concurrent with construction	ingestion, dermal absorption	residents in vicinity
Transport	atmospheric suspension, sediment transport	inhalation, ingestion, dermal absorption	transport crew, residents in vicinity of transportation routes
Storage/ Handling	atmospheric suspension, sediment transport	inhalation, ingestion, dermal absorption	storage crew, residents in vicinity
Disposal	potential release to solvent-ground-water systems at a controlled landfill	ingestion	residents in vicinity who utilize ground water and surface

[a]Population size and resultant environmental concentrations were not determined.

mentation of the recommended on-site plan because this plan involves (1) the transport of nonhazardous materials to the site and (2) the grading of uncontaminated materials.

Since the adoption of the off-site plan would result in potentially greater adverse effects to human health, as well as costing approximately 17 times as much, the on-site plan is the more cost-effective.

Conclusions

An exposure assessment is the determination or estimation of the magnitude, frequency, duration, and routes of exposure by humans to a chemical. The extent of exposure is an important factor that determines which remedial action at sites containing contaminated soil and groundwater are cost-effective in abating potential adverse effects.

References

[1] Schneiter, R. W., Dragun, J., and Erler, T. G., "Groundwater Contamination: III. Remedial Action," *Chemical Engineering*, Vol. 91, 26 Nov. 1984, pp. 73–78.

[2] Block, R. M., Dragun, J., and Kalinowski, T. W., "Groundwater Contamination: II. Health and Environmental Aspects of Setting Cleanup Criteria," *Chemical Engineering*, Vol. 91, 26 Nov. 1984, pp. 70–73.

[3] Dragun, J., Schneiter, R. W., and Kuffner, A. C., "Groundwater Contamination: I. Transport and Transformations of Organic Chemicals," *Chemical Engineering*, Vol. 91, 26 Nov. 1984, pp. 65–70.

[4] Doull, J., Klassen, C. D., and Amdur, M. O., *Toxicology*, MacMillan, New York, 1980, pp. 28–38.

[5] Loomas, T. A., *Essentials of Toxicology*, 3rd ed., Lea and Febiger, Philadelphia, 1978, pp. 67–79.

[6] Bonazountas, M., in *Fate of Chemicals in the Environment*, R. L. Swann and A. Eschenroeder, Eds., American Chemical Society, Washington, DC, 1983, pp. 41–65.

[7] Burns, L. A., in *Fate of Chemicals in the Environment*, R. L. Swann and A. Eschenroeder, Eds, American Chemical Society, Washington, DC, 1983, pp. 25–40.

[8] Erler, T. G., III, Dragun, J., and Weiden, D. R., *Proceedings*, 38th Annual Purdue Industrial Waste Conference, Ann Arbor Science, 1984, pp. 369–375.

[9] Kenega, E. E., *Ecotoxicology and Environmental Safety*, Vol. 4, No. 1, Jan. 1980, pp. 26–38.

[10] Haque, R., Schmedding D. W., and Freed, V. H., *Environmental Science and Technology*, Vol. 8, No. 1, Jan. 1974, pp. 139–142.

[11] Moein, G. J., "Study of the Distribution and Fate of PCBs and Benzenes After Spill of Transformer Fluid," EPA Report 904/9-76-014, U.S. Environmental Protection Agency, Washington, DC, 1976.

[12] Moein, G. L., Smith, A. J., Jr., and Stewart, P. L., *Proceedings*, 1976 National Conference on Control of Hazardous Material Spills, Information Transfer, Inc., Rockville, MD, 1976.

[13] Oloffs, P. C., Albright, L. S., Szeto, S. Y., and Lau, J., *Canadian Journal of Fisheries Research Board*, Vol. 30, No. 12, Dec. 1973, p. 1619.

[14] Paris, D. F., Steen, W. C., and Baughman, G. L., *Chemosphere*, Vol. 7, No. 4, April 1978, pp. 319–325.

[15] Clark, R. R., Chian, E. S. K., and Griffin, R. A., *Applied and Environmental Microbiology*, Vol. 37, No. 4, April 1979, pp. 680–685.

[16] Furukawa, K., Tonomura, K., and Kamibayashi, A., *Applied and Environmental Microbiology*, Vol. 35, No. 2, Feb. 1978, pp. 223–227.

[17] Griffin, R. A., Au, A. K., and Chian, E. S. K., Illinois State Geological Survey, Urbana, IL, 1979, Reprint 1979G.

[18] Liu, D., *Water Research*, Vol. 14, No. 10, Nov. 1980, pp. 1467–1475.

[19] "Principles for Evaluating Chemicals in the Environment," ISBN 0-309-02248-7, National Academy of Sciences, Washington, DC, 1975, pp. 33–44.

John E. Matthews[1] and Anthony A. Bulich[2]

A Toxicity Reduction Test System to Assist in Predicting Land Treatability of Hazardous Organic Wastes

REFERENCE: Matthews, J. E. and Bulich, A. A., "**A Toxicity Reduction Test System to Assist in Predicting Land Treatability of Hazardous Organic Wastes,**" *Hazardous and Industrial Solid Waste Testing: Fourth Symposium, ASTM STP 886*, J. K. Petros, Jr., W. J. Lacy, and R. A. Conway, Eds., American Society for Testing and Materials, Philadelphia, 1986, pp. 176–191.

ABSTRACT: Migration of toxic organics contained in the water-soluble fraction (WSF) of land-applied hazardous wastes poses the most serious threat to groundwater resources. Wastes containing toxic organics that are potentially leachable must be applied to the land at a rate that will allow attenuation of such constituents before their migration from the treatment zone can occur.

A toxicity reduction (TR) test system is being proposed that will serve to determine whether and to what extent attenuation of WSF organic constituents, including both parent compounds and intermediate transformation products, will occur in a well-managed land treatment system.

The test system employs reduction of acute toxicity exerted by WSF organics over time as the measurement criteria. Four sample sets of selected waste-soil combinations are prepared in duplicate for extraction with deionized water (DW) at 14-day intervals during a 42-day experimental period. Acute toxicity of each DW extract is determined by using a bacterial bioluminescence assay. Dose-response curves for each loading rate and subsequent time interval are compared with those of previous sample sets. The waste is predicted to be a candidate for land treatment if a significant reduction of acute toxicity exists during the experimental period at any of the loading rates tested. The maximum acceptable initial loading rate (MAIL) for waste application is predicted from these data. The predicted MAIL rate is the highest rate tested in which toxicity reduction for the waste-soil DW extract progresses steadily during the experimental period with the Day-42 EC_{50} approaching or exceeding 100 percent. (The EC_{50} is defined as the effective concentration of the DW extract that causes a 50% decrease in bacterial light output.) The higher the EC_{50} the lower the toxic effect; therefore, a 42-day EC_{50} approaching 100% signifies toxicity reduction in a waste-soil DW extract.

[1]Research biologist, U.S. Environmental Protection Agency, Kerr Environmental Research Laboratory, Ada, OK 74820.
[2]Microbiologist, Clinical Diagnostic Division, Beckman Instruments, Inc., Carlsbad, CA 92008.

The TR procedure currently is being used by the EPA's Hazardous Waste Land Treatment Research Program to assist in evaluating land treatability potential for a variety of organic hazardous wastes. The most extensive use of the procedure to date has involved evaluation of land treatability for oily wastes from different sources within a refinery: i.e., lagoon bottom sludge, API separator sludge, slop oil emulsion solids, and dissolved air flotation unit skimmings. Additional evaluation tests have been conducted by using wastes from the wood preserving and paint industries. Results from these evaluation tests indicate the potential usefulness of the TR procedure as one of the initial tests conducted in a laboratory screening test sequence for predicting the land treatability potential of organic hazardous wastes.

KEY WORDS: toxicity, tests, bioassay, treatability, land treatment, hazardous wastes, screening

Regulations governing treatment of hazardous waste in land treatment (HWLT) units [1] stipulate that hazardous constituents contained in the applied waste must be detoxified or otherwise rendered innocuous within the defined soil treatment zone. A demonstration of treatment is required to obtain a HWLT facility operating permit. This demonstration must focus on maximizing attenuation of hazardous constituents via natural soil processes and interactions and minimizing their escape to groundwater, surface water, and air. It is recommended that the treatability studies address, at a minimum, degradation/transformation, mobility, and toxicity of hazardous constituents [2].

One promising approach to generating the diverse data needed to make initial land treatability decisions is through the use of a standardized battery of relatively rapid laboratory screening tests, each of which addresses at least one of the aforementioned areas of concern. Following an initial waste and soil characterization phase, this approach involves subjecting the waste in question to a prescribed sequence of screening procedures in which the proposed site soil is used as the treatment medium. Integrated data from the different areas of testing are used to categorize land treatability potential of a specific waste if applied to a specific soil. The major objective of laboratory screening is to identify any treatability problem rapidly and at a minimum of cost. Bench or pilot-scale verification studies are conducted to complete the land treatment demonstration. This phased approach to conducting a land treatment demonstration program is depicted in Fig. 1.

Screening tests that require extraction and analyses for specific chemical constituents also require the largest resource allocation. Therefore, the Phase 2 laboratory screening sequence should begin with procedures that use gross parameters relative to the total pollutant matrix as the measurement criteria. Results from this first tier of screening tests can be used to make decisions pertaining to which subsequent screening tests should be conducted and to define the highest waste loading rate to be used in these tests.

A major concern associated with land treatment of hazardous wastes is adequate protection of groundwater resources. The water-soluble fraction

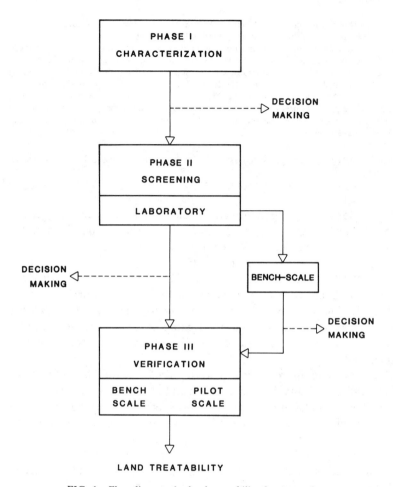

FIG. 1—*Flow diagram for land treatability demonstration.*

(WSF) of land-applied wastes possesses the greatest potential for leaching and subsequent groundwater contamination [3,4]. Any waste that contains organic hazardous constituents in the WSF must be applied at a rate that will allow for detoxification of such constituents within the treatment zone before leaching into the unsaturated zone can occur. Therefore, one of the first tests conducted in the screening sequence should address leaching and detoxification potentials of these WSF organic constituents.

This paper presents a toxicity reduction (TR) test system that can be used as one of the initial screening tests conducted during the Phase 2 land treatability laboratory test sequence (Fig. 2). Its results can be used to predict whether and to what extent detoxification of the WSF organic pollutant matrix will occur in the site soil. Coupled with results from microbial activity

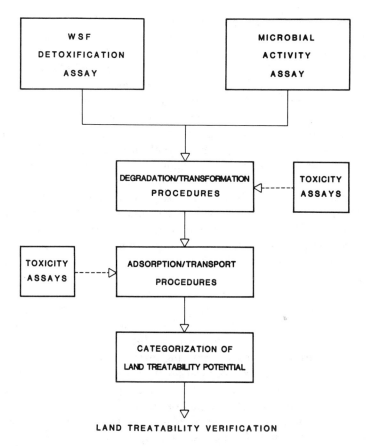

FIG. 2—*Laboratory screening test sequence for predicting land treatability potential of hazardous wastes.*

assays, comparing the detoxification of the WSF of several waste-soil combinations allows the maximum waste loading rate for subsequent screening tests to be defined.

Experimental Method

If the WSF of any hazardous waste exhibits a significant level of acute toxicity using a specific bioassay system, then reduction of the toxicity as measured by that bioassay becomes a measure of treatability. The basic design concept of the proposed test system assumes that for a given waste-soil loading rate, significant toxicity reduction of the waste-soil WSF pollutant matrix during the experimental period defines a positive land treatability potential at that loading rate.

The relative acute toxicity of the WSF is determined by conducting a bacterial bioluminescence assay (designated as Microtox®) on the deionized water (DW) extract of the waste and specific waste-soil combinations. The Microtox system uses a suspension of marine luminescent bacteria (*Photobacterium phosphoreum*) as bioassay organisms. Suspensions with approximately 10^6 bioluminescent organisms in each are "challenged" by addition of serial dilutions of the DW extract. A reduction of light output reflects physiological inhibition, thereby signifying the presence of toxic constituents in the sample.

This assay was selected for use because of the small volume of sample required, the ease of determining the end point, and its reported effectiveness in determining relative acute toxicity of complex wastewaters containing toxic organic constituents [5,6]. A recent report published by the Canadian Petroleum Association [7] concluded that the bacterial bioluminescence assay is a viable method of screening for apparent toxicity in complex waste drilling fluids. Results correlated closely with those from rainbow trout bioassays. The report also recommends that this assay be used as a tool in evaluating the effects of these drilling fluids on soils.

Acute toxicity reduction of WSF organics as depicted by bacterial bioluminescence assays is used as the measurement criteria for predicting land treatability potential. The effective concentration of the DW extract which causes a 50% decrease in light output under the conditions of the test (EC_{50}) is used as the primary unit for comparing toxicity reduction.

A TR experiment requires preparation of 32 test units. Each unit consists of a 500-mL flask containing 50 g of site soil that has been crushed and sieved to 2 mm. Three appropriate weights or volumes of waste are thoroughly mixed with the soil to achieve the desired loading rates. A routine experiment employs four sample sets of test units. Each set consists of eight units, including duplicates for three loading rates and a control. Results of preliminary toxicity tests run on extracts of the waste, soil, and four waste-soil loadings (1, 5, 10, and 15 weight-percent) are used to select loading rates for a toxicity reduction experiment. Loading rates generally are selected in an arithmetic or geometric series that will allow the Day 0 EC_{50} to fall within a range of 20 to 60%. Sample sets are sacrificed for DW extraction and toxicity testing on Day 0 and at 14-day intervals during a routine 42-day experiment.

Test units are maintained at room temperature ($22 \pm 2°C$). Soil moisture is maintained between 40 and 70% of the soil's moisture holding capacity (MHC). These values are well within the 30 to 90% range presented by Dibble and Bartha [8] as being favorable for biodegradation of oily sludges in soil. When the soil reaches 40% of its MHC, the waste-soil combination is mixed and DW added to return the moisture level to 70%.

The extraction procedure consists of adding 400 mL of DW to each test unit at the specified time interval, tightly sealing the unit, and mixing for 22 ± 2 h in a rotary tumbler mixer at approximately 30 rpm. This procedure provides for sufficient agitation of the mixture so that sample surfaces are continually brought into contact with the extracting medium. Following this mixing period, the resulting slurry for each test unit is filtered under pressure through 0.45-μm filter paper. Osmotic adjustment of the resulting filtrate is then made because the test organism is a marine species.

Bacterial bioluminescence assays are conducted by using serial dilutions of each osmotically adjusted filtrate. Dose-response curves, EC_{50} values, and 95% confidence limits are calculated for each loading rate at each time interval by using data reduction procedures described in the manufacturer's operating manual [9] or other appropriate methods for reducing acute toxicity data [10]. The extent of toxicity reduction is determined by comparing subsequent EC_{50} values for a given loading rate with those for the Day 0 sample. Control soil samples are extracted and tested to determine if the soil matrix itself exhibits a residual toxicity.

Experimental results are used to predict the maximum acceptable initial loading (MAIL) rate for applying the tested waste to the tested soil. The MAIL rate is the highest at which acute toxicity reduction for WSF organic compounds contained in the DW extract progresses steadily during the experimental period. Since the higher the EC_{50} the lower the toxicity, if the EC_{50} for Day 42 approaches or exceeds 100% significant toxicity reduction in a waste-soil extract is indicated. The calculated EC_{50} can exceed 100% if the undiluted (100% dilution) DW extract does not elicit a 50% decrease in light output from the luminescent bacteria. This is a typical response for many control soils. The highest loading rate that should be considered in subsequent screening tests is considered to be twice the MAIL rate.

A loading rate is not considered to be acceptable for initial loading if the Day 14 EC_{50} for that rate does not reflect an initiation of toxicity reduction, regardless of what happens in the subsequent sample sets. An initiation of toxicity reduction is indicated if the Day 14 EC_{50} is significantly higher than the Day 0 EC_{50}. A significant increase exists when 95% confidence intervals for the EC_{50} values of the Day 0 and Day 14 sample sets do not overlap. The basis for this rejection criterion is the high potential for migration of toxic water-soluble organic compounds below the treatment zone during such a long lag period. Significant toxicity reduction in subsequent sample sets, however, might indicate the potential for the use of this higher loading rate once microbial acclimation to the waste has occurred.

A loading rate at which the EC_{50} of the Day 28 sample decreases significantly from the Day 14 level may also be unacceptable for initial loading. Such a condition may reflect the formation of a significant quantity of toxic intermediates that are soluble or increased solubility of toxic parent compounds. Under the test conditions, the former is more likely to occur. The potential exists for migration of these toxics below the treatment zone if a subsequent lag period in their toxicity reduction occurs. This possibility is being investigated in ongoing land treatability studies.

Land treatability potential for a variety of hazardous wastes is currently being evaluated at the Robert S. Kerr Environmental Research Laboratory as part of the U.S. Environmental Protection Agency (EPA) Hazardous Waste Land Treatment Research Program. Treatability predictions for selected organic wastes are made based on results from a prescribed battery of laboratory screening tests that address all potential areas of concerns: degradation, migration, and toxicity. The TR procedure is included in this screening test battery. Bench-scale or field-plot studies are then used to verify these predictions.

Results and Discussion

The TR procedure described is the first test conducted in the laboratory screening test sequence (Fig. 2). Results are used to predict an acceptable initial loading rate and to define the highest rate to be used in subsequent screening tests. Wastes for which TR experiments have been completed are presented in Table 1.

The TR experimental results indicate that each of these wastes is a potential candidate for land treatment; however, environmentally acceptable loading rates predicted vary considerably. Predicted acceptable loading rates for each waste type are presented in Table 2.

Results from TR experiments for the lagoon bottom sludge compared favorably with those generated during subsequent field plot experiments. The EC_{50} profiles for the lagoon bottom sludge TR and field plot experiments are presented in Figs. 3 and 4, respectively.

Laboratory TR results indicated that six weight-percent of waste would be an environmentally acceptable loading rate for initial application of the lagoon bottom sludge to the loam soil used in this experiment. Toxicity reduction at this loading rate progressed steadily during the experiment. There also was no indication that a significant quantity of toxic intermediates would be formed or the solubility of toxic parent compounds would be increased if this loading rate were used.

TABLE 1—*Hazardous waste types subjected to TR experiments.*

	Soil Type		
Waste Types	Loam	Silty Loam	Clay Loam
Oily lagoon bottom sludge	X	X	
API separator bottoms		X	
Slop oil emulsion solids			X
DAF[a] skimmings			X
Wood preserving treatment sludge			X
Paint waste			X

[a]Dissolved air flotation unit.

TABLE 2—*Predicted loading rates for hazardous wastes subjected to TR experiments.*

Waste Type	Predicted Loading Rate[a]	
	MAIL[b]	HLR[c]
Oil lagoon bottom sludge	6	12
API separator sludge	4	8
Slop oil emulsion solids	5	10
DAF[d] skimmings	6	12
Wood preserving treatment sludge	0.1	0.2
Paint waste	4	8

[a]Grams of waste applied and incorporated in 100 g of dry soil.
[b]Maximum acceptable initial loading rate predicted for application of tested waste in tested soil.
[c]Highest loading rate that should be considered for use in subsequent treatability testing.
[d]Dissolved air flotation unit.

Soil-core samples from field plots to which the lagoon bottom sludge was applied were subjected to DW extraction and toxicity testing in the same manner as laboratory TR experimental units. The EC_{50} profiles for these samples again indicated that 6 weight-percent waste would be an environmentally acceptable loading rate for initial application of this waste to the loam soil. The increased toxicity observed ten weeks following waste reapplication most probably reflected the extended period of hot, dry weather that occurred in late summer. Since supplemental moisture was not added during that period, microbial activity would have decreased, and further transformation of any toxic intermediates formed would have been minimized. In addition, the solubility of some toxic parent compounds could have increased under these conditions because solubility generally increases with increasing temperature. After cessation of the hot, dry conditions, toxicity reduction appeared to progress in a normal fashion.

Results from a TR experiment using the lagoon bottom sludge and another soil type (silty loam) gave essentially the same results as those using the loam soil. Six weight-percent waste was indicated to be an environmentally acceptable loading rate for initial application to the silty loam soil.

Results from the TR experiment for the API separator sludge were used for comparison with results generated during an extensive field plot research project conducted at the University of Oklahoma [11]. The EC_{50} profiles for this experiment are presented in Fig. 5. The 4 weight-percent MAIL rate predicted from the TR results agreed favorably with the field plot data.

The TR experiments for the remaining four wastes listed in Table 1 have been conducted as part of a hazardous waste land treatability screening program currently under evaluation at the Kerr Laboratory. The EC_{50} profiles for these four wastes are presented in Figs. 6 to 9.

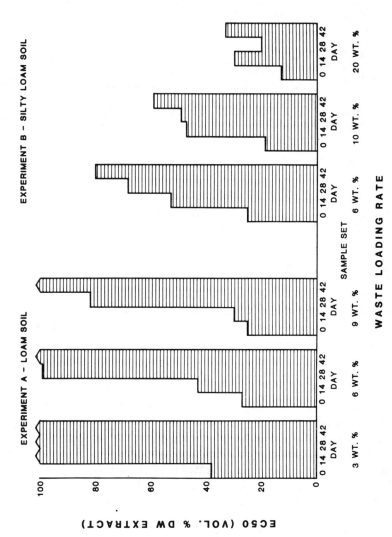

FIG. 3—EC_{50} profiles for lagoon bottom sludge, TR experiments.

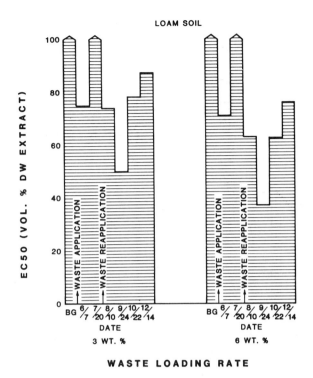

FIG. 4—EC_{50} profiles for lagoon bottom sludge, field plot core samples.

Comparable oil contents for the four oily wastes tested were 25, 51, 15 and 11 weight-percent respectively for the lagoon bottom, API separator, slop oil, and dissolved air floatation (DAF) samples. Predicted MAIL rates for each waste are similar when compared by weight-percent of waste (6, 4, 5, and 6 weight-percent waste). When the rates are converted to oil in soil content, however, there is a significant difference in predicted MAIL rates for the lagoon bottom and API samples (1.5 and 2.0 weight-percent oil) when compared to those for the slop oil and DAF samples (0.75 and 0.65 weight-percent oil).

The current practice at many operating oily waste land treatment facilities is to use oil in soil content as the primary loading parameter. Variations in predicted MAIL rates for the four oily wastes indicate that there is not a universal oil in soil content that will be environmentally acceptable for initial loading at every land treatment facility. The oil in soil content acceptable for initial loading is both soil- and waste-specific, with the waste WSF pollutant matrix being the dominant factor of the two. This lends credence to the current requirement that land treatability studies be conducted using the waste to be treated and the soil to which that waste will be applied.

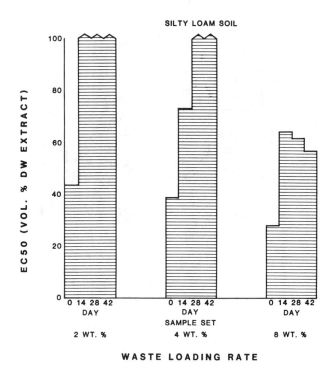

FIG. 5—*EC₅₀ profiles for API separator bottoms, TR experiment.*

Two TR experiments have been conducted for the wood preserving treatment sludge. Each of the three loading rates used in the first experiment was considered to be environmentally unacceptable (Fig. 8). Although the Day 0 EC_{50} was noticeably higher at the lowest loading rate tested (0.4 weight-percent waste), there was no apparent toxicity reduction between any of the sample sets at any of the three rates tested. A second experiment was conducted using loading rates of 0.1, 0.2 and 0.3 weight-percent to try to predict a MAIL rate that would be compatible with the environment. The predicted MAIL rate for this waste was the lowest of the six rates tested, 0.1 weight-percent waste.

The most significant toxic organic identified in the wood preserving waste was pentachlorophenol (PCP). The concentration of PCP was much higher than any of the other potential toxicants. The TR experimental data indicate that wood preserving wastes containing PCP can be land-treated in an environmentally acceptable manner; however, such a practice will require careful management to control the PCP content in the soil. There appears to be a sharp demarcation between an initial PCP loading rate which will be environmentally acceptable and one which will be unacceptable.

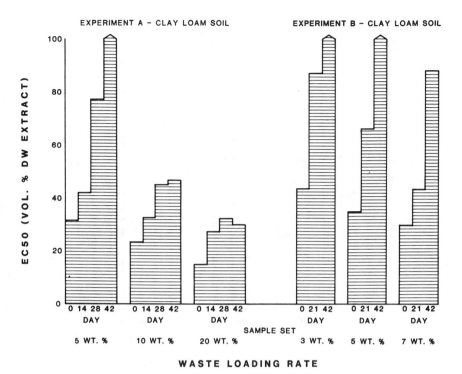

FIG. 6—*EC₅₀ profiles for slop oil emulsion solids, TR experiment.*

The paint waste exhibited the most atypical toxicity reduction pattern of those tested to date. Day 0 samples for each of the experimental loading rates exhibited a high WSF acute toxicity level; however, Day 14 samples reflected at least a sixfold decrease in this toxicity. The Day 14 EC_{50} for the WSF of the low–loading rate sample (4 weight-percent waste) exceeded 100%. This indicates the presence of a high quantity of toxic volatile organics in the WSF of the paint waste. The Day 14 EC_{50} values for the 8 and 12 weight-percent waste loading rates, however, indicate that a significant quantity of semivolatile toxic organics also are contained in this waste. Since none of the volatile constituents identified in this waste appear on the list of hazardous constituents presented in the Resource Conservation and Recovery Act (Appendix VIII, Part 261, Section 3001, Subpart D), this volatilization does not preclude the use of land treatment as a management alternative. Air-stripping of volatile components, however, should be considered in the final determination of an acceptable waste loading rate.

The EC_{50} profiles for each of the six wastes tested to date have exhibited the same general trend. Acute toxicity, as defined by the bacterial bioluminescence assay, increased with increased loading rate within a given sample set at

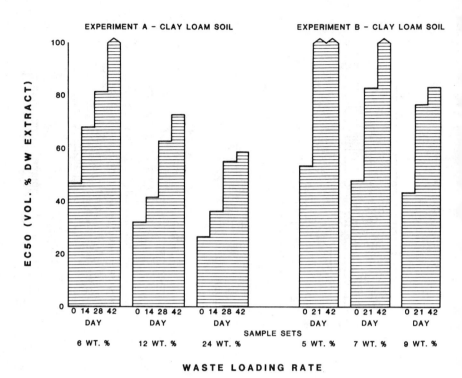

FIG. 7—*EC50 profiles for DAF skimmings, TR experiment.*

any point in time and decreased for a given loading rate with each subsequent sample set as a function of time. In addition, as the loading rate increased for a given waste, significant lag periods or reversals in toxicity reduction were common occurrences, as indicated by comparisons of EC_{50} 95% confidence intervals.

Initial sharp reductions in toxicity depicted in some EC_{50} profiles indicate that volatilization of toxic organics may have contributed to toxicity reduction during these experiments. In such cases, the major impact most likely occurred within a short time following mixing of the waste and soil. A simple experiment was conducted to check for volatilization potential of WSF toxic organics. Fifty grams of an oily sludge were added to three flasks and extracted with DW on Days 0, 14, and 28. The EC_{50} values for the DW extracts were 18.0, 24.0, and 24.5, respectively, indicating the potential for loss of volatile toxicants through air-stripping following waste applications to soil.

The significance of volatilization as a factor in toxicity reduction is dependent on both the organic pollutant matrix and the waste loading rate. Of the wastes tested to data, volatilization of toxic organics was most apparent in the paint waste TR experiment (Fig. 9). In contrast to results of previous experi-

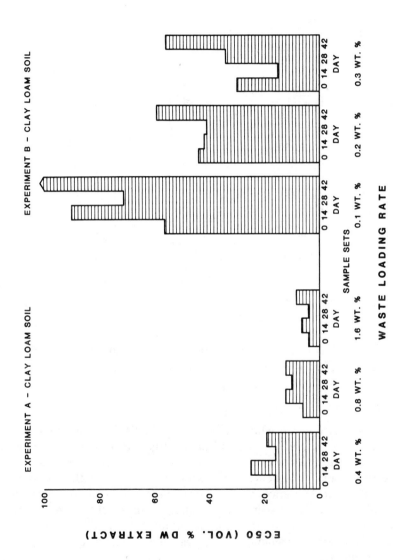

FIG. 8—EC_{50} profiles for wood preserving treatment sludge, TR experiment.

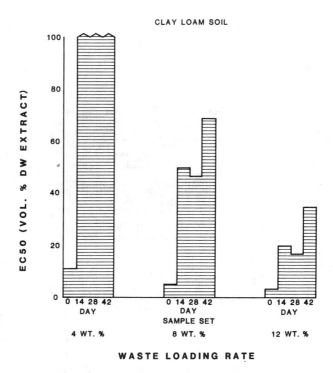

FIG. 9—EC_{50} profiles for paint waste, TR experiment.

ments, sharp reductions in toxicity during the 14 days following waste application were exhibited at each of the three loading rates tested.

Summary

A toxicity reduction procedure employing a bacterial bioluminescence assay is currently being used as one of the initial tests conducted in a laboratory screening test battery for determining land treatability potential of hazardous waste. Experimental results obtained thus far demonstrate that the TR procedure can be used to determine the detoxification potential of the WSF organic compounds that provide the greatest threat to groundwater resources. This allows prediction of an environmentally acceptable loading rate for the waste as well as definition of the highest loading rate that should be used in subsequent screening tests.

Disclaimer

Although the research described in this paper has been funded wholly or in part by the U.S. EPA, it has not been subjected to the agency's peer and administrative review and therefore may not necessarily reflect the views of the agency and no official endorsement should be inferred. Reference to equipment brand names or suppliers in this paper is not to be interpreted as an endorsement of the products or suppliers by the U.S. EPA.

References

[1] U.S. Environmental Protection Agency, "Hazardous Waste Management System; Permitting Requirements for Land Disposal Facilities," CFR 47-143, Subpart M—Land Treatment, 26 July 1982, pp. 32361–32367.

[2] U.S. Environmental Protection Agency, "RCRA Guidance Document: Land Treatment Units," 1st ed., U.S. EPA, Office of Solid Waste, Washington, DC, May 1983.

[3] Beak Consultants, Ltd., "Landspreading of Sludges at Canadian Petroleum Facilities," Report 81-5A, Petroleum Association for Conservation of the Canadian Environment, Ottawa, Dec. 1981.

[4] Brown, K. W., Deuel, L. E., Jr., and Thomas, J. C., "Land Treatability of Refinery and Petrochemical Sludges," final report, Grant R805474013, PB 83-247 148, U.S. Environmental Protection Agency, Office of Research and Development, Cincinnati, OH, Nov. 1983.

[5] Qureshi, A. A., Flood, K. W., Thomson, S. R., Janhurst, S. M., Inniss, C. S., and Rokosh, D. A., "Comparison of a Luminescent Bacterial Test with Other Bioassays for Determining Toxicity of Pure Compounds and Complex Effluents," *Aquatic Toxicology and Hazard Assessment: Fifth Conference, ASTM STP 766,* J. G. Pearson, R. B. Foster, and W. E. Bishop, Eds., American Society for Testing and Materials, Philadelphia, 1982, pp. 179–195.

[6] Vasseur, P., Ferod, J. F., Rost. C., and Larbaigt, G., "Luminescent Marine Bacteria in Ecotoxicity Screening Tests of Complex Effluents," *Toxicity Screening Procedures Using Bacterial Systems,* D. L. Lui and B. J. Dutka, Eds., Marcel Dekker, New York, 1984, pp. 23–36.

[7] Strosker, M. T., "A Comparison of Biological Testing Methods in Association with Chemical Analysis to Evaluate Toxicity of Waste Drilling Fluids in Alberta," Vol. 1, Canadian Petroleum Association, Calgary, Alta., Feb. 1984.

[8] Dibble J. T. and Bartha, R., "Effects of Environmental Parameters on the Biodegradation of Oil Sludge," *Applied and Environmental Microbiology,* Vol. 37, No. 4, April 1979, pp. 729–743.

[9] "Microtox® System Operating Manual," Section 11, Beckman Instruments, Inc., Microbics Operations, Carlsbad, CA, 1982.

[10] Peltier, W. and Weber, C. I., Eds., "Methods for Measuring the Acute Toxicity of Effluents to Aquatic Organisms," 3d ed., U.S. Environmental Protection Agency, Office of Research and Development, Environmental Monitoring and Support Laboratory, Cincinnati, OH, March 1985.

[11] Streebin, L. E., Robertson, J. M., and Schornick, H. M., "Land Treatment of Petroleum Refinery Sludges," final report, EPA Cooperative Agreement CR807957810, U.S. Environmental Protection Agency, Office of Research and Development, Ada, OK, Oct. 1984.

Edward F. Neuhauser,[1] Raymond C. Loehr,[2] and Michael R. Malecki[1]

Contact and Artificial Soil Tests Using Earthworms to Evaluate the Impact of Wastes in Soil

REFERENCE: Neuhauser, E. F., Loehr, R. C., and Malecki, M. R., "**Contact and Artificial Soil Tests Using Earthworms to Evaluate the Impact of Wastes in Soil,**" *Hazardous and Industrial Solid Waste Testing: Fourth Symposium, ASTM STP 886,* J. K. Petros, Jr., W. J. Lacy, and R. A. Conway, Eds., American Society for Testing and Materials, Philadelphia, 1986, pp. 192-203.

ABSTRACT: This study was designed to evaluate two methods using earthworms that can be used to estimate the biological impact of organic and inorganic compounds that may be in wastes applied to land for treatment and disposal. The two methods were the contact test and the artificial soil test. The contact test is a 48-h test using an adult worm, a small glass vial, and filter paper to which the test chemical or waste is applied. The test is designed to provide close contact between the worm and a chemical, similar to the situation in soils. The method provides a rapid estimate of the relative toxicity of chemicals and industrial wastes. The artificial soil test uses a mixture of sand, kaolin, peat, and calcium carbonate as a representative soil. Different concentrations of the test material are added to the artificial soil, adult worms are added, and worm survival is evaluated after two weeks.

These studies have shown that (1) earthworms can be used to distinguish among a wide variety of chemicals with a high degree of accuracy, (2) earthworms are a suitable biomonitoring tool to measure the impact of chemicals in wastes added to soils, (3) the contact and artificial soil tests can measure the biological impact of chemicals applied to soils, (4) earthworms can be used to differentiate the impact of chemical groups and specific chemicals, and (5) the contact and artificial soil tests appear to identify the same relative toxicity of chemicals as rat toxicity data.

KEY WORDS: land treatment, contact test, artificial soil test, industrial wastes, hazardous wastes, earthworms, biological testing, heavy metals, organic chemicals

Land application has been used successfully to treat many industrial wastes. However, laboratory and field studies of this treatment method rarely

[1]Research associate and research support specialist, respectively, Department of Agricultural Engineering, Cornell University, Ithaca, NY 14853.

[2]Professor of civil engineering, University of Texas, Austin, TX 78712; formerly, professor of agricultural engineering and environmental engineering, Cornell University, Ithaca, NY 14853.

consider the impact of potentially hazardous waste materials on the soil eco-system. This paper evaluates methods that measure the impact of waste materials on earthworms, an important component of the soil biota. Previous work with this approach was reported at an earlier ASTM symposium [1].

The complexity of soil fauna is a complicating factor when biological assays are used to evaluate the impact of wastes on the soil ecosystem. Soil microorganisms and soil arthropods are very diverse, difficult to identify, and difficult to use in biological testing procedures. Earthworms, a major biological component of soils, are relatively easy to obtain in large numbers and biomass and can be a suitable soil test organism.

This paper discusses the applicability of two methods originally developed by the European Economic Community (EEC) [2] to estimate the biological impact of wastes applied to soils and wastes leaking from landfills and impoundments. These methods are designed to evaluate the short-term impact of materials on earthworm survival through skin absorption (contact test) and the longer term impact on earthworm survival through skin absorption and ingestion (artificial soil test).

Procedures

Earthworms

Eisenia fetida (Savigny) was the earthworm species used in these test methods. Large numbers of this species of known age and size can be obtained easily for testing. *Eisenia fetida* occurs in specific habitats in temperate climates; considerable information has accumulated concerning its growth [3], reproduction [4], physical requirements [5], use in waste management systems [6], and response to metals [7,8]. This body of information and the ease of using this species under laboratory conditions has made it appropriate to use *E. fetida* in biological testing methods.

The worms used in these experiments were obtained from stock cultures in the laboratory of the authors. The cultures are maintained at room temperature (21 to 24°C), with cocoons being harvested every two weeks. Aerobically digested domestic sludge was used as the maintenance organic food substrate for the worms.

Chemicals

Both organic and inorganic chemicals were used to evaluate the potential impact of wastes applied to soils. The water soluble salts (acetate, chloride, nitrate and sulfate) of five metals (cadmium, copper, nickel, lead, and zinc) were selected on the basis of their potential toxicity and presence in municipal and industrial wastes that may be applied to soils. All of the metal salts used were Fisher-certified chemicals. The organic chemicals used in these studies were organic priority pollutants [9] and were at least 98% pure.

Contact Test

The contact test procedures described by the EEC [2] were followed. In this test, glass vials are lined with filter paper (Whatman No. 1) to which the chemical is added. The chemical is distributed over the surface of the paper, one worm is added to each vial, and the vials are incubated at 20°C. The worm is in contact with the chemical and can absorb the chemical through its skin. After incubation for a defined time period (two days at 20°C), worm survival is determined. A range of chemical concentrations is evaluated and a large number of replicates are used for each concentration. The results are evaluated to determine a median lethal concentration (LC_{50}) value for each chemical.

Glass vials [9 mL (3 oz)] were lined with 12 by 6.5 cm of filter paper to give an area of 80 cm^2 (Fig. 1). Fresh aqueous metal salt solutions were made for each series of tests.

The testing of organic chemicals used distilled water as the solvent if possible. If the test chemical was not soluble in water, other solvents (acetone or chloroform) were used. The organic solvent used to add the chemical should volatilize readily and be evaporated from the vial before the 1 mL of distilled water and the worm are added to the vial. The loss of organic solvents was evaluated and did occur in all cases before the worms were added. In addition, experimentation determined that none of the chemicals added to the vials with the organic solvents volatilized with the organic solvent. The 1 mL of water was added to provide enough moisture for earthworm survival. One adult earthworm (300 to 500 mg) was added per vial and the vials were kept on their side at 20°C in a darkened incubator. Only one worm is added per vial because the death of one worm in the vial could adversely affect the well-being of any other worms present.

After 48 h, the viability of the earthworms was determined. Earthworms were considered dead if they did not respond to a gentle mechanical stimulus.

1 worm/vial

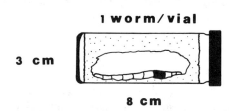

3 cm

8 cm

FIG. 1—*Cutaway view of glass vial lined with filter paper (12 cm by 6.7 cm, or 80 cm^2 per vial) used for the two-day contact test at 20° C.*

The number of dead worms was related to the concentration of the chemical to which the worms were exposed.

The LC_{50} value for each chemical tested was obtained by first using a series of screening tests to determine the range of chemical concentrations that had an adverse impact. This permitted the use of a narrow, more definitive range to determine the more specific LC_{50} value. The range finding tests included concentrations that resulted in mortalities from 0 to 100%. This was accomplished by using chemical concentrations that varied over several logs.

At least five concentrations were evaluated in the definitive test, with ten or more replicates included for each concentration tested. Controls, which included vials that contained everything except the test chemical, were included in each series of experiments.

The LC_{50} value for each chemical tested was calculated using the method of Litchfield and Wilcoxon [10]. In the contact test, the LC_{50} values are reported in terms of micrograms of test chemical per square centimetre of filter paper.

Artificial Soil Test

The EEC tentative artificial soil test method [2] was followed exactly. This test uses a medium which consists by weight of 10% finely ground sphagnum peat, 20% colloidal kaolin clay, 69% fine sand, and 1% pulverized calcium carbonate. The final pH was adjusted to 6.0 ± 0.5. The water content was raised to 35% of the dry weight of the four components. The test chemical was dissolved in water and then added to the artificial soil.

If the test chemical was not water soluble, it was mixed with a small volume of suitable organic solvent (acetone or chloroform) and then mixed with the artificial soil. Tests were run to determine the amount of time needed to completely evaporate the organic solvent from the artificial soil. After the organic solvent evaporated, the artificial soil was brought to a moisture content of 35%. Controls, which consisted of all components except the test chemical, were included in each set of experiments.

Ten adult *E. fetida* (300 to 500 mg) were added to 400 g (dry weight) of the test mixture in round glass dishes 6.5 cm high and 12.5 cm in diameter (Fig. 2). The dishes were placed in an incubator at 20°C and mortality was assessed after two weeks. Earthworms were considered dead when they did not respond to a gentle mechanical stimulus.

Concentrations causing mortalities of 0 to 100% were determined by using range-finding tests that included chemical concentrations that varied over several logs. After the range of chemicals had been narrowed, a definitive test was performed. This test had at least five chemical concentrations, with four repetitions for each test concentration and four control dishes, all containing ten worms per dish.

The LC_{50} value for each chemical was calculated by using the method of

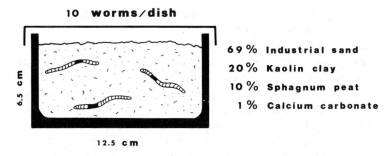

FIG. 2—*Cutaway view of glass dish filled with artificial soil (400 g) used for the 14-day artificial soil test at 20° C.*

Litchfield and Wilcoxon [*10*] and is reported in terms of milligram chemical per kilogram dry weight of artificial soil.

Results

Metals

Four salts of each metal were used in the contact test to determine if the metal salts exhibited different toxicities (Table 1). Statistical analyses determined that the four salts of each metal (acetate, chloride, nitrate, and sulfate) had similar LC$_{50}$ values. The different water-soluble salts of each metal did not show any difference in toxicity to the worms. Of the metals tested, copper was the most toxic, followed by zinc, nickel, cadmium, and lead in order of decreasing toxicity.

The artificial soil test requires more work than the contact test. Because none of the metal salts showed any differences, only one salt (nitrate) was

TABLE 1—*The earthworm contact test LC$_{50}$ values for the metals tested.*

	Metal Salt LC$_{50}$, $\mu g/cm^2$ (95% CI)[a]			
Metal	Acetate	Chloride	Nitrate	Sulfate
Cadmium	20 (16–25)*[b]	18 (12–25)*	24 (12–45)*	26 (17–39)*
Copper	6.7 (5.3–8.5)[†]	4.9 (4.0–5.9)[†]	7.4 (3.8–14.4)[†]	6.3 (4.6–8.6)[†]
Lead	50 (30–83)[‡]	NT[c]	64 (45–92)[‡]	NT[c]
Nickel	17 (13–22)[§]	17 (12–24)[§]	25 (19–34)[§]	20 16–25)[§]
Zinc	13 (10–16)[‖]	12 (6–24)[‖]	10 (8–12)[‖]	13 (7–26)[‖]

[a]CI = confidence interval.
[b]Means with a common superscript symbol are not significantly different, $P < 0.05$.
[c]Not tested.

used in the artificial soil test. The LC_{50} values obtained for the nitrate salts of the five metals tested in the artificial soil test are noted in Table 2. The artificial soil test results also indicated that copper, nickel, and zinc were the most toxic metals, with no statistical difference among them. Cadmium was the next most toxic metal tested, followed by lead.

Organics

The results of the 44 organics tested with the earthworm contact test are summarized in Table 3. The phenols were the most toxic chemical group tested, followed closely by the amines. The major remaining chemical groups, in order of toxicity from most toxic to least toxic, were substituted aromatics, halogenated aliphatics, polycyclic aromatic hydrocarbons, and phthalates. The individual contact test results for the 44 chemicals are given in Table 4. The individual LC_{50} values for one of the chemical groups studied, the substituted aromatics, are summarized in Table 5. There was no statistical dif-

TABLE 2—*Artificial soil test LC_{50} values for the nitrate salts of the metals tested.*

Metal	Artificial Soil Test LC_{50}, mg/kg (95% CI)
Cadmium	1843 (1660-2045)
Copper	643 (549-753)[a]
Lead	5941 (5292-6670)
Nickel	757 (661-867)[a]
Zinc	662 (574-674)[a]

[a]Means footnoted are not significantly different, $P < 0.5$.

TABLE 3—*Summary of means and ranges for the six chemical groups in the contact test and the means and ranges for the rat oral LD_{50} values for the same chemical groups.*

Chemical Group[a]	Worm LC_{50}, $\mu g/cm^2$ (range)		Rat Oral LD_{50}, mg/kg (range)	
Phenols ($n = 9$)	2.9	(0.6-5.9)	1 000	(30-3 200)
Amines ($n = 3$)	5.0	(2.0-11.0)	740	(70-1 650)
Substituted aromatics ($n = 9$)	41	(14-98)	2 100	(250-5 000)
Halogenated aliphatics ($n = 14$)	93	(10-304)	2 300	(75-10 000)
Polycyclic aromatic hydrocarbons ($n = 4$)	1 470	(49-4 670)	4 560	(1 780-10 000)
Phthalates ($n = 5$)	1 760	(550-3 140)	20 225	(6 900-31 000)

[a]n = number of chemicals tested with the contact test.

TABLE 4—*Earthworm LC_{50} contact test values, amounts of chemicals used, water solubilities, and vapor pressures for the 44 chemicals tested with the contact test.*

Chemical	Contact Test LC_{50}, $\mu g/cm^2$ (95% CI)[a]	Chemical[b] Solubility at 20°C[c] μg per Vial	$\mu g/mL$	Vapor Pressure, mm^c
Phenols				
2,4-Dinitrophenol	0.6 (0.5–0.7)*[d]	48	5 620	NF[e]
4-Nitrophenol	0.7 (0.6–0.8)*	56	16 000	2.2
2-Chlorophenol	2.2 (1.9–2.4)[t]	176	28 500	40
2,4-Dimethylphenol	2.2 (1.7–2.8)[t]	176	4 200	0.06
Pentachlorophenol[f]	2.4 (1.8–3.0)[t]	192	10^{-3}	14
2,4-Dichlorophenol	2.5 (2.3–2.8)[t]	200	4 600	0.12
2,4,6-Trichlorophenol	5.0 (4.1–6.1)[‡]	400	800	1.0
Phenol	5.0 (4.2–5.8)[t]	400	82 000	0.2
2-Nitrophenol	5.9 (5.5–6.3)[‡]	472	2 100	1.0
Amines				
Chloroacetamide	2.0 (1.9–2.1)*	160	>10 000	NF
Diphenyl nitrosamine[f]	2.4 (2.2–2.6)[t]	192	<2.4	NF
Di-n-propyl nitrosamine	11.0 (10.0–12.0)[‡]	880	9 900	NF
Substituted aromatics				
Carbaryl[f]	14 (10–19)*	1 120	40	0.15
Nitrobenzene	16 (15–17)*	1 280	1 900	0.15
1,2-Dichlorobenzene[f]	21 (19–23)[t]	1 680	145	1.0
1,2,4-Trichlorobenzene[f]	27 (24–32)[‡]	2 160	19	0.42
Chlorobenzene[f]	29 (22–38)[‡]	2 320	500	8.8
Ethylbenzene[f]	47 (42–53)[§]	3 760	150	7.0
Toluene[f]	75 (68–83)[‖]	6 000	5.0	10
Benzene[f]	98 (78–123)[‖]	7 840	1 780	76
Hexachlorobenzene[f]	nontoxic up to 1 000[#]	80 000	0.1	10^{-5}
Halogenated aliphatics				
Hexachlorobutadiene[f]	10 (9–11)*	800	2.0	0.15
Tetrachloroethane	14 (11–17)*[t]	1 120	2 900	5
Hexachloroethane[f]	19 (16–23)[t‡]	1 520	50	0.4
Bis(2-chloroethyl)ether	19 (18–20)[‡]	1 520	10 200	0.7
2-Chloroethyl vinyl ether	33 (31–36)[§]	2 640	15 000	26
1,1,2-Trichloroethane	42 (35–49)[§]	3 360	4 500	19
1,2-Dichloroethane	60 (54–68)[‖]	4 800	8 700	61
1,2-Dichloropropane[f]	64 (59–70)[‖]	5 120	2 700	42
1,1,1-Trichloroethane[f]	83 (76–91)[#]	6 640	4 400	100
Trichloroethene[f]	105 (91–122)**	8 400	1 100	60
Trichloromethane[f]	111 (103–122)**	8 880	8 000	160
Tetrachloromethane[f]	160 (142–179)	12 800	800	90
1,2-$trans$-Dichloroethene[f]	286 (264–308)[tt]	22 880	800	200
Dichloromethane[f]	304 (258–358)[tt]	24 320	20 000	349
Polycyclic aromatic hydrocarbons				
Acenaphthene[f]	49 (34–69)*	3 920	3.4	10^{-3}
Fluorene[f]	171 (128–224)[t]	13 680	1.8	10^{-3}
Fluoranthene[f]	2 160 (1 550–3 000)[‡]	172 800	0.2	10^{-5}
Naphthalene[f]	4 670 (2 730–7 980)[§]	373 600	3.4	0.5

TABLE 4—*Continued.*

Chemical	Contact Test LC$_{50}$, $\mu g/cm^2$ (95% CI)^a	Chemical^b Solubility at 20°C^c		Vapor Pressure, mm^c
		μg per Vial	$\mu g/mL$	
Phthalates				
Dimethyl phthalate	550 (440–690)*	44 000	5 000 000	<0.01
Diethyl phthalate	850 (660–1 090)†	68 000	1 000 000	14
n-Butyl phthalate^f	1 360 (1 050–1 750)‡	108 800	13 000	0.1
Dioctyl phthalate^f	3 140 (2 270–4 330)§	251 200	3 000	1.2
Bis(2-ethylhexyl)phthalate^f	nontoxic up to 25 000	2 × 10^6	400	<0.01

^{a}Numbers in parentheses are the 95% confidence intervals (CI).
^{b}Indicates the total amount of chemical present per vial at the LC$_{50}$ concentration.
^{c}Data from Ref 9.
^{d}Means with a common superscript symbol are not significantly different ($P < 0.05$) within a chemical group.
^{e}NF—not found.
^{f}The amount of chemical in the vial at the LC$_{50}$ concentration exceeds the solubility of the chemical in 1 mL of water.

TABLE 5—*Earthworm LC$_{50}$ contact test values for the substituted aromatics tested.*

Chemical	LC$_{50}$, $\mu g/cm^2$ (95% CI)
Carbaryl	14 (10–19)*^{a}
Nitrobenzene	16 (15–17)*
1,2-Dichlorobenzene	21 (19–23)
1,2,4-Trichlorobenzene	27 (24–32)†
Chlorobenzene	29 (22–38)†
Ethylbenzene	47 (42–53)
Toluene	75 (68–83)‡
Benzene	98 (78–123)‡
Hexachlorobenzene	nontoxic up to 1 000 $\mu g/cm^2$

^{a}Means with a common superscript symbol are not statistically different, $P < 0.05$.

ference in toxicity between carbaryl and nitrobenzene, the most toxic substituted aromatics tested. The chlorine-substituted aromatics were all more toxic than the alkyl-substituted aromatics. Hexachlorobenzene, perhaps due to its extremely low solubility, 0.1 $\mu g/mL$, was the least toxic of all the substituted aromatics tested. The two-day contact test may not have permitted enough time for the worms to be affected by the hexachlorobenzene.

The large amount of effort involved in conducting the artificial soil tests limited the number of chemicals that could be evaluated. Table 6 lists the

TABLE 6—*Impact of chemicals in the artificial soil test and contact test.*

Chemical	Artificial Soil Test LC_{50}, mg/kg (95% CI)		Contact Test LC_{50}, $\mu g/cm^2$ (95% CI)	
Chloroacetamide	26	(22–31)	2.0	(1.9–2.1)
4-Nitrophenol	38	(32–45)	0.7	(0.6–0.8)
2,4,6-Trichlorophenol	58	(54–62)	5.0	(4.1–6.1)
Carbaryl	106	(81–138)	14	(10–19)
Diphenyl nitrosamine	151	(136–167)	2.4	(2.2–2.6)
Fluorene	173	(150–201)	171	(128–224)
Phenol	401	(347–463)	5.0	(4.2–5.8)
2-Chloroethyl vinyl ether	740	(570–970)	33	(31–36)
Dimethyl phthalate	3 160	(2 450–3 920)	550	(440–690)
1,2-Dichloropropane	4 240	(3 830–4 680)	64	(59–70)

LC_{50} values found for the organics tested with the artificial soil test in order of decreasing toxicity. The contact test LC_{50} values for the organics tested with the artificial soil test are included for comparative purposes. In general, the chemicals that were the most toxic to the worms in the contact test were the most toxic in the artificial soil test.

Discussion

The development of any bioassay or toxicity index requires that the data obtained with the new methods be compared to existing toxicity data. One of the most readily available toxicity data bases is rat oral lethal dose for 50% (LD_{50}) data. Table 3 lists the mean earthworm LC_{50} values and ranges and the mean rat oral LD_{50} values and range for the six chemical classes tested. The two chemical groups that were the most toxic to the earthworms, phenols and amines, were also the most toxic to the rat. The third most toxic group to the earthworms, substituted aromatics, was also the third most toxic chemical group to the rat. The other three chemical groups, halogenated aliphatics, polycyclic aromatic hydrocarbons and phthalates, were less toxic to the earthworms and the rats but had the same relative toxicity. This comparison suggests that chemicals toxic to the earthworms have the same relative toxicity to rats and vice versa.

Comparable data for specific organic chemicals, in terms of earthworm LC_{50} and rat LD_{50} values, are presented in Table 7. This comparison also shows the same relative toxicity pattern. The substituted aromatics that were the most toxic to the rat also were the most toxic to the worms. Carbaryl, nitrobenzene, and 1,2-dichlorobenzene were the three most toxic substituted aromatics and toluene, benzene, and hexachlorobenzene were the three least toxic substituted aromatics tested.

The LC_{50} data from the artificial soil test also were compared to the rat

TABLE 7—*Earthworm LC_{50} contact test values and rat oral LD_{50} values for the substituted aromatics tested.*

Chemical	Contact Test LC_{50}, $\mu g/cm^2$	Rat Oral LD_{50}, mg/kg
Carbaryl	14[*][a]	250
Nitrobenzene	16[*]	640
1,2-Dichlorobenzene	21	500
1,2,4-Trichlorobenzene	27[†]	756
Chlorobenzene	29[†]	2 910
Ethylbenzene	47	3 500
Toluene	75[‡]	5 000
Benzene	98[‡]	3 800
Hexachlorobenzene	nontoxic up to 1 000 $\mu g/cm^2$	10 000

[a]Means with a common superscript symbol are not significantly different, $P < 0.05$.

LD_{50} data (Table 8). Again, in general, the chemicals that were most toxic to the worms also were the most toxic to the rat.

A comparison of the relevant factors associated with the contact and artificial soil tests (Table 9) identifies the advantages and disadvantages of both methods. The contact test is more rapid and less expensive to run than the artificial soil test. The artificial soil test may be more representative of what really happens to earthworms in soil ecosystems since this test permits both skin contact and ingestion of the test chemical. Neither test lasts long enough to evaluate growth and reproductive effects.

The comparisons of relative order of toxicity (Tables 6, 7, and 8) indicated that nitro-substituted compounds were generally more toxic than chloro-

TABLE 8—*Impact of the chemicals in the artificial test and rat oral LD_{50} values.*

Chemical	Artificial Soil Test LC_{50}, mg/kg	Rat oral LD_{50}, mg/kg
Chloroacetamide	26	70
4-Nitrophenol	38	350
2,4,6-Trichlorophenol	58	820
Carbaryl	106	250
N-Nitrosodiphenylamine	151	1 650
Fluorene	173	NF[a]
Phenol	401	414
2-Chloroethyl vinyl ether	740	250
Dimethyl phthalate	3 160	6 900
1,2-Dichloropropane	4 240	1 900

[a]Not found.

TABLE 9—*Comparison of the parameters involved with the contact and artificial soil tests.*

Parameter	Contact Test	Artificial Soil Test
Time	short—2 days	long—14 days
Cost	lower	higher
Mode of impact	topical	topical and ingestion
Usefulness	for compounds that pass through skin membranes	for low-volatility compounds
Complexity of test	fewer components to procedure	more components to procedure

substituted compounds, which in turn were more toxic than alkyl-substituted compounds. All of the substituted compounds were more toxic than the parent unsubstituted compound.

The contact test and the artificial soil test are biological assay methods that can be used to estimate the biological impact of wastes and chemicals added to the soil. Both are relatively short-term methods: the contact test takes two days and the artificial soil test takes 14 days. The length of a biological assay can affect the toxicity of a chemical. If the chemical takes a long time to affect the biological system of a worm, the contact and artificial soil tests may not identify the true toxicity. An example could be cadmium, which in both the contact and artificial soil tests was not as toxic as copper, zinc, or nickel. However, in our other studies [7], in which the same five metals were analyzed for their effect on growth and reproduction, cadmium was the most toxic of the five metals tested.

Summary

The contact test and artificial soil tests are designed to screen the biological impact of potentially toxic chemicals in wastes applied to soils. Four salts (acetate, chloride, nitrate and sulfate) of five metals (cadmium, copper, nickel, lead, and zinc) were evaluated by using the contact test. The different water-soluble salts of each metal did not indicate any difference in toxicity to the worms. The order of toxicity for the metals in the contact test, from most toxic to least toxic, was Cu · Zn · Ni ≃ Cd · Pb. The order of toxicity for the metals in the artificial soil test, from most toxic to least toxic, was Cu ≃ Zn ≃ Ni · Cd · Pb.

A number of organic chemicals were tested for potential toxicity by using both the contact test and artificial soil test. The results were compared to available information on the rat oral LD_{50} values for the same compounds. The comparison indicated similarity between the relative toxicity of the chemicals to the worm and the rat. The chemicals that were most toxic to the worm were most toxic to the rat and those that were least toxic to the worm were least toxic to the rat.

The same relative toxicity was apparent for the six major chemical groups that were tested. The two most toxic chemical groups to the worms, phenols and amines, were also the most toxic to the rat, and the two least toxic chemical groups to the worms, polycyclic aromatic hydrocarbons and phthalates, were the two least toxic chemical groups to the rat.

The earthworm contact and artificial soil tests are useful in predicting the potential biological impact of organics and inorganics in wastes that may be applied to land for treatment and disposal.

Acknowledgements

This work was supported by grants from the National Science Foundation, "Waste Management Using Earthworms—Engineering and Scientific Relationships," Project 80-16764, and the U.S. Environmental Protection Agency, "Earthworms as a Bioassessment Tool," cooperative agreement CR-810006.

References

[1] Neuhauser, E. F., Malecki, M. R., and Loehr, R. C., in *Hazardous and Industrial Solid Waste Testing (Second Conference), ASTM STP 805,* American Society for Testing and Materials, Philadelphia, 1983, pp. 313-320.
[2] *EEC Guidelines for the Testing of Chemicals,* "The Contact and Artificial Soil Tests," EEC 79/831, Rev. 3., European Economic Community, Brussels, Belgium, 1982.
[3] Neuhauser, E. F., Hartenstein, R., and Kaplan, D. L., *Oikos,* Vol. 35, 1980, pp. 93-98.
[4] Hartenstein, R., Neuhauser, E. F., and Kaplan, D. L., *Oecologia,* Vol. 43, 1979, pp. 329-340.
[5] Kaplan, D. L., Hartenstein, R., Neuhauser, E. F., and Malecki, M. R., *Soil Biology and Biochemistry,* Vol. 12, 1980, pp. 347-352.
[6] Neuhauser, E. F., Kaplan, D. L., Malecki, M. R., and Hartenstein, R., *Agricultural Wastes,* Vol. 2, 1980, pp. 43-60.
[7] Malecki, M. R., Neuhauser, E. F., and Loehr, R. C., *Pedobiologia,* Vol. 24, 1982, pp. 129-137.
[8] Neuhauser, E. F., Malecki, M. R., and Loehr, R. C., *Proceedings,* 37th Purdue Industrial Waste Conference, Ann Arbor Science Publishers, Ann Arbor, MI, 1983, pp. 253-258.
[9] "Water-Related Environmental Fate of 129 Priority Pollutants," EPA-440/4-79-029, U.S. Environmental Protection Agency, Washington, DC, 1979.
[10] Litchfield, J. T. and Wilcoxon, F., *Journal of Pharmacology and Experimental Therapeutics,* Vol. 96, 1949, pp. 99-113.

Barbara Andon,[1] *Marcus Jackson,*[1] *Virginia Houk,*[2] *and*
Larry Claxton[2]

Evaluation of Chemical and Biological Methods for the Identification of Mutagenic and Cytotoxic Hazardous Waste Samples

REFERENCE: Andon, B., Jackson, M., Houk, V., and Claxton, L., "**Evaluation of Chemical and Biological Methods for the Identification of Mutagenic and Cytotoxic Hazardous Waste Samples,**" *Hazardous and Industrial Solid Waste Testing: Fourth Symposium, ASTM STP 886,* J. K. Petros, Jr., W. J. Lacy, and R. A. Conway, Eds., American Society for Testing and Materials, Philadelphia, 1986, pp. 204–215.

ABSTRACT: To assist in the development of methods for identifying potentially hazardous wastes, we have conducted studies on the extraction of toxicants from several solid waste samples. The extracts were tested for toxicity by the Chinese hamster ovary (CHO) cytotoxicity test and for mutagenic potential in the *Salmonella* histidine reversion assay. A new technique was also employed that measured the mutagenicity of neat waste samples by coupling thin layer chromatography (TLC) with the *Salmonella* histidine reversion assay.

The wastes selected for study were coke plant waste, herbicide manufacturing acetone-water effluent, and oil refining waste. Three extraction solvents—ethanol, dichloromethane (DCM), and dimethylsulfoxide (DMSO)—were chosen based on their solubility and compatibility with bioassay procedures. Each sample was divided into three parts and extracted with each of the three solvents separately.

All extracts were tested in the *Salmonella* assay at five dose levels with five Ames tester strains in the presence and in the absence of an exogenous metabolizing system. The DMSO and DCM extracts were utilized for CHO cytotoxicity evaluations. The three neat waste samples and two extracts were assayed by the *Salmonella*/TLC technique. In addition to the biological assessments, the gross chemical parameters for each sample were determined.

Results showed that coke plant waste and herbicide manufacturing acetone-water were mutagenic to *S. typhimurium* with the standard plate test. With the *Salmonella*/TLC

[1]Research associates, Environmental Health Research and Testing, Inc., Research Triangle Park, NC 27709.
[2]Research toxicologist and research biologist, respectively, U.S. Environmental Protection Agency, Health Effects Research Laboratory, Research Triangle Park, NC 27711.

technique, the coke plant waste was mutagenic and oil refining waste was both toxic and mutagenic. Oil refining waste was also toxic to CHO cells. Results of the gross chemistry determinations showed the three samples to have a wide range of solid content, total organic content, and extractables. Evaluation of the chemical extraction methods demonstrated few differences in extraction capabilities with respect to mutagenic activity.

KEY WORDS: *Salmonella typhimurium*, Chinese hamster ovary cells, hazardous wastes, mutagenicity, cytotoxicity, thin layer chromatography

Short-term in-vitro cytotoxicity and mutagenicity assays have been used to predict the impact of pollutants on the environment and to support risk assessment procedures. Few in-vitro studies, however, have evaluated the biological activity that may be associated with industrial wastes. In the past the major research effort concerning wastes has been directed toward elucidating the toxicity associated with leachates [1]. Biological and chemical test methods have seldom been evaluated for use in detecting biologically active solid wastes.

In this study, the genotoxic and cytotoxic effects associated with three industrial waste samples have been determined. The samples were coke plant waste, herbicide manufacturing waste acetone-water, and oil refining waste. The biological test methods employed were the *Salmonella* histidine reversion assay and the Chinese hamster ovary cytotoxicity assay (CHO). Chemical methods involving the extraction of the three wastes with ethanol (EtOH), dichloromethane (DCM), and dimethylsulfoxide (DMSO) were evaluated. An additional technique, which coupled the chemical separation of wastes by thin layer chromatography (TLC) with the *Salmonella* histidine reversion assay, was also used.

This paper reviews the above methods for their ability to identify biologically active waste samples.

Materials and Methods

Waste Samples

The hazardous waste samples (Table 1) evaluated in this study were obtained from Battelle Columbus Laboratories, Columbus, Ohio, courtesy of Dr. M. McKown. Samples were stored in the dark at 5°C in brown glass bottles with Teflon®-lined caps.

Gross Chemical and Physical Characterization of Waste Samples

The following gross chemical and physical characteristics were determined for the three wastes: pH, density, total organic content, total solids, volatiles, and semivolatile content. Selected methods are presented below.

TABLE 1—*Gross chemical and physical characteristics of waste samples.*

Waste	Description	pH	Density	TOC, mg/L $(mg/g)^a$	Total Solids, %	Volatiles, mg/mL	Extract-ables, %
Coke plant waste	light brown fluid with suspended solids, 95% DMSO-soluble, water-insoluble	8.0	1.0	189	4.0	0.78	0.27
Herbicide manufacturing acetone-water	red-brown clear liquid, 100% DMSO-soluble	6.5	1.1	20 044	20.5	1.1	0.06
Oil refining waste	dark liquid with brown-black flocculent and oil droplets, DMSO- and water-soluble	7.0	NA^b	(7.3)	58.5	NA	4.22

^{a}Total organic content.
^{b}Not applicable.

Organic Extractables

The organic extractables were determined gravimetrically from the organic residue weight of a sample extract. A homogenized filtered sample was transferred to a separatory funnel and extracted three times with methylene chloride. The extracts were combined and transferred to a Kuderna-Danish apparatus with a tared receiver. The extract was concentrated to 2 mL and then taken to dryness under a nitrogen stream. Each sample was subjected to aspirator vacuum for approximately 2 min to ensure methylene chloride removal. The containers were reweighed to determine the absolute amount of organic extractables, which was reported as percent extractables (based on total sample weight).

Volatiles Content

The approximate volatiles content of aqueous hazardous waste samples was determined by using the Purgeable Halocarbons Method (modified Method 601) of the U.S. Environmental Protection Agency (purge-and-trap analysis). The samples were analyzed with a Fischer-Victoreen gas chromatograph (GC) equipped with flame ionization and Hall electrolytic conductivity detectors. A 1% SP-1000 packed column was used for analysis. Each sample (100 μL) was introduced into the purge vessel, which contained 5 mL of pre-purged distilled, deionized water. The sample was then purged following Method 601 procedures. The GC column effluent was split, with a portion of the stream passing to the flame ionization detector and a portion passing to the Hall detector. All GC peaks were electronically integrated. Approximate quantitation was achieved by comparing the sample peak areas with peak areas of external standards in n-nonane and toluene for the flame ionization detector and methyl chloroform for the Hall electrolytic conductivity detector. Because the individual components (and hence response factors) of the samples were not known, the levels determined must be viewed as approximate.

Extraction of Waste Samples for Bioassay

Dichloromethane Extracts—The sample was placed in a preweighed screw cap culture tube. After the sample weight was determined, a known volume of DCM was added to the tube. The tube was shaken to break up clumps, vortexed for 10 min, and centrifuged for 30 min. The DCM phase was withdrawn and filtered through a 5-μM Teflon filter into a capped vial. An aliquot (0.50 mL) of the extract was removed and blown to dryness under nitrogen gas, and then the sample extract mass and concentration were determined. A known volume of sample extract was transferred to a clean vial and the solvent exchanged into 1.0 mL DMSO for bioassay.

Ethanol Extracts—The above procedure was also employed for the EtOH extracts. However, no solvent exchange was performed since the solvent was compatible with bioassay procedures.

Dimethylsulfoxide Extracts—The same procedure was used as for the DCM extracts, except no solvent exchange was performed and no extract concentration was determined.

Bioassay methods

Salmonella *Histidine Reversion Assay*—The *S. typhimurium* plate incorporation assay was performed as described by Ames et al [2]. Most test samples were assayed twice with each of the five tester strains (TA1535, TA100, TA1537, TA1538, and TA98), both with and without an Aroclor-1254-induced rat liver S9 homogenate at a concentration of 1.05 mg protein per plate. Triplicate plates at five test concentrations were evaluated. The five test concentrations employed resulted in a 4.5, 3.6, 2.7, 1.8, and 0.9% vol/vol supplement of the neat extract to the bacterial cell suspension in agar. The 4.5% vol/vol supplement represented the maximum concentration of solvent that could be exposed to the bacteria before solvent toxicity was observed. Criteria for a positive mutagenic response were based on the presence of a dose-response curve and the number of revertants for a treated culture equal to two times the background spontaneous value. In addition, the program of Stead et al [3] for statistical evaluation of Ames test results was employed.

Thin Layer Chromatography/Ames Technique—The TLC/Ames technique was a modification of the procedure developed by Bjørseth et al [4,5].

All samples were evaluated twice with tester strains TA98 and TA100, both with and without metabolic activation.

The TLC separations were performed on commercially available glass backed cellulose plates (100 by 100 mm). Neat samples or extracts were spotted on the plates at an origin 15 mm from the bottom edge and 20 mm apart. Dose-response relationships were determined by making four individual applications of increasing volumes per plate. Spot diameters were kept as small as possible to enhance resolution. The plates were developed at room temperature in chloroform for 15 to 20 min, at which point the solvent front had ascended to approximately 10 mm from the top edge of the plate. After the plates were removed from the developing chamber and the mobile phase had evaporated, the plates were placed into petri dishes (150 by 15 mm). A mixture containing agar, minimal growth media, the bacterial strain, and the optional metabolizing system was poured over the TLC plates. The petri dishes were incubated for 72 h at 37°C. After incubation, the plates were examined for toxicity, as evidenced by (1) a reduction in the overall number of revertants when compared to spontaneous values or (2) a discrete toxic zone devoid of bacterial growth. A mutagenic event was indicated by the appearance of one or more localized clusters of histidine revertants or as a twofold or greater increase in revertant counts (for the total plate) over spontaneous background values.

The CHO Cytotoxicity Test—The samples were evaluated for cytotoxicity by using the CHO cell system. Detailed aspects of analysis, sample handling, and culture and treatment of cells were performed as previously reported [6]. Briefly, CHO cells were added to individual wells of cluster dishes containing the waste extracts. Each extract was evaluated twice in triplicate at the maximum concentration obtainable with an 0.5% vol/vol supplement of extract. After a 20-h exposure, cells were harvested and cell number and viability were determined simultaneously by exclusion of trypan blue using a hemacytometer. Adenosine-5-triphosphate (ATP) was assayed by luminescence, using the luciferin-luciferase reaction [7]. Results from each determination were expressed as a percent of the control cell data, except for percent viability, and combined from replicate experiments. The viability index was determined by multiplying the percent viability by the surviving fraction of cells for each treated culture. The viability index is an expression of cell viability that adjusts for any reduction in cell number resulting from toxic treatment. Any sample found to affect cell viability by depressing ATP generation below 50% of control was reevaluated in an effort to establish a dose-response relationship.

Results

The hazardous waste samples with their gross chemical parameters and physical descriptions are shown in Table 1.

The results of the mutagenic evaluation of the wastes with both the standard plate test and the TLC technique demonstrate that coke plant waste, herbicide manufacturing waste acetone-water, and oil refining waste were all mutagenic under one or more test conditions (Table 2). The model slope value is given for positive responses with the standard plate test as an indication of the sample's relative activity. Since slope values cannot be determined with the TLC technique, a positive response is defined as a (+) for the TLC technique. A negative response is designated by a (−) for both assays. The concentrations of the EtOH and DCM extracts used in the experiments are also listed. Because of the low vapor pressure of the solvent DMSO, no extract weights could be determined. As a result, the maximum volume of extract per plate used is listed. Where possible, density values were used to calculate the exposure concentration of neat samples in the TLC assay.

Coke plant waste demonstrated the greatest mutagenic activity, with slope values ranging from 0.1 to 30. A positive mutagenic response with the coke sample was also detected with the TLC technique. The herbicide sample was the next most active sample, with two slope values of 20. Oil refining waste was not mutagenic with the standard plate test, but mutagenicity was detected with the TLC technique using the neat sample. This sample also demonstrated toxicity with the TLC technique in the absence of microsomal activation. In general, however, the addition of a rat liver metabolizing system enhanced the detection of mutagenic samples.

The evaluation of a maximum (0.5% vol/vol) concentration of the waste extracts with CHO cells showed only the oil refining waste to be cytotoxic (Table 3). This sample produced a greater than 50% reduction in the viability index and cellular ATP content. A more definitive evaluation for dose-response relationships, however, shows no parameter to be depressed greater than 50% (Fig. 1). This may be a result of the sample's oily nature, which caused some difficulty in delivering reproducible sample replicates and in its solubility.

Only the DMSO and DCM extracts are reported in Table 3, since the ethanol solvent proved to be cytotoxic at an 0.5% vol/vol concentration. Lowering the exposure volume to a nontoxic concentration of 0.01% vol/vol resulted in too low a sample concentration to warrant performing the experiments. In general, evaluation of the chemical extraction methods demonstrated few differences in extraction capabilities with respect to mutagenic activity.

Discussion

Of the three samples tested, the coke plant waste was the most mutagenic. The activity of this sample resulted in plate counts that were 20 times greater than the spontaneous his^+ background values under certain test conditions (raw data not given). This observation is supported by the model slope values and the relatively low exposure concentrations. It was not surprising that this

TABLE 2—Results of the mutagenic activity of waste samples using the S. typhimurium histidine reversion assay.

Waste	Extract	Concentration, μg (volume) per plate[a]	Standard Plate Test[b] TA100 +S9	TA100 -S9	TA1535 +S9	TA1535 -S9	TA1537 +S9	TA1537 -S9	TA1538 +S9	TA1538 -S9	TA98 +S9	TA98 -S9	Concentration, μg (volume) per plate[c]	TLC Technique TA98 +S9	TA98 -S9	TA100 +S9	TA100 -S9
Coke plant	neat	...	NT[d]	NT	NT	NT	NT	NT	NT	NT	NT	NT	10	—	+[f]	—	—
	EtOH	50	10.2[e]	—[f]	—	—	0.9	0.1	6.9	0.3	30.0	—	3	+	—	—	—
	DCM	60	1.3	—	—	—	1.1	—	3.6	—	2.8	—	...	NT	NT	NT	NT
	DMSO	(100 μL)	1.4	—	—	—	0.8	1.4	—	—	15.0	—	(8 μL)	—	—	+	+
Herbicide manufacturing acetone-water	neat	...	NT	NT	NT	NT	NT	NT	NT	NT	NT	NT	4	—	—	—	—
	EtOH	4000	—	—	—	—	—	—	0.1	0.01	0.006	—	...	NT	NT	NT	NT
	DCM	10	—	—	—	—	20.7	—	11.6	2.4	19.9	—	...	NT	NT	NT	NT
	DMSO	(100 μL)	—	—	—	—	—	—	—	—	—	—	...	NT	NT	NT	NT
Oil refining	neat	...	NT	NT	NT	NT	NT	NT	NT	NT	NT	NT	4	+	T[f]	+	—
	EtOH	2000	—	—	—	—	—	—	—	—	—	—	...	NT	NT	NT	NT
	DCM	16	—	—	—	—	—	—	—	—	—	—	...	NT	NT	NT	NT
	DMSO	(100 μL)	—	—	—	—	—	—	—	—	—	—	...	NT	NT	NT	NT

[a] Approximate concentration per plate; maximum volume per plate where weight of extract could not be determined.
[b] +S9, experiment performed in the presence of Aroclor-induced S9; −S9, experiment performed in the absence of Aroclor-induced S9.
[c] Approximate concentrations derived from density values for neat samples.
[d] NT, not tested.
[e] Model slope values for positive responses with the standard plate test.
[f] −, negative response; +, positive response with the TLC technique; T, toxic response.

TABLE 3—*Results of the toxic potential of three hazardous waste samples using the CHO cytotoxicity test.*

Waste	Extract	Concentration, μg/mL (% vol/vol)	Percent Viability	Viability Index[a]	Cell, n/mL[a]	ATP, fg/mL[a]
Coke plant	DCM	6	97	107	111	87
	DMSO	(0.5)	99	99	100	108
Herbicide	DCM	1	98	111	114	93
manufacturing	DMSO	(0.5)	99	93	94	95
acetone-water						
Oil refining	DCM	1560	68	43	63	24
	DMSO	(0.5)	78	74	78	78

[a]Expressed as a percentage of controls.

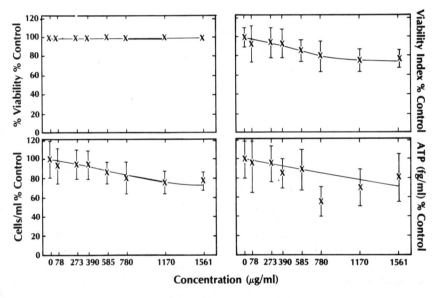

FIG. 1—*Toxicity of oil refining waste in the CHO assay DCM extracts (mean of six replicates). Concentrations shown are approximate.*

sample was mutagenic, since gas chromatography/mass spectrometry (GC/MS) analysis (Table 4) performed by Battelle Columbus Laboratories showed it to contain 1.12 μg/g benzo(*a*)pyrene, 0.69 μg/g benzo(*b*)- and benzo(*k*)fluoranthene, and 0.50 μg/g chrysene and benzo(*a*)anthracene [8]. Other investigators also have shown coke effluents to be mutagenic [9].

Since identity of the actual herbicide(s) contained in the herbicide manufacturing sample is unknown, little can be said about the activity of this sam-

TABLE 4—*Results of GC-MS analysis of coke plant waste.*

Compound	Coke Plant Waste, $\mu g/g$
Purgeable organics	
Methylene chloride	0.09
Trichloroethylene	0.00
Benzene	0.00
1,1,2,2-Tetrachloroethene	0.00
1,1,2,2-Tetrachloroethene	0.00
Toluene	0.00
Chlorobenzene	0.00
Ethylbenzene	0.00
Base/neutral extractable semivolatile organics	
Naphthalene	0.00
Diethyl phthalate	0.00
Di-n-butyl phthalate	0.00
Bis(2-ethylhexyl)phthalate	1.49
Chrysene/benzo(a)anthracene	<0.50
Benzo(b)/benzo(k)fluoranthenes	0.69
Benzo(a)pyrene	1.12
Acid Extractable Semivolatile Organics	
2-Chlorophenol	0.00
Phenol	<0.50
Pentachlorophenol	2.17

ple. The GC/MS analysis (Table 5), again performed by Battelle Columbus Laboratories, revealed that 75 $\mu g/g$ of the sample was an alkylamine(s) ($C_8H_{19}N$) [10]. Chemical compounds containing this mass distribution are generally considered nonhazardous; however, the conversion of the amine group to a more active species is a possibility.

A limited amount of mutagenic activity was demonstrated for oil refining waste by using the *Salmonella*/TLC method. This observation was not unexpected, since a variety of organic bacterial mutagens, especially the polycyclic aromatic hydrocarbons (PAH) and alkylated PAH, are known to be associ-

TABLE 5—*Results of GC-MS analysis of herbicide manufacturing acetone-water.*

Compound	$\mu g/g$
Alkylamine ($C_8H_{19}N$)	75
Alkylamine	9
Alkyl ketone (MW=128)	8
Unknown	20
Unknown	30
Unknown	9

ated with fossil fuels and their combustion effluents [11]. In addition, several reports defining the mutagenic activity of oil shale retort process water have been published in the literature [12,13].

Oil refining waste is currently subject to Resource Conservation and Recovery Act (RCRA) regulation because it has been shown to contain significant levels (in waste streams) of the toxic metals lead and chromium [14]. The *Salmonella* assay with the standard plate test or with the TLC technique does not detect the mutagenic or cytotoxic potential of metals [15]. Bioassay modifications or additional strains are required to enhance the sensitivity of the test for this class of chemicals. Only one metal, chromium, has been shown not to require alterations in assay parameters. As a result, the *Salmonella* assay may not be suitable for the routine mutagenic analysis of metal-containing wastes unless alternative experimental conditions are employed. For these reasons, it may be possible that the low levels of mutagenic and toxic activities observed with the neat sample may be some small percentage of the cumulative effects of these metal components under nonoptimal assay conditions.

The CHO cytotoxicity test, unlike the *Salmonella* assay, is sensitive to a variety of metallic compounds, including lead and chromium salts [16]. This may explain the observed cytotoxic effects with oil refining waste in CHO. Unfortunately, no GC/MS data are available to aid in the analysis of the observed mutagenic and cytotoxic effects.

Each of the three assays employed in this study were capable of detecting the biological activity of the hazardous waste samples. There are, however, advantages and disadvantages associated with each system. The standard plate test, as a rule, is not capable of using the neat complex mixture for routine analysis. Because of the incompatibility of the neat sample with the test system, extracts are employed. The TLC technique has the advantage of using neat samples. This technique is rapid, inexpensive, and sensitive and may provide a convenient method for prioritizing samples for additional testing. One disadvantage is that the separation and detection of mutagenic chemical components depends greatly on the chemical characteristics of the test sample (solubility, polarity, charge) and the mobile phase selected. An additional disadvantage is that the applied sample volumes must remain quite small, thus limiting the possibility of detecting weak mutagens present in low concentrations. The CHO system, which is a more rapid technique than the bacterial test systems, may be useful in prioritizing samples for additional testing as well as aiding in identifying ranges of toxic potential. This system also has the advantage of being able to evaluate neat samples as well as extracts or even leachates. The major disadvantage of the CHO system is that it does not identify mixtures or compounds capable of causing heritable genetic damage.

Effective and inexpensive methods for evaluating the potential health effects of industrial process residuals are needed. Most of these bioassay methods must use extraction techniques that reliably remove a representative population of chemical components from the waste sample. The extraction solvents must also be compatible with short-term in-vitro assays capable of detecting genetic damage and cytotoxicity. The results presented here demonstrate that EtOH, DCM, and DMSO extraction of a variety of solid wastes does remove and allow detection of mutagenic and cytotoxic components. A useful addition to the screening procedure assay would be the use of bacterial strains capable of identifying wastes containing mutagenic metals. Future work will involve the characterization of additional classes of waste samples to further discriminate between test systems and chemical methods. Attempts will be made to identify the mutagenic and cytotoxic components of selected samples by coupling high-performance liquid chromatography with suitable bioassays.

Acknowledgments

The authors would like to thank S. Warren, R. Thomas, and J. Scott for their technical assistance. We also acknowledge L. Scearce and S. Staton for the preparation of this manuscript.

The research described in this paper has been reviewed by the Health Effects Research Laboratory, U.S. Environmental Protection Agency, and approved for publication. Mention of trade names or commercial products does not constitute an endorsement by the agency or recommendation for use.

References

[1] Epler, J. L., "Toxicity of Leachates," final report, EPA 600/280-057, U.S. Environmental Protection Agency, Municipal Environmental Research Laboratory, Cincinnati, OH, March 1980.

[2] Ames, B. N., McCann, J., and Yamasaki, E., "Methods for Detecting Carcinogens and Mutagens with the *Salmonella*/Mammalian-Microsome Mutagenicity Test," *Mutation Research*, Vol. 31, Dec. 1975, pp. 347-364.

[3] Stead, A. G., Hasselblad, V., Creason, J. P., and Claxton, L., "Modeling the Ames test," *Mutation Research*, Vol. 85, Feb. 1981, pp. 13-27.

[4] Bjørseth, A., Eidsa, G., Gether, J., Landmark, L., and Møller, M., "Detection of Mutagens in Complex Samples by the *Salmonella* Assay Applied Directly on Thin Layer Chromatography Plates," *Science*, Vol. 215, Jan. 1982, pp. 87-89.

[5] Houk, V., "Screening Complex Hazardous Wastes for Mutagenic Activity Using the TLC/Ames Assay," M.S. thesis, University of North Carolina, Chapel Hill, North Carolina, 1984.

[6] Garrett, N. E., Campbell, J. A., Stack, H. F., Waters, M. D., and Lewtas, J., "The Utilization of the Rabbit Alveolar Macrophage and Chinese Hamster Ovary Cell for Evaluation of the Toxicity of Particulate Materials," *Environmental Research*, Vol. 24, No. 4, April 1981, pp. 366-376.

[7] Waters, M. D., Vaughan, T. O., Abernathy, D. J., Garland, H. R., Cox, C. C., and Coffin, D. L., "Toxicity of Platinum (IV) Salts for Cells of Pulmonary Origin," *Environmental Health Perspectives*, Vol. 12, 1975, pp. 45-46.

[8] Miller, H. C., James, R. H., Dickson, W. K., Neptune, M. D., and Carter, M. H., "Evaluation of Methodology for the Survey Analysis of Solid Wastes," *Hazardous Solid Waste Testing: First Conference, STP 760*, R. A. Conway and B. C. Malloy, Eds., American Society for Testing and Materials, Philadelphia, 1982, pp. 240-266.

[9] Lewtas, J., Bradow, R., Jungers, R., Harris, B., Zweidinger, R., et al, "Mutagenic and Carcinogenic Potency of Extracts of Diesel and Related Environmental Emissions: Study Design, Sample Generation, Collection, and Preparation," *Environment International*, Vol. 5, April 1981, pp. 383-387.

[10] Battelle Columbus Laboratories, "Collaborative Study for the Evaluation of a Selected Method for Hazardous Waste Analysis," final report, EPA Contract 68-02-3169, Battelle Columbus Laboratory, Columbus, OH, 8 Dec. 1981.

[11] Barfknecht, T. R., Andon, B. M., Thilly, W. G., and Hites, R. A., "Soot and Mutation in Bacteria and Human Cells, *Chemical Analysis and Biological Fate: Polynuclear Aromatic Hydrocarbons*, M. Cooke and A. F. Dennis, Eds., Battelle Press, Columbus, OH, 1981, pp. 231-242.

[12] Smith-Sonneborn, J., McCann, E. A., and Palizzi, R. A., "Bioassay of Oil Shale Process Waters in *Paramecium* and *Salmonella, Short-Term Bioassays in the Analysis of Complex Environmental Mixtures III*, M. D. Waters et al, Eds., Plenum, New York, 1983, pp. 197-210.

[13] Strinste, G. F., Bingham, J. M., Spall, W. D., Nickols, J. W., Okinaka, R. T., and Chen, D. J., "Fractionation of Oil Shale Retort Process Water: Isolation of Photoactive Genotoxic Components," *Short-Term Bioassays in the Analysis of Complex Environmental Mixtures III*, M. D. Waters et al, Eds., Plenum, New York, 1983, pp. 139-152.

[14] U.S. Environmental Protection Agency, Resource Conservation and Recovery Act, Subtitle C, Background Document, "Petroleum Refining," U.S. Government Printing Office, Washington, DC, 19 May 1980.

[15] Sirover, M., "Effects of Metals in *in-Vitro* Bioassays," *Environmental Health Perspectives*, Vol. 40, Aug. 1981, pp. 163-172.

[16] Garrett, N. E. and Lewtas, J., "Cellular Toxicity in Chinese Hamster Ovary Cell Cultures. 1. Analysis of Cytotoxicity Endpoints for Twenty-Nine Priority Pollutants, "*Environmental Research*, Vol. 32, No. 2, Dec. 1983, pp. 455-465.

Joellen Lewtas[1] *and Barbara Andon*[2]

A Proposed In-Vivo/In-Vitro Approach to the Toxicological Assessment of Hazardous Waste

REFERENCE: Lewtas, J. and Andon, B., **"A Proposed In-Vivo/In-Vitro Approach to the Toxicological Assessment of Hazardous Waste,"** *Hazardous and Industrial Solid Waste Testing: Fourth Symposium, ASTM STP 886,* J. K. Petros, Jr., W. J. Lacy, and R. A. Conway, Eds., American Society for Testing and Materials, Philadelphia, 1986, pp. 216–229.

ABSTRACT: An in-vivo/in-vitro toxicological screen has been developed to evaluate potentially hazardous waste samples and process stream residuals. The biological methods employed in the screen are designed to be rapid, cost-effective, and capable of screening large numbers of wastes. Emphasis is placed on identifying a wide range of potential toxic responses associated with each waste by employing diverse test methods. The toxic end points identified by the screen include mutagenesis/carcinogenesis, general toxicology, neurotoxicology, reproductive toxicology, teratology, and immunotoxicology.

The protocol, called the toxicological screen, involves the oral administration of waste material to male rodents for ten consecutive days. At the end of the ten-day period the whole animal, body tissues, and fluids are evaluated for toxicity. The toxicological screen also involves the oral treatment of pregnant rodents for ten consecutive days to assess transplacental effects and neonatal survival. A major goal of the toxicological screen is to maximize the amount and type of potential health effects information that can be obtained by exposing metabolically competent intact animals to complex waste mixtures.

The toxicological screen is being validated by using a series of positive (and negative) control compounds of known and defined toxicity. The validation study ensures that the protocol will be capable of detecting biologically active wastes and will allow assessment of the protocol's use as a predictive tool for known chronic effects. An additional goal of the validation study is to identify a subset of the assays from the toxicological screen that most readily detect the toxic potential of the positive control compounds. These selected assays will constitute a prescreen for the entire protocol. The prescreen can then be used as a rapid means of prioritizing waste samples to be evaluated in the full toxicological screen.

Initial trials of the validation study with four positive control compounds demonstrate the toxicological screen's ability to detect biologically active chemicals and to generate dose-response data.

[1]Chief, Genetic Bioassay Branch, Health Effects Research Laboratory, U.S. Environmental Protection Agency, Research Triangle Park, NC 27711.

[2]Hazardous Waste Program coordinator, Environmental Health Research and Testing, Research Triangle Park, NC 27709.

KEY WORDS: hazardous wastes, industrial wastes, environmental pollutants, toxic wastes, toxicity testing, hazardous waste assessment, solid waste, mutagenesis/carcinogenesis, general toxicology, immunotoxicology, neurotoxicology, reproductive toxicology, teratology

The disposal of the vast quantity of wastes generated in the United States each year is of great economic, public, and environmental concern. The development of suitable waste management methods requires a means of identifying the "hazardous" portion of the nations's total waste load. It is only with the identification of hazardous wastes that proper storage, transportation, and disposal methods can be utilized to ensure public health and safety and integrity to the environment.

Over the past ten years much research has been directed toward the identification and hazard assessment of a variety of complex mixtures and environmental pollutants [1]. Many programs have relied totally on an in-vitro battery of tests of increasing complexity to identify complex mixtures with the potential for chronic toxicity (such as mutation or cancer) [2]. Others have estimated the hazard associated with pollutants by identifying the chemical components of a complex mixture. This information is then used in combination with computer programs designed to estimate the biological activity (of the mixture). Although both approaches provide valuable information concerning sample activity, they do not estimate the potential synergistic or antagonistic (greater than or less than additive) effects a pollutant may exhibit in the intact animal [3,4]. An in-vivo or combined in-vivo/in-vitro toxicological test system may best approximate this situation.

The Health Effects Research Laboratory (HERL) at Research Triangle Park, NC, has developed an in-vivo/in-vitro toxicological screen for the identification of biologically active hazardous waste mixtures. The screen involves the short-term repeated exposure of rodents to waste materials followed by toxicological assessment of the whole animal, body tissues, and fluids. Emphasis is placed on identifying the type of biological activity associated with wastes in addition to the degree of activity.

The protocol, called the toxicological screen, has been developed to provide the informational basis for regulatory judgments regarding the classification of wastes under the Resource Conservation and Recovery Act (RCRA). Any waste classified (or listed) as hazardous would be subject to stringent disposal methods as specified under Subtitle C of RCRA.

The toxicological screen's protocol has been reviewed by an external committee of scientists [5]. Their recommendations and comments have been incorporated into the toxicological screen's design. Additional review will be made during the validation of the toxicological screen's protocol in an effort to ensure the quality of the test methods developed.

This paper presents the general toxicological screen protocol and validation study design.

Methods

The protocol for the toxicological screen can be divided into two sections. The first is the treatment protocol, which defines the animal exposure regimen. The second is the biological screens, which include the toxicological assessments performed with the whole animal or various body tissues and fluids (Figs. 1 and 2). Only the logistics of performing the screen as a single protocol are described here. Specific protocols concerning the biological endpoints (procedural details, equipment, reagents, and the like) are beyond the scope of this paper.

The Treatment Protocol

Seventy-day-old Fischer 344 male rats are exposed by gavage to the waste mixture for ten consecutive days. Five concentrations of the test material and two negative control concentrations (cage and vehicle) are used (Fig. 1). The maximum concentration delivered to the animal is 1 g/kg per day. The remaining four doses are reduced usually by factors of two or three. Exposure of the test material is adjusted according to the weight gain or loss of the animal in an effort to maintain a constant ratio of dose to animal weight. The animals are weighed on Days 0, 3, 6, and 10 of the ten-day study. The dosing volume is calculated for each rat based on its latest weight. The neat waste material and extracts of the wastes are employed for in-vivo exposure.

Under this protocol, seven groups of animals are exposed to the wastes. The cage control and two of the five treated groups each have ten animals. The vehicle control group and three of the five treated groups each have 14 animals per group, bringing the total number to 86 animals. The extra four animals per group are used for sister chromatid exchange analysis.

Biological End points

During or at the end of the ten-day treatment the whole animal or its body tissues and fluids are evaluated in a series of toxicological assays. These assays are listed in Figs. 1 and 2. A time-course description of the order in which these assays are performed is given below and in Fig. 3.

Mutagens in the Urine

The mutagenic evaluation of urine from rats gavaged with waste samples will assist in identifying metabolic and physiological events which occur in the intact organism and which can not be duplicated in vitro. Body fluid analysis will be performed in a attempt to identify the in-vivo formation of mutagens and will be compared with in-vitro studies using wastes with rat liver preparations [6]. If successful, these analyses could play a critical role in determining the relative risk of genetic damage resulting from exposure to waste samples.

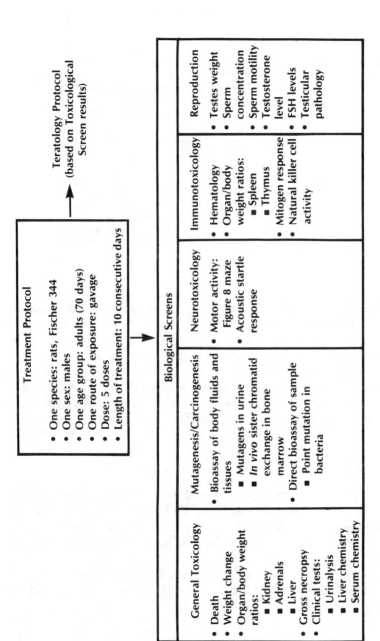

FIG. 1—*Toxicological screen.*

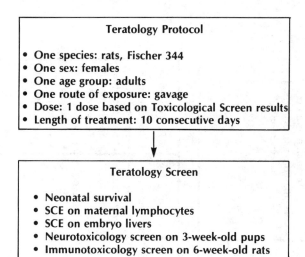

FIG. 2—*Teratology portion of toxicological screen.*

Immediately following the administration of the fifth consecutive dose (Day 5), 24-h urine samples are collected from one animal in each treatment group. The seven animals are randomly selected and placed in seven Nalgene® metabolism cages. Each cage is suspended over dry ice to ensure the integrity of the urine samples. Immediately following collection (Day 6) the raw urine is evaluated for the presence of mutagens using the Ames Test [7]. Each urine sample is evaluated once using strains TA98 and TA100 at three to five concentrations (depending on sample size) in triplicate, both with and without a rat liver microsomal preparation.

The In-Vitro Salmonella *Histidine Reversion Assay*

Each complex mixture to be evaluated in the toxicological screen will be assayed for its mutagenic potential using the in-vitro *Salmonella* histidine reversion assay (Ames Test) [7]. Samples will be tested with two to five tester strains at three to five concentrations, both with and without an exogeneous metabolizing system.

The generation of an in-vitro dose-response curve will allow (1) the identification of mutagenic samples and (2) the comparison of in-vitro and in-vivo generation of mutagenic metabolites.

In-Vivo Sister Chromatid Exchange (SCE)

Sister chromatid exchange analysis has become widely used as a sensitive measure of induced chromosomal alterations [8]. In the present study, in-

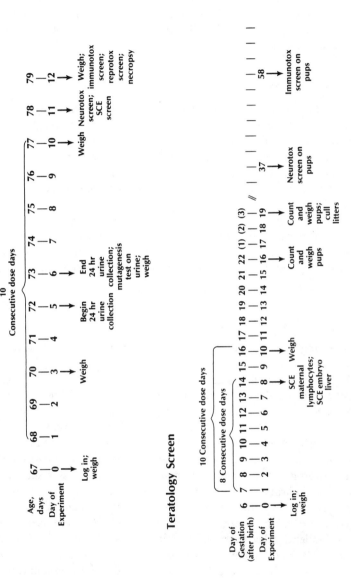

FIG. 3—*Schematic overview of the toxicological screen.*

vivo methodological approaches were used to measure SCE frequencies in treated rat bone marrow cells.

On the morning following the tenth consecutive dose, four animals designated for SCE studies from three of the five treated groups and the vehicle control group are pretreated with bromodeoxyuridine. These 16 animals are then given colchicine on Day 11, followed by the harvesting of bone marrow cells for SCE analysis.

Neurotoxicology Screen

Because behavorial and neurological complaints are reported by people exposed to industrial wastes, it is necessary to employ neurotoxicological tests to provide an indication of the hazard posed by specific waste samples. The neurotoxic potential of waste samples will be assessed by routine short-term behavorial tests, including motor activity and acoustic startle response [9,10]. These behavorial tests are among the most rapid and reliable tests of neurotoxicity.

Twenty-four hours after the tenth consecutive dose (Day 11 of the study), ten animals from each concentration group are evaluated in the neurotoxicological screen. This involves the assessment of motor activity using the figure-eight maze [10] and response to auditory stimuli using the acoustic startle response test [9].

Remaining Screens

The following determinations are performed on each animal unless otherwise stated:

On Day 12 the same ten animals evaluated in the neurotoxicological screen are anesthetized, weighed, and exsanguinated through the abdominal aorta. A small portion of whole blood (0.5 mL) is placed in a test tube for leukocyte count determinations. A blood smear is also made for differential leukocyte counts. Serum is prepared from the remaining blood for liver function and other serum chemistry tests, as well as testosterone and follicle-stimulating hormone determinations. These parameters are used to assess specific and general somatic normality. In order to determine the immunotoxic potential of wastes, the thymus and spleen are removed aseptically and weighed. Spleen cells from each concentration group are utilized for mitogen response and natural killer (NK) cell assays. The mitogen assay is a good in-vitro correlate of in-vivo T- and B-lymphocyte function [11]. Natural killer cells are a nonspecific first line of defense against tumors, certain viruses, and certain infectious agents. The 4-h ^{51}C release assay employed here for NK activity correlates very well with in-vivo tumor model assays [12].

For an initial assessment of reproductive toxicity, the testes and epididymides are removed and weighed. The right testis from each animal is trimmed and placed in Bouin's fixative for pathological examination. The epididymal fluid from the left cauda epididymis is utilized for studies of sperm concentration, motility, and abnormal spermatozoa. Sperm head counts are made with the left testis. These physiological measurements are common for assessment of risk in the male rat [13].

As part of the necropsy performed on Day 12, urine is collected directly from the bladder with a sterile syringe and applied to a diagnostic reagent strip for urinalysis. Each animal is then examined for gross thoracic, abdominal, and external pathology, including examination of the heart, lung, liver, kidney, and adrenal. The liver, kidney, and adrenals are weighed and organ-to-body-weight ratios are calculated to assess specific organ toxicity and stress response [14]. Except for the spleen, testes, and epididymides all animals and organs are discarded at the end of experimental Day 12.

Teratology Screen (Neonatal Survival)

The neonatal survival screen, which was developed at HERL, is a simple, cost-effective test [15]. Its results are highly correlated with those obtained in the standard teratology protocol. It relies upon the well-known observation that rodent pups born with birth defects are generally cannibalized by the dam, and those that do survive show significant growth deficits. These end points—death and weight loss during the first three days postpartum—are extremely easy to measure. This screen, therefore, provides a simple way to identify agents having the potential to affect the embryo/fetus adversely.

Forty-eight (24 treated, 24 controls) pregnant Fischer 344 female rats are exposed by gavage to one concentration of the waste material beginning on Day 7 of gestation and ending on Day 16 (10 consecutive days). The selection of the single exposure concentration for neonatal survival is made near or at the maximum tolerable dose level observed for the males in the toxicological screen. The dosing volume is calculated based on the initial weight of the female on Day 6 of gestation. This dosing volume is used for 10 consecutive days. The change in maternal weight during the treatment period is determined by weighing the dams after the tenth dose. The animals are allowed to give birth (Day 22 of gestation) and the litters are counted and weighed at birth and at three days of age. The resulting litters are randomly reduced to one male and one female. At three weeks of age the pups are evaluated in the neurotoxicological screen; at six weeks the pups are killed for the immunotoxicological screen. This sequence is shown in Figs. 2 and 3.

In order to determine if transplacental SCE inductions occur with treatment of the dams with wastes, SCE frequencies in maternal lymphocytes and embryo liver cells are compared to control levels.

After the eighth consecutive dose, maternal blood is collected from the eight animals designated for SCE analysis (four treated, four control). The blood samples are cultured for lymphocyte SCE analysis. Immediately following blood removal, the dams are administered 5-bromodeoxyuridine. The following day the animals are killed and embryo livers are assessed for SCE frequencies.

Validation of the Toxicological Screen

Before the toxicological screen can be used with complex waste samples, its ability to detect biologically active compounds and complex waste mixtures must be determined. This first goal of the validation study is being accomplished by using a series of positive and negative control compounds with the entire protocol. The positive control compounds selected for study are well-characterized chronic toxins of two basic types. The first type causes a wide variety of chronic toxic effects (mutagenic, immunotoxic, teratogenic, and so on). The second type has a very narrow and specific activity (for example, neurotoxic only). After a prescribed number of individual compounds have been evaluated, additional validation will be accomplished with the use of well-characterized complex waste mixtures.

The use of control compounds will allow assessment of the ten-day screen's ability as a predictive tool for known chronic effects. The evaluation of compounds of known chronic effects in the toxicological screen will allow for a comparison of the predictability of results obtained in these short-term tests for known chronic effects. The potential identification of chronic effects with waste samples would be of great value when considering alternatives for disposal of the wastes.

A second goal of the validation study is to identify which biological end points in the toxicological screen are most predictive of the toxic potential of the pure control compounds. Control compounds which are identified as toxic for end points other than the anticipated end points provide data concerning the sensitivity of the tests (Fig. 4). Those compounds possessing a wide variety of activity will identify tests responding in agreement with known

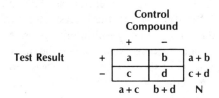

Several useful scores for each test may be defined as follows:

Sensitivity = a/(a + c)
Specificity = d/(b + d)
Accuracy = (a + d)/N

FIG. 4—*Definition of scores useful in the prescreen test selection.*

toxic effects. With a combination of both sets of information it may be possible to identify one or several tests from the toxicological screen that have the greatest potential for identifying biologically active test materials. The identification of those tests in the toxicological screen that are most sensitive plays an important role in the validation study. Many compounds must be evaluated in the toxicological screen before such assays will be statistically apparent. A minimum of 20 pure compounds have been suggested. So far, four compounds have been used for the validation of the toxicological screen: acrylamide, chlordecone, cyclophosphamide, and diethylstilbestrol. Additional compounds are still under consideration. If such tests are evident they will be used to prescreen samples for (1) the identification of grossly biologically active samples and (2) the prioritization of wastes for evaluation in the full toxicological screen.

An example of a prescreen protocol with a *hypothetical* selection of biological end points is given in Fig. 5. The results of the validation study may demonstrate that changes in body weight, the Ames test with wastes, the neurotoxicological assessments of the figure-eight maze, and the acoustic startle response are the most sensitive tests in identifying biologically active samples. If so, they will be used in combination with an abbreviated form of the toxicological screen's treatment protocol. Although the biological end points presented in Fig. 5 are hypothetical, the treatment protocol is not. In the prescreen, the treatment protocol consists of ten animals exposed by gavage to one concentration of wastes (≤ 1 g/kg) for ten consecutive days instead of five concentrations, as in the toxicological screen. If it is possible to select a prescreen protocol, this approach will facilitate the identification of biologically active wastes. Should any waste evaluated at one concentration in the prescreen demonstrate activity or toxicity, it will not be reevaluated in the prescreen at a different single concentration. Instead, it will become an immediate candidate for the toxicological screen.

It is important to remember that the prescreen is designed to identify grossly active samples and to prioritize waste samples for testing in the toxicological screen. The prescreen neither characterizes activity nor provides any dose-response data for waste samples. For these reasons information obtained from the prescreen can not aid in regulatory or disposal decisions. In addition, any sample that demonstrates no activity in the prescreen remains a candidate for the toxicological screen but at a low priority level. Failure to respond in the prescreen will not constitute a biologically "inactive" waste sample.

An integral part of the toxicological screen is to develop dose-response relationships for waste samples. The prescreen tests a single concentration and does not meet this criteria. For this reason any test selected for the prescreen will not be eliminated from the toxicological screen. The prescreen tests will remain part of the toxicological screen and will be performed using five concentrations of the waste sample in an effort to generate dose-response rela-

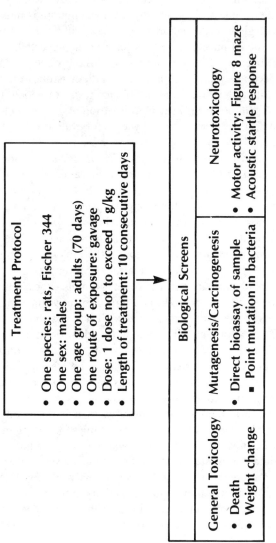

FIG. 5—*Hypothetical prescreen protocol.*

tionships. The only exception to this would be for the Ames Test, which is a totally in-vitro test and uses several concentrations of the neat wastes or extracts.

Examination of the validation study results may indicate the prescreen concept to be totally inadequate for prioritizing wastes. Should this be the case, the prescreen concept will be eliminated. The screening of hazardous wastes would then commence with the toxicological screen.

Currently, the toxicological screen is designed to include several tests from six areas of toxicology. Such a design is intended to maximize the probability of detecting biologically active wastes. It is possible, however, that the validation study may prove one or several of the toxicological screen's tests to be relatively uninformative, insensitive, or redundant as to the type and degree of information it provides. The results of the validation study will allow assessment of each test with respect to the above points. This assessment may result in the elimination of any test(s) demonstrating little utility and merit. This approach will help assure the toxicological screen's efficiency for screening waste samples both across and within the six categories of toxicology.

Criteria for the Prescreen Test Selection

Each biological screen employed in the toxicological screen will be evaluated with positive and negative control compounds. The test results will be scored in comparison to the known toxicity of control compounds. The criteria for selection of tests to be used in the prescreen will be based on the sensitivity, specificity, and accuracy scores for each test as shown in Fig. 4. These scores will be determined individually for each biological screen as judged against the positive and negative compounds selected to evaluate that screen (for example, the immunotoxicological screens will be scored against known positive and negative immunotoxins).

In order to evaluate each biological screen as a candidate for the prescreen test, the sensitivity of each test in predicting other toxicities will be determined (for example, the immunotoxicological screens will be scored against known positive and negative carcinogens, teratogens, and so on). Sensitivity will be optimized so that a minimum number of biological screens could be selected to detect any potential toxicity. Accuracy will be maximized to the extent possible; however, specificity will not be a critical consideration in selection of prescreen tests.

Evaluation of the entire toxicological screen's utility in accurately confirming the toxicity of a waste mixture and in determining the toxic end points affected by the waste mixture will rely more heavily on the specificity score than on the sensitivity score.

Discussion

The toxicological screen has been developed to identify potentially hazardous waste samples resulting from varying industrial processes. The experimental design of the screen was chosen to meet three goals.

The first goal of the toxicological screen was to maximize the amount and type of information obtained under a given protocol by combining the in-vivo exposure with in-vitro methods in six areas of toxicology. Using a single group of animals (except for teratology) maximizes the amount of information obtained by minimizing its loss. The type of toxicological activity each sample possesses is defined with evaluation in the many biological end points employed. The ranking of toxic end points affected as a function of dose for the complex wastes will aid in selecting health end points (for example, neurotoxicity) that require more definitive and elaborate toxicological studies for risk assessment.

The second goal of the toxicological screen was to determine at what concentration a toxic effect can be observed by exposing the animals to five concentrations of the test material. This establishment of a dose-response data base will aid in determining the relative risk of exposure to the waste. In addition, the use of pure chemicals whose chronic toxic effects are known (as part of the validation study) will allow a correlation of results in this short-term test with long-term chronic toxicity studies. By employing both the hazardous waste and pure chemical data bases, it should be possible to determine the relative toxicity of various wastes, support risk assessment, and perhaps predict or estimate human response.

The third goal of the toxicological screen was to develop its methodology to provide adequate data for use in waste regulation by the U.S. Environmental Protection Agency (EPA) Office of Solid Waste. The development of technologies to control hazardous waste disposal is critical to EPA's mission to guarantee the protection of public health and the environment from the adverse effects of hazardous wastes. The search for effective technologies is complicated by the large number of hazardous waste streams resulting from numerous industrial processes. The toxicological screen presented here is currently being validated for its ability to detect a wide variety of biologically active wastes in a timely, cost-effective manner. Evaluation of wastes in six areas of toxicology will increase the likelihood of identifying toxic wastes, while providing a basis for waste regulation and risk assessment.

Acknowledgments

The scientists responsible for the program's design and the areas of evaluation included in the study are E. Berman for general toxicology, J. Allen, L. Claxton, J. Lewtas, and T. K. Rao for genetic toxicology, L. Reiter for neurotoxicology, R. Smialowicz for immunotoxicology, J. Laskey for reproductive toxicology, N. Chernoff for teratology, and J. Creason and B. Most for statistical evaluation and consultation. L. Hall is responsible for the selection of the doses for control and test compounds. The overall study design, coordination, and management is under the direction of J. Lewtas and B. Andon. The authors would like to thank A. Hunter and J. Finley for the preparation of this manuscript.

Disclaimer

The research described in this paper has been reviewed by the Health Effects Research Laboratory, U.S. Environmental Protection Agency, and approved for publication. Mention of trade names or commercial products does not constitute an endorsement by the agency or recommendation for use.

References

[1] Hoffmann, G. R., "Mutagenicity Testing in Environmental Toxicology," *Environmental Science and Technology*, Vol. 16, No. 10, Oct. 1982, pp. 560A-574A.

[2] Waters, M. D., Simmon, V. F., Mitchell, A. B., Jorgenson, T. A., and Valencia, R., "A Phased Approach to the Evaluation of Environmental Chemicals for Mutagens and Presumptive Carcinogenesis," *In Vitro Toxicity Testing of Environmental Agents*, Part B, Plenum Press, New York, 1983, pp. 417-441.

[3] Murphy, S. D., "Pesticides," *Toxicology: The Basic Science of Poisons*, Macmillan, New York, 1980, pp. 398-400.

[4] Luckey, T. D. and Venugopal, B., *Metal Toxicity in Mammals 1: Physiologic and Chemical Basis for Metal Toxicity*, Plenum Press, New York, 1977, p. 130.

[5] Miya, T. S., Turner, A. G., Little, L. W., Weil, C. S., Boreiko, C. J., et al, "Hazardous Waste Health Effects Research Program: Peer Review Report," HERL-0408, U.S. EPA, Research Triangle Park, NC, 1983.

[6] Legator, M. S., Bueding, E., Batzinger, R., Connor, T. H., Eisenstadt, E., et al, "An Evaluation of the Host Mediated Assay and Body Fluid Analysis: A Report of the U.S. EPA Gene-Tox Program," *Mutation Research*, Vol. 98, No. 3, May 1982, pp. 319-374.

[7] Ames, B. N., McCann, J., and Yamasaki, E., "Methods for Detecting Carcinogens and Mutagens with the *Salmonella*/Mammalian Microsome Mutagenicity Test," *Mutation Research*, Vol. 31, No. 6, Dec. 1975, pp. 347-364.

[8] Latt, S. A., Allen, J., Bloom, S. E., Carrano, A., Flake, E., et al, "Sister-Chromatid Exchanges: A Report of the Gene-Tox Program," *Mutation Research*, Vol. 87, No. 1, July 1981, pp. 17-62.

[9] Ruppert, P. H., Dean, K. F., and Reiter, L. W., "Development and Behavorial Toxicity Following Acute Postnatal Exposure of Rat Pups to Trimethyltin," *Neurobehavioral Toxicology and Teratology*, Vol. 5, No. 4, July-Aug. 1983, pp. 421-429.

[10] Reiter, L. W., "Chemical Exposures and Animal Activity: Utility of the Figure-Eight Maze," *Developments in the Science and Practice of Toxicology*, Elsevier, New York, 1983, pp. 73-84.

[11] Anderson, A., Moller, G., and Sjorberg, O., "Selective Induction of DNA Synthesis in T- and B-Lymphocytes," *Cellular Immunology*, Vol. 4, No. 4, 1972, pp. 381-393.

[12] Hanna, N. and Burton, R. C., "Definitive Evidence that Natural Killer (NK) Cells Inhibit Experimental Tumor Metastases In Vivo," *Journal of Immunology*, Vol. 127, No. 5, Nov. 1981, pp. 1745-1758.

[13] Christian, M., *Advances in Modern Environmental Toxicology. Vol. III. Assessment of Reproductive and Teratogenic Hazards*, Princeton Scientific Publishers, Princeton, NJ, 1983, pp. 69-92.

[14] Weil, C. S. and McCollister, D. D., "Safety Evaluation of Chemicals: Relationship Between Short- and Long-Term Feeding Studies in Designing an Effective Toxicity Test," *Journal of Agricultural and Food Chemistry*, Vol. 11, No. 6, 1963, p. 486.

[15] Chernoff, N. and Kavlock, R. J., "An in vivo Teratology Screen Utilizing Pregnant Mice," *Journal of Toxicology and Environmental Health*, Vol. 10, Nos. 4 and 5, Oct.-Nov. 1982, pp. 541-550.

Land Treatment and Disposal
Test Methods

John J. Bowders,[1] *David E. Daniel,*[1] *Gregory P. Broderick,*[1]
and Howard M. Liljestrand[1]

Methods for Testing the Compatibility of Clay Liners with Landfill Leachate

REFERENCE: Bowders, J. J., Daniel, D. E., Broderick, G. P., and Liljestrand, H. M.,
"Methods for Testing the Compatibility of Clay Liners with Landfill Leachate," *Hazardous and Industrial Solid Waste Testing: Fourth Symposium, ASTM STP 886,*
J. K. Petros, Jr., W. J. Lacy, and R. A. Conway, Eds., American Society for Testing and
Materials, Philadelphia, 1986, pp. 233–250.

ABSTRACT: The compatibility of compacted clay liners with landfill leachate may be
evaluated with laboratory permeability tests. Compaction mold permeameters are recommended for laboratory-compacted specimens of clay subjected to low overburden pressure. Flexible wall permeameters are recommended for laboratory-compacted specimens
subjected to relatively high overburden pressure and for undisturbed specimens of clay
obtained from the field. Before permeability tests are performed, the index properties of
the soil should be determined by mixing the soil with water and by mixing the soil with the
leachate. If the leachate does not affect the index properties of the soil, it is not likely to
affect the permeability. The solubility of the soil in the leachate should be checked if the
soil is acidic or basic. Also, for dilute leachates, batch equilibrium tests are recommended
before a permeability test is initiated so that the tendency for the soil to adsorb materials
in the leachate can be evaluated. The effluent liquid should be monitored to determine the
degree of breakthrough of key constituents. It is best to continue the permeability test
until full breakthrough is achieved.

KEY WORDS: permeability, hydraulic conductivity, soil, clay liner, compatibility,
leachate, hazardous waste, adsorption, index test, chemicals

Solid waste landfills and liquid waste impoundments are frequently constructed at sites underlain by natural clay or man-made clay liners. The compatibility of the clay with any leachate that might emanate from the waste
requires investigation. It is generally assumed that if the leachate does not

[1]Graduate research assistants (Bowders and Broderick) and assistant professors (Daniel and
Liljestrand), Department of Civil Engineering, University of Texas, Austin, TX 78712.

adversely affect the permeability (hydraulic conductivity) of the clay, then the clay and leachate are compatible.

Unfortunately, no standardized method of permeability testing exists. Some investigators have measured large increases in the permeability of clay to certain chemicals [1–8], while other investigators have tested similar, or in some instances identical, liquids and soils and have found no increase in permeability [9–15]. Differences in testing procedures are thought to be at least partly responsible for the discrepancies.

The purpose of this paper is to describe the available methods of permeability testing using landfill leachate and clay liner materials, to discuss the advantages and disadvantages of each method, and to recommend the most suitable methods. It is also the authors' aim to present a decision tree to be used as a guide in selecting an appropriate testing program for a given site.

Methods of Permeability Testing

Compaction Mold Permeameters

Compaction mold permeameters usually consist of a cylindrical compaction mold, two end plates, and a collar placed above the mold to contain the permeant liquid (Fig. 1). The most common type of mold is described in the ASTM Test for Moisture-Density Relations of Soils and Soil-Aggregate Mixtures, using 5.5-lb (2.49-kg) Rammer and 12-in. (304.8-mm) Drop (D 698) and the ASTM Test for Density of Soil in Place by the Sand-Cone Method (D 1556), Method A. Such molds have a diameter of 10.2 cm and a height of 11.6 cm; the mold into which the soil is compacted also serves as the confining ring in the permeameter. The permeant liquid is usually pressurized with compressed air to achieve the desired hydraulic gradient (usually 20 to 300). The rate of flow is determined by measuring the quantity of inflow or outflow over a known interval of time. The permeability is then computed from Darcy's law. Aliquots of effluent liquid may be obtained for chemical analysis or other testing. Further details of this type of permeameter may be found in Refs 16 to 18.

The compaction mold permeameter is the simplest and most economical device available for testing samples of clay that are compacted in the laboratory. The disadvantages of the device are that complete saturation of the void spaces of the soil with permeant liquid is difficult, there is no means for measuring the amount of shrinkage or swelling that might take place within the soil sample, there is little to no control over the stresses that are applied to the soil, and there is a potential for spurious leakage of permeant liquid along the sidewall of the test specimen. Incomplete saturation is important because permeability decreases as the quantity of air in the test specimen increases [18]. Inability to measure deformations of the test specimen is a serious drawback for research purposes but is insignificant for most practical applica-

PRESSURE   VENT

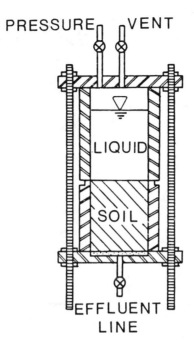

EFFLUENT
LINE

FIG. 1—*Compaction mold permeameter with a reservoir of permeant liquid contained within a collar located directly above the soil specimen.*

tions. Lack of control over stress is important in cases where the soil will be subjected to large stresses in the field (the applied vertical stress is zero in laboratory permeability tests performed with the typical compaction mold permeameter). In the field, there will usually be some vertical stress acting on the clay from the weight of overlying soil or solid waste. Tests performed with a consolidation cell permeameter (to be discussed later), which is similar to a compaction mold device except that known vertical stresses can be applied to the test specimen, illustrate the effect that stresses can have on the permeability of clays. Data are shown in Fig. 2 for specimens of kaolinite that were permeated with methanol at various stress levels. The results show order-of-magnitude differences in the measured permeability when the vertical stress was changed by an amount equivalent to the weight of just a few meters of soil overburden. At low stress, methanol appears to cause flocculation of the clay and an increase in the size of the largest pores within the test specimen. When a sufficiently large vertical stress is applied, there seems to be little structural alteration of the soil and no adverse impact on permeability.

A potential problem with the compaction mold permeameter is that leakage may occur along the contact between the soil and the rigid wall of the mold [19]. Sidewall leakage is probably not important with compacted clays

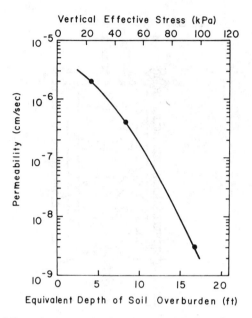

FIG. 2—*Permeability of kaolinite to methanol in consolidation cell permeameter at different effective stresses.*

that are permeated with water and that tend to swell when soaked [20], but can be important when clays are permeated with concentrated organic chemicals. Research indicates that results from compaction mold cells compare well with results from field tests on prototype liners subjected to relatively low overburden pressure and permeated with concentrated organic solvents [21].

Double-Ring Compaction Mold Permeameters

Anderson et al [22] have developed a modified compaction mold device known as a double-ring permeameter. As indicated in Fig. 3, a 15.2-cm-diameter compaction mold contains a specimen of soil while a ring built into the base plate separates the outflow that occurs through the central portion of the soil from the outflow that occurs near or along the sidewall. If there is significant sidewall leakage, the rate of flow into the outer collection ring will be much greater than the rate of flow into the inner ring. If the rates of flow, adjusted to take into account the differences in cross-sectional area of the test specimen that is intercepted, are unequal, one could reject the test and set up a new one.

Although there is relatively little experience with this device, the cell has worked well in a few tests and shows excellent promise. The disadvantages are

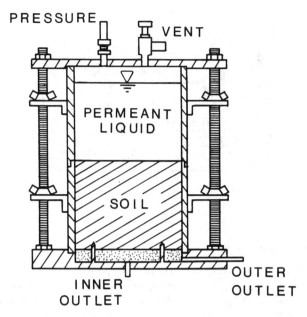

FIG. 3—*Double-ring compaction mold permeameter developed by Anderson et al [22].*

the same as those indicated previously for the compaction mold permeameter, except that one has some indication of the magnitude of sidewall leakage.

Consolidation Cell Permeameters

A diagram of a consolidation cell permeameter is shown in Fig. 4. The soil
is trimmed into a ring (typical diameter of 5 to 8 cm and height of 13 to 25
mm), and the ring is clamped into position. Water is introduced into the reservoir that surrounds the sample, and the soil is consolidated to a predetermined vertical stress. Permeation is initiated by flowing leachate upward
through the test specimen. Further details may be found in Ref *18.*

The advantages of consolidation cell permeameters are the convenient dimensions of the ring for trimming undisturbed samples into the ring, the ability to apply a known vertical stress (which not only simulates the effect of
overburden but also helps to squeeze the soil back against the wall of the
consolidation ring and thus minimize sidewall leakage), and the ability to
measure vertical deformations of the soil. The main disadvantages are higher
cost than compaction mold devices, potential for sidewall leakage, and the
thinness of the test specimen. In addition, most consolidation cell permeameters have no provisions for ensuring complete saturation of the soil with
liquid.

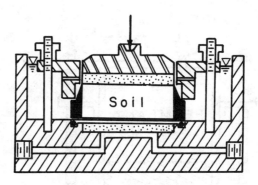

FIG. 4—*Consolidation cell permeameter.*

Flexible Wall Permeameters

Flexible wall permeability tests are performed in cells that are designed to confine the sides of the test specimen with a flexible membrane rather than a rigid wall. Pressure is applied to the outside of the membrane to press it against the soil specimen and thereby minimize sidewall leakage. The membrane conforms to the irregular sidewalls that typify many test specimens. Several designs are described in the literature [18,23-25]. A schematic drawing of the cell used by the authors is shown in Fig. 5, and a schematic diagram of the hydraulic system is shown in Fig. 6. The test method is described in detail in Ref 25.

Flexible wall cells have several advantages. Undisturbed samples are easily tested because minimal trimming is required and irregular surfaces on the test specimen are easily accommodated. Back pressure [25] is normally used, which helps to saturate the soil fully and eliminate any gas that might tend to cause one to measure a permeability that is too low. The deformation of the test specimen can be measured. One can control the stresses that are applied to the soil. The chances of sidewall leakage are nil.

There are three principal disadvantages of flexible wall cells. Firstly, the membranes used to confine the soil can be attacked and destroyed by certain chemicals. Organic chemicals with dielectric constants below ten are particularly destructive [26]. The problem can be eliminated by wrapping the soil specimen with Teflon® tape and placing the membrane over the tape [25]. Diffusion of organics through the membrane and into the cell water has also been reported as a problem [9]. A second disadvantage of flexible wall cells is a significantly higher cost than the other types of permeameters.

The third and most important disadvantage is related to the effect of the confining pressure itself, which tends to close any cracks that may be present in the soil. If the confining pressures used in the permeability test signifi-

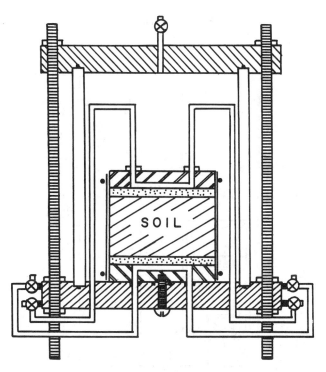

FIG. 5—*Flexible wall permeameter.*

cantly exceed the values expected in the field, erroneous results may be obtained [*24*].

Recommended Permeameters

Two types of permeameters are recommended for the majority of testing: the compaction mold permeater and the flexible wall permeameter. The double-ring compaction mold permeameter may also be useful.

For laboratory-compacted soils, the compaction mold permeameter is the simplest device to use and might be used for preliminary testing. If the leachate causes a significant increase in permeability with compaction mold permeameters, the tests might be repeated using double-ring permeameters or flexible wall permeameters to ensure that sidewall leakage is not causing erroneous readings. If the soil in the field will be subjected to relatively large overburden pressures, one may elect to skip the phase of testing with compaction mold cells and go directly to tests with flexible wall cells.

For undisturbed samples of clay, the flexible wall cell is the device that would normally be recommended. However, with relatively soft clays, the

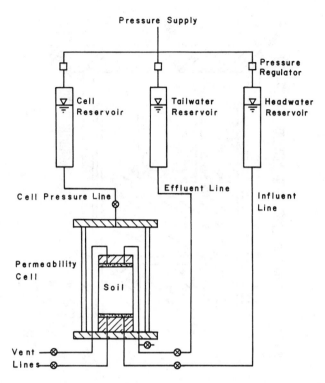

FIG. 6—*Schematic diagram of the hydraulic system employed for flexible wall permeability tests.*

consolidation cell may work well because the applied vertical stress tends to squeeze soft clays against the rigid confining ring and thereby minimize sidewall leakage.

Presentation of Permeability Test Results

Measurements are normally obtained of the permeability k of the soil and the normalized effluent concentration c/c_0, where c is the concentration of a chemical in the effluent liquid and c_0 is the concentration in the leachate. The cumulative quantity of flow is also determined, and whenever measurements are obtained the number of pore volumes of flow is computed. A pore volume is defined as the volume of void space in a test specimen. Permeability tests should preferably be continued until at least two pore volumes of flow are achieved, to ensure that the original soil water is flushed out and the void spaces are filled with leachate. It is also best to continue the test until c/c_0 approaches unity.

Typical results are shown in Fig. 7. For this test, the leachate was an actual

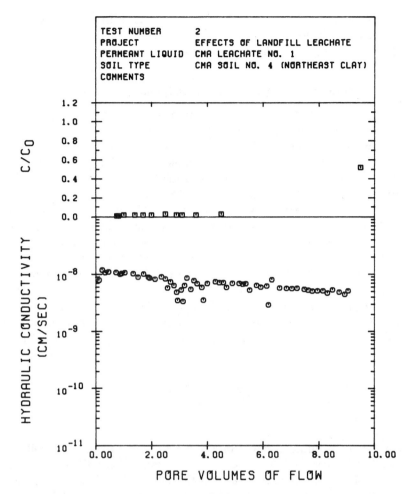

FIG. 7—*Results of 12-month permeability test on northeastern clay. The permeant liquid was a landfill leachate.*

landfill leachate that contained 1400 mg/L of total organic carbon (TOC) and a variety of organic and inorganic compounds. The c/c_0 values are based upon TOC.

Sampling of Effluent Liquid

The method used to obtain effluent samples depends on the type of permeameter. With the compaction mold cell, the effluent line is placed in a suitable container and a sample is collected directly. With flexible wall permeameters, the effluent line is pressurized. A tee connection in the effluent line is used as a sampling port. One branch of the tee is fitted with a septum. A syringe with a hypodermic needle is used to obtain samples of effluent liquid through the septum. The usual sample size in our tests is 25 μL.

Batch Equilibrium Tests to Determine Adsorption Characteristics

The procedure that has been used by the authors for batch equilibrium tests is as follows. The clay is air-dried and ground with a mortar and pestle until a uniform powdery texture is obtained. Variable amounts of soil, ranging from 0 to 10 g, are placed in glass sample bottles. A constant volume of leachate is then added to each bottle and the bottle is capped tightly. The bottles are then placed in a rotary shaker and are kept at a constant temperature of 20°C. The bottles are shaken for at least 48 h to ensure equilibrium.

At the end of the shaking period, the samples are centrifuged to separate the clay from the liquid. One of the bottles, however, contains just the leachate and no clay. The supernatant liquid from the bottles is filtered, and the equilibrium concentration in the liquid phase of a constituent of interest (c, expressed in units of mass of constituent per unit volume of liquid) is measured using the appropriate analytical method. The concentration of the constituent in the leachate itself is determined from the bottle with no soil and is denoted c_0. The adsorption mass ratio q is computed for each bottle as follows:

$$q = \frac{(c_0 - c) V}{M} \tag{1}$$

where V is the volume of liquid in a bottle and M is the mass of soil in a bottle. The numerator in Eq 1 represents the mass of constituent adsorbed onto the solid phase, and it is divided by the mass of the soil to obtain a measure of the relative mass of the constituent adsorbed onto the solid phase.

The values of q are plotted as a function of the equilibrium concentration. The resulting plot is often linear, in which case the plot is called a "linear isotherm." Other shapes are also possible. For a linear isotherm, the slope of the plot is termed the "distribution coefficient" and is denoted K_d. The distribution coefficient has units of volume of liquid per mass of soil.

The purpose of performing batch equilibrium tests is to obtain an estimate of how many pore volumes of flow will be necessary to achieve breakthrough of a constituent into the effluent liquid. If the number of pore volumes of flow is denoted R, then the number of pore volumes of flow required to achieve breakthrough is given by the expression:

$$R = 1 + \frac{\rho_d}{\eta} K_d \tag{2}$$

where ρ_d is the dry mass density (mass of dry solids divided by the total volume of the soil specimen) of the test specimen and η is the porosity of the test specimen. The parameter R is also called the "retardation factor" and is the ratio of the breakthrough time of an adsorbed chemical relative to the time of elution of a nonadsorbed tracer. In applying Eq 2, it is assumed that breakthrough of a nonadsorbed tracer would occur at one pore volume of flow. If there is no molecular diffusion or hydrodunamic dispersion, then breakthrough curves for adsorbed and nonadsorbed chemicals would be assumed to have the form shown in Fig. 8. If diffusion and dispersion occur, breakthrough would not occur suddenly (Fig. 8); "breakthrough" in this case means $c/c_0 = 0.5$.

Further theoretical background is provided in Ref 27; some limitations are discussed in Refs 28 and 29. The degree to which batch equilibrium tests correctly predict breakthrough characteristics is not known for clay. In the batch equilibrium tests, the full surface area of the soil particles comes into contact with the leachate. In column tests, the leachate may move preferentially through certain flow channels and come into contact with only a fraction of the total area of the soil particles. Nevertheless, the batch equilibrium test can provide valuable insight into sorption characteristics.

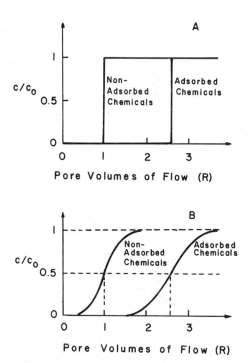

FIG. 8—*Breakthrough curves if* (a) *there is no diffusion or dispersion and* (b) *diffusion and dispersion do occur.*

To evaluate the applicability of batch equilibrium tests, the same soil and leachate used for the permeability test for which results are reported in Fig. 7 were used for a batch equilibrium test. The isotherm is plotted in Fig. 9 and the measured breakthrough curve is shown in Fig. 7. The predicted number of pore volumes to achieve breakthrough is 9.7. As seen in Fig. 7, breakthrough at $c/c_0 = 0.5$ has occurred at nine to ten pore volumes of flow, although there is only one data point in the region of most interest. The test is still in progress, and attempts will be made to carry the test to full breakthrough (which may take several years).

The batch equilibrium test is best applied to individual compounds or groups of compounds of similar hydrophobic character. If a mixture of compounds contains a wide range of hydrophilic to hydrophobic species, then the average isotherm determined by an overall measurement such as TOC is really a weighted mean of the individual isotherms.

Index Tests

The authors have evaluated the use of soil index tests to predict the potential for a leachate to react with a clay liner in a manner that is significantly different from the reaction of the clay liner with water. The index tests are performed first by mixing the soil with water and then performing the relevant tests; second by mixing the soil with the leachate of interest and then repeating the tests. The index tests that have been investigated are Atterberg limits tests and sedimentation analyses. For acidic leachates, solubility tests are also recommended. The purpose of such tests is to provide a simple, fast, and economical means for preliminary screening of a leachate/soil combination to determine if there is a significant potential for the leachate to degrade the permeability of the liner.

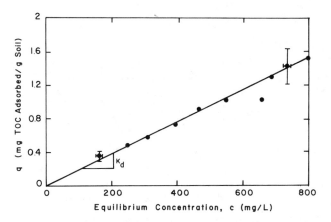

FIG. 9—*Isotherm for northeastern clay and landfill leachate.*

The following data are presented to illustrate how Atterberg limits can be used. Liquid and plastic limits were measured on kaolinite using various concentrations of methanol in an aqueous solution and following the ASTM Test for Liquid Limit, Plastic Limit, and Plasticity Index of Soils (D 4318). The results are summarized in Table 1. The change in liquid limit was negligible, with the exception of the solution containing pure methanol, which caused a 28% increase in the liquid limit compared to the value for water. The plastic limit with 100% methanol is 32% higher than for water. The results of permeability tests using the same range of solutions are shown in Fig. 10. The trends seen in the Atterberg limits tests are the same as those observed in the permeability tests—namely, the aqueous solution must contain more than 80% methanol for either the Atterberg limits to change significantly or the permeability to increase to a value greater than the permeability to water.

TABLE 1—*Effect of aqueous solutions of methanol on the Atterberg limits of kaolinite.*

Methanol, %	Liquid Limit, %	Plastic Limit, %
0	58	34
20	60	32
40	60	31
60	59	32
80	58	33
100	74	45

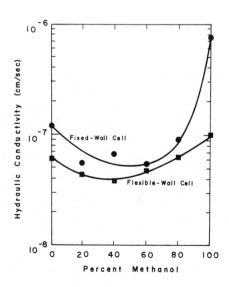

FIG. 10—*Permeability of compacted kaolinite to aqueous solutions containing methanol.*

Sedimentation tests may be performed according to ASTM Particle-Size Analysis of Soils (D 422). It is possible to sediment the clay not only in water but also in the leachate. Hydrometer readings are normally taken at various times, and the hydrometer/time data are used to evaluate the grain size distribution curve. However, hydrometer analysis is impossible if the hydrometer is too heavy to be bouyed by the liquid, as is the case with methanol, for example. Rather than using a hydrometer, it has been found to be convenient to record the distance from the top of the liquid in a sedimentation test to the top of the soil suspension. This works well in differentiating liquids which will allow the soil to remain dispersed in suspension (such as water) and liquids which cause flocculation and rapid sedimentation (such as concentrated organic chemicals). The distance is recorded at various times, although the effects of the leachate, if they are significant, will probably be obvious within an hour.

The results of sedimentation tests for kaolinite mixed with various methanol-water solutions are plotted in Fig. 11. A dispersing agent was used as outlined in ASTM D 422. With pure water and 20% methanol, there was virtually no sedimentation of soil particles after 24 h. The sedimentation rate was much faster with 40, 60, 80, and 100% methanol. There is a range in methanol concentration (40 to 80%) in which sedimentation velocities were different than the velocities with a kaolinite-water slurry but in which the permeability is the same as that to water (Fig. 10). Sedimentation analysis appears to be useful in identifying soil/leachate combinations in which permeability might be adversely affected, but leachates that affect sedimentation velocity do not always affect permeability.

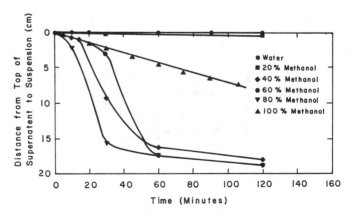

FIG. 11—*Sedimentation pattern for 50 g of kaolinite suspended in 1 L of aqueous solutions containing methanol.*

Selecting a Program of Testing

The authors offer the decision tree outlined in Fig. 12 as a guide to establishing a program of testing. The basic assumption in the decision tree is that simple index property tests can be used to identify soil/leachate combinations that do not require extensive testing. In the decision tree, the liquids are divided into two categories: neutral liquids and liquids that are strongly acidic or basic. Those in the latter group are capable of dissolving the clay and require special consideration. One could exit from the decision tree at any point if it were clear that the clay is incompatible with a particular liquid.

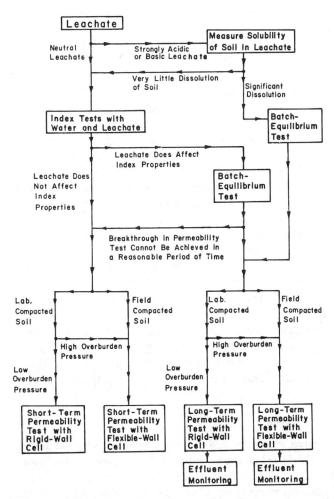

FIG. 12—*Suggested decision tree to be used as an aid in selecting a program of testing.*

Conclusions

Several types of permeameters are available for testing the compatibility of clay liner materials with landfill leachates. The compaction mold permeameter and the flexible wall permeameter are the most commonly used devices and are recommended for most applications. For laboratory-compacted specimens of clay, the compaction mold device is suggested for cases where the clay liner will be subjected to little or no overburden pressure; in cases where there will be significant overburden pressure from overlying soil or solid waste, flexible wall permeameters are recommended. Flexible wall cells are recommended for testing undisturbed samples of soil, although the tests will be difficult to perform at low overburden pressure. The double-ring permeameter shows promise as an alternative to the normal compaction mold permeameter.

It is usually best to monitor the chemistry of the effluent liquid to ensure that the influent liquid appears in essentially full concentration in the effluent line before dismantling a test. For nonadsorbed chemicals, the permeability test should be carried out to a cumulative quantity of flow of at least two pore volumes. Even more flow may be required with leachates that contain chemicals that will be adsorbed by the soil. The solubility of the clay in the leachate should also be measured if the leachate is strongly acidic or basic. To aid in predicting the breakthrough characteristics of any particular constituent in the leachate, batch equilibrium tests may be performed before permeability testing. One sometimes finds that many pore volumes of flow would be needed to obtain full breakthrough and that the required testing time might stretch into many months or even years.

Index property tests, performed by mixing soil with water and with leachate in separate batches, can be useful in identifying whether the leachate is likely to affect the permeability of the soil. If the index properties are unaffected by the leachate, it is unlikely that the permeability will be affected. By performing index property tests at the beginning, one may be able to reduce the required number of permeability tests.

A decision tree to serve as an aid in selecting a program of testing is presented in Fig. 12. By following a logical sequence, one can frequently keep the permeability testing to a minimum and have some knowledge of what will happen in the long-term permeability tests before the tests are even started.

Acknowledgments

This research was sponsored by grants from the National Science Foundation (CEE-8204967) and the Chemical Manufacturers Association. The technical input provided by J. Matey, J. Petros, C. Knowles, D. Caputo, S. Daniels, and other members of the CMA tesk group that coordinated the research is appreciated. The help of S. Trautwein, S. Boynton, and D. Foreman in developing and refining the testing equipment is acknowledged. F. Hulsey helped with the equipment used for chemical analyses.

References

[1] Anderson, D., Brown, K. W., and Green, J., "Effect of Organic Fluids on the Permeability of Clay Soil Liners," *Land Disposal of Hazardous Waste,* Report EPA-600/9-82-002, U.S. Environmental Protection Agency, Cincinnati, OH, 1982, pp. 178–190.

[2] Crim, R. G., Shepherd, T. A., and Nelson, J. D., "Stability of Natural Clay Liners in a Low pH Environment," *Second Symposium on Uranium Mill Tailings Management,* Colorado State University, Fort Collins, CO, 1979, pp. 41–53.

[3] Fireman, M., "Permeability Measurements on Disturbed Soil Sample," *Soil Science,* Vol. 58, No. 5, 1944, pp. 337–355.

[4] Gee, G. W., Campbell, A. C., and Optiz, B. E., "Interactions of Uranium Mill Tailings Leachate with Morton Ranch Clay Liner Soil Material," *Third Symposium on Uranium Mill Tailings Management,* Colorado State University, Fort Collins, CO, 1980, pp. 333–354.

[5] Laguros, J. G. and Robertson, J. M., "Problems of Interaction Between Industrial Residues and Clays," *Third Annual Conference on Treatment and Disposal of Industrial Wastes, Wastewaters, and Residues,* Houston, TX, Hazardous Materials Control Research Institute, 1978, pp. 289–292.

[6] Mesri, G. and Olson, R. E., "Mechanisms Controlling the Permeability of Clays," *Clays and Clay Minerals,* Vol. 19, No. 3, 1971, pp. 151–158.

[7] Michaels, A. S. and Lin, C. S., "Permeability of Kaolinite," *Industrial and Engineering Chemistry,* Vol. 46, No. 6, June 1954, pp. 1239–1246.

[8] Wilkinson, W. B., Discussion, *In-Situ Investigation in Soils and Rocks,* British Geological Society, Institute of Civil Engineers, London, 1969, pp. 311–313.

[9] Acar, Y. B. and Field, S. D., "Organic Leachate Effects to Hydraulic Conductivity in Fine-Grained Soil," Report GE-82/01, Louisiana State University, Department of Civil Engineering, Baton Rouge, LA, 1982.

[10] Daniel, D. E. and Liljestrand, H. M., "Effects of Landfill Leachates on Natural Liner Systems," Report to Chemical Manufacturers Association, Report GR83-6, University of Texas Geotechnical Engineering Center, Austin, 1984.

[11] Dunn, R. J., "Hydraulic Conductivity of Soils in Relation to the Subsurface Movement of Hazardous Wastes," Ph.D. dissertation, University of California, Berkeley, 1983.

[12] Gordon, B. B. and Forrest, M., "Permeability of Soil Using Contaminated Permeant," *Permeability and Groundwater Contaminant Transport, ASTM STP 746,* American Society for Testing and Materials, Philadelphia, pp. 101–120.

[13] Green, J. W., Lee, G. F., and Jones, R., "Clay-Soils Permeability and Hazardous Waste Storage," *Journal of the Water Pollution Control Federation,* Vol. 53, No. 8, August 1981, pp. 1347–1354.

[14] Griffin, R. A. and Shimp, N. F., "Attenuation of Pollutants in Municipal Landfill Leachate by Clay Minerals," Report EPA-600/2-78-157, U.S. Environmental Protection Agency, Washington, DC, 1978.

[15] Nasiatka, D. M., Shepherd, T. A., and Nelson, J. D., "Clay Liner Permeability in Low pH Environments," *Fourth Symposium on Uranium Mill Tailings Management,* Colorado State University, Fort Collins, CO, 1981, pp. 627–645.

[16] Brown, K. W. and Anderson, D. C., "Effect of Organic Chemicals on Clay Liner Permeability," *Proceedings of the Sixth Annual Research Symposium on Land Disposal of Hazardous Waste,* Report EPA-600/9-80-010, U.S. Environmental Protection Agency, Cincinnati, OH, 1980, pp. 123–134.

[17] Anderson, D. C., "Organic Leachate Effects on the Permeability of Clay Soils," M.S. thesis, Texas A&M University, College Station, 1981.

[18] Olson, R. E. and Daniel, D. E., "Measurement of the Hydraulic Conductivity of Fine-Grained Soil," *Permeability and Groundwater Contaminant Transport, ASTM STP 746,* American Society for Testing and Materials, Philadelphia, 1981, pp. 18–64.

[19] Zimmie, T. F., "Geotechnical Testing Considerations in the Determination of Laboratory Permeability for Hazardous Waste Disposal Siting," *Hazardous Solid Waste Testing: First Conference, ASTM STP 760,* American Society for Testing and Materials, Philadelphia, 1981, pp. 293–304.

[20] Foreman, D. E. and Daniel, D. E., "Effects of Hydraulic Gradient and Method of Testing on the Hydraulic Conductivity of Kaolinite to Water, Methanol, and Heptane," *Proceedings of the Tenth Annual Research Symposium on Land Disposal of Hazardous Waste,* EPA 600/9-84-001 U.S. Environmental Protection Agency, Cincinnati, OH, 1984, pp. 138–144.

[21] Brown, K. W., Green, J. W., and Thomas, J. C., "The Influence of Selected Organic Liquids on the Permeability of Clay Liners," *Proceedings of the Ninth Annual Research Symposium on Land Disposal of Hazardous Waste,* U.S. Environmental Protection Agency, Cincinnati, OH, 1983, pp. 114–125.

[22] Anderson, D. C., Crawley, W., and Zabick, D., "Effect of Various Liquids on Clay Soil-Bentonite Slurry Mixes," *Hydraulic Barriers in Soil and Rock, ASTM STP 874,* American Society for Testing and Materials, Philadelphia, 1985, in press.

[23] Carpenter, G. W., "Assessment of the Triaxial Falling Head Permeability Testing Technique," Ph.D. dissertation, University of Missouri, Rolla, 1982.

[24] Boynton, S. S., "An Investigation of Selected Factors Affecting the Hydraulic Conductivity of Compacted Clay," M.S. thesis, University of Texas, Austin, 1983.

[25] Daniel, D. E., Trautwein, S. J., Boynton, S. S., and Foreman, D. E., "Permeability Testing with Flexible-Wall Permeameters," *Geotechnical Testing Journal,* Vol. 7, No. 3, September 1984, pp. 113–122.

[26] Rad, N. S. and Acar, Y. B., "A Study of Membrane-Permeant Compatibility," *Geotechnical Testing Journal,* Vol. 7, No. 2, June 1984, pp. 104–106.

[27] Ogata, A. and Banks, R. B., "A Solution of the Differential Equation of Longitudinal Dispersion in Porous Media," Professional Paper 411-A, U.S. Geological Survey, Washington, DC, 1961.

[28] Cameron, D. R. and Klute, A., "Convective-Dispersive Solute Transport with a Combined Equilibrium and Kinetic Model," *Water Resources Research,* Vol. 13, No. 1, February 1977, pp. 183–188.

[29] Connolly, J. P., Armstrong, N. E., and Miksad, R. W., "Adsorption of Hydrophobic Pollutants in Estuaries," *Journal of Environmental Engineering ASCE,* Vol. 109, No. 1, February 1983, pp. 17–135.

Norbert B. Schomaker[1]

U.S. EPA Program in Hazardous Waste Landfill Research

REFERENCE: Schomaker, N. B., **"U.S. EPA Program in Hazardous Waste Landfill Research,"** *Hazardous and Industrial Solid Waste Testing: Fourth Symposium, ASTM STP 886*, J. K. Petros, Jr., W. J. Lacy, and R. A. Conway, Eds., American Society for Testing and Materials, Philadelphia, 1986, pp. 251–262.

ABSTRACT: The U.S. Environmental Protection Agency's hazardous waste land disposal research program is collecting data necessary to support implementation of disposal guidelines mandated by the Resource Conservation and Recovery Act of 1976 (RCRA— PL 94-580). This program relating to the categorical areas of landfills, surface impoundments, and underground mines encompasses state-of-the-art documents, laboratory analysis, economic assessment, bench and pilot studies, and full-scale field verification studies. Over the next five years the research will be reported as Technical Resource Documents in support of the RCRA Guidance Documents. These documents will be used to provide guidance for conducting the review and evaluation of land disposal permit applications. This paper will present an overview of this program and will report the current status of work in the areas of landfills, surface impoundments, underground mines, and dioxin engineering.

KEY WORDS: hazardous wastes, hazardous materials, pollution, environmental surveys, waste disposal, landfills, environmental engineering, environmental surveys, environmental tests, waste treatment, decontamination, containment, hydrogeology, leaching, linings

Congress, through the Solid Waste Disposal Act, also known as the Resource Conservation and Recovery Act (RCRA—PL 94-580), directed the U.S. Environmental Protection Agency (EPA) to develop programs for the management of hazardous and nonhazardous solid wastes. To effect the degree of coordination required of such an endeavor, a committee was formed to prepare and update an agency-wide research strategy. The Solid and Hazardous Waste Research Committee coordinates the efforts of the Office of Research and Development (ORD) with those of the Office of Solid Waste

[1]Chief, Disposal Branch, U.S. Environmental Protection Agency, Hazardous Waste Engineering Research Laboratory, Land Pollution Control Division, Cincinnati, OH 45268.

(OSW), the Office of Enforcement (OE), the Office of Water Programs (OWP), the Office of Emergency and Remedial Response (OERR), and the EPA regional offices.

The fiscal year (FY) 1980 solid and hazardous waste research strategy was based on the anticipated needs of the EPA as it implemented the mandates of RCRA. The major emphasis was placed on hazardous waste while efforts continued in the municipal solid waste area. The FY 1981–1984 strategies focused entirely on hazardous wastes. In FY 1984, the Solid and Hazardous Waste Research Strategy Committee was combined with the Superfund Research Strategy Committee to form the Hazardous Waste/Superfund Research Strategy Committee.

Agency responsibilities under RCRA are reasonably specific, and future activities to fulfill those responsibilities have been scheduled according to a step-by-step strategy. Within this RCRA hazardous waste strategy, seven research categories have been established:

• Waste identification
• Waste characterization (assessment)
• Hazardous releases (spills)
• High hazards (dioxins)
• Incineration
• Land disposal
• Alternative technology (treatment)

The above seven categories of research activities relating to RCRA are being investigated by various research groups within the EPA. The hazardous waste landfill research program, which is the focus of this paper, is oriented toward meeting the needs addressed in the land disposal category of this research committee. This research program is conducted by the Land Pollution Control Division, Hazardous Waste Engineering Laboratory, Environmental Research Center in Cincinnati, Ohio.

Land Disposal

The hazardous waste land disposal research program, encompassing state-of-the-art documents, laboratory analysis, bench and pilot studies, and full-scale field verification studies, is at various stages of implementation. The objective of this program is to compile a more adequate data base of information so that current hazardous waste disposal technology to the land may be upgraded by (1) developing proper site selection design and operational criteria for the establishment of new waste disposal sites and (2) developing proper control technology for upgrading existing waste disposal sites. The procedures thus developed will consider and describe those specific functions required to eliminate the pollution potential from landfills, surface impoundments, and underground mines used as hazardous waste storage facilities.

Specific functions include the ability to predict or provide guidance for leachate quality and quantity, quick indicator tests for the selection of lining materials, quick indicator tests for chemical fixation performance and compatibility with various generic waste streams, and a methodology for minimizing moisture infiltration.

Over the next five years the research will be reported as criteria and guidance documents for user communities. The land disposal research program is currently developing and compiling a data base for use in the development of guidelines and standards for waste residual disposal to the land as mandated by the RCRA. RCRA Guidance Documents, which provide guidance for conducting the review and evaluation of permit applications, are currently being prepared by the OSW in Washington, DC. Technical Resource Documents (TRD) in support of the Guidance Documents are also being prepared by the ORD in specific areas to provide the best engineering technologies and methods for evaluating the performance of the applicant's design. The information and guidance presented in these documents will constitute a suggested approach for review and evaluation and should be useful references for design and operation.

The current FY 1984 hazardous waste land disposal research program has been divided into four general areas: (1) landfills, (2) surface impoundments, (3) underground storage, and (4) dioxin engineering.

Landfills

The landfill program examines components of a hazardous waste landfill and optimizes their performance. The goal is to control or be able to predict the movement of liquids and gases in and around a landfill. The program develops basic information for each component and prepares reports of research and user documents for permit writers, design engineers, and operators.

The components and unit operations of a landfill to be addressed in this paper include:

- Multilayered covered systems
- Waste leaching
- Clay soil liners
- Flexible membrane liners
- Waste modification

Multilayered Cover Systems

The objective of this activity is to develop and evaluate the effectiveness of various cover systems in relation to their functional requirements for actual field application. Validation efforts are being performed in the laboratory

and field with model work development being pursued for eventual incorporation into a TRD.

The laboratory effort currently being pursued is evaluating alternatives to
soil materials for use as landfill covers. Such materials as geomembranes,
flexible membrane liners, and waste materials are being investigated.

The model efforts relate to the assessment of time settlement/subsidence
effects on cover systems and to the development and field verification of the
two-dimensional subsurface drainage model entitled "Hydrologic Evaluation
of Landfill Performance (HELP)." The time settlement/subsidence effort will
utilize a theoretical approach incorporating classical consolidation theory
and documented with practical field observations at existing sites. Computer
models have been run to determine amounts of settlement based upon different scenarios. The HELP model has been developed into user-friendly technical support information for establishment and evaluation of landfill performance. Field verification is currently being conducted for covers and leachate
collection systems with large-scale physical models.

The field efforts being investigated relate to evaluating maintenance-free
vegetative cover species and a three-layered surface cover system. The maintenance-free vegetative effort is evaluating deep-rooted vegetative growth species, which require minimum maintenance care, in relation to their effects on
cover integrity. The three-layered surface cover system effort has resulted in
plots being constructed where surface runoff, subsurface runoff, leachate formation, and soil moisture distribution will be measured under applied rainfall. A controlled level of subsidence will also be generated and its effect on
the cover integrity will be measured.

The TRD efforts are focusing on methods for control of infiltration to improve cover design and on the development of construction, maintenance,
and inspection manuals for clay liners/caps and leachate collection/cap
drainage. The infiltration control effort is basically developing and compiling
various methods to control infiltration into hazardous waste landfills. This
information will be incorporated as updated material in the current TRD entitled "Evaluating Cover Systems for Solid and Hazardous Waste" (SW-867).
This document can be ordered from the U.S. Government Printing Office
(GPO); the GPO stock number is 055-000-00228-2. Another TRD document
entitled "Hydrologic Simulation of Solid Waste Disposal Sites" (SW-868) describing a one-dimensional model of flow through a cover system has been
printed by GPO (Stock Number 055-000-00225-8). The construction manual
efforts are being pursued to provide guidance to ensure proper installation
and quality control during the construction of these multilayered cover systems for hazardous waste landfills.

Waste Leaching

The objective of this activity is to develop and evaluate laboratory techniques for working with a sample of a waste or a mixture of waste to predict

the composition of actual leachates obtained under field conditions. Results from laboratory and model predictions are being compared with results from pilot scale and field scale work to develop better procedures and an updated TRD on waste leaching.

The laboratory effort is evaluating and comparing the effectiveness of batch extraction versus column leaching in generating leachate from four representative hazardous waste samples with diverse physicochemical characteristics. The model efforts relate to transit flow times and field capacity determination for water movement through wastes. The transit flow time effort is evaluating predictive methods for determining liquid flow rates through the landfilled wastes; the conclusions are that deterministic predictive methods would be difficult to apply. The field capacity effort relates to developing a model for predicting saturation levels in landfills to determine the effectiveness of systems for controlling the production of leachate with time.

The field effort is determining the rate of release of pollutants from an actual field site and comparing the results with the model for predictive validation.

The TRD effort has produced a draft version on batch leaching procedures entitled "Solid Waste Leaching Procedure" (SW-924). The report was issued for public comment in March 1984: appropriate comments were incorporated into final publication in the fall of 1985. Our previous TRD, entitled "Landfill and Surface Impoundment Performance Evaluation" (SW-869), which relates to leachate generation, has been published and can be obtained from the GPO (Stock Number 055-000-00233-9).

Clay Soil Liners

The objective of this activity is to evaluate the effectiveness of clay soils as liners and surface caps to contain or minimize leachate movement and infiltration and to predict performance with time. Laboratory and field studies are being performed to develop tools for predicting and evaluating performance of soil liners. The information produced will be published in TRDs and construction manuals.

The laboratory efforts involve evaluating the effects of organic chemicals and inorganic leachates on clay soil liners. The organic chemical effort is assessing organic solvent effects on clay soil permeability and developing a matrix to predict soil/waste interaction. The inorganic leachate effort is evaluating the adverse impact of shrinkage on clay liner performance when selected inorganic salt solutions are placed in contact with the liner.

The methodology efforts are attempting to develop tools for predicting the retention and rate of movement of pollutants through soil liners. One effort relates to determining the effective porosity of soil and whether it is usable in a regulatory mode. Another effort is attempting to validate the precision and reliability of permeability (hydraulic conductivity) measurements as determined in the laboratory and in the field. A third effort relates to correlating the laboratory porosity measurements of both disturbed and undisturbed soils with similar soil samples in place in the field.

The field efforts involve investigating the field testing of hydraulic conductivity for in situ soils and evaluating clay soil liner failures. The hydraulic conductivity effort is evaluating the effects of hydraulic gradient, gas diffusion, sample volume change, and confinement on the measurements to determine permeability. The clay liner failure effort will evaluate actual in-place or recompacted clay liners where organic chemicals have migrated through the clay liner and it has failed. Failure mechanisms will be identified.

The TRDs are focusing on the evaluation of clay soils for use as liners and on installation procedures. Also a manual is being developed to assist with proper construction, maintenance, and inspection for clay soil liner installations. The initial evaluation effort of clay soil liners has been published by the GPO in a TRD entitled "Lining of Waste Impoundment and Disposal Facilities" (SW-870) (Stock Number 055-000-00231-2). The clay liner installation TRD will incorporate previous work on liquid control, waste leaching, pollutant movement, and field quality assurance with the existing liner TRD report (SW-870) to produce a new guidance document. Two new TRD reports are currently published in draft form for review by the public. These reports are entitled "Soil Properties, Classification, and Hydraulic Conductivity Testing" (SW-925) and "Procedures for Modeling Flow Through Clay Liners to Determine Required Liner Thickness" (EPA 530-SW-84-001). A third new TRD, entitled "Batch Soil Procedure to Design Clay Liners for Pollutant Removal," was completed in September 1985. The construction manual effort is being pursued to provide guidance to ensure proper control during the construction of the clay soil liner control measure at hazardous waste landfills.

Flexible Membrane Liners

The objective of this activity is to evaluate the effectiveness of synthetic membrane or flexible membrane as liners and caps to contain leachates/ moisture infiltration and to predict their performance with time. Efforts are being performed in the laboratory and the field for developing tools to establish flexible membrane liner performance criteria. Information produced will be published in TRDs and construction manuals.

The laboratory efforts are evaluating the adequacy of available laboratory testing procedures to determine hydraulic integrity of flexible membrane liners (FMLs). Also various line/waste compatibility efforts are being pursued to determine the relationship between resistance to chemical modification and change in liquid and gas permeability of FMLs over a variety of time spans. The liquid and gas permeability of FMLs is being determined by the methods described in the ASTM Tests for Water Vapor Transmission of Materials (E 96) and the ASTM Test for Determining Gas Permeability Characteristics of Plastic Film and Sheeting to Gases (D 1434), Procedure V, respectively. One effort is compiling chemical resistance data from manufacturers and literature exposure data in conjunction with utilizing the cohesive energy density numbers to predict service life. A second laboratory effort is attempting to quantify leak rates through holes in FML.

The field efforts are investigating the strength and durability of seaming techniques and methods to repair damaged liners. The strength and durability effort is comparing the results obtained from laboratory studies of the fabricated material as supplied by the manufacturer to those results obtained from actual field applications. The repair effort is evaluating current repair techniques of patching and grouting FML materials and to identify other new repair techniques to improve the long-term integrity of FML materials. One major field effort is an attempt to obtain access to an operating hazardous waste disposal site for the purpose of pursuing various research tasks with respect not only to liners, but also to leachate collection, waste characterization, and subsidence.

The TRDs are focusing on updating existing TRDs by pursuing efforts to expand the data base on FMLs and on developing liner compatibility guidance. The effort to expand the data base is correlating available information on existing FML types. These materials are also being subject to additional hazardous wastes and are being compared to overall performance. In the compatibility guidance effort, a liner acceptability ranking system is being developed from comparability of liner to waste data. The initial evaluation effort of FML has been published by the GPO in a TRD entitled "Lining of Waste Impoundment and Disposal Facilities" (SW-870; Stock Number 055-000-00231-2). Also in conjunction with the publication of this TRD, a leachate treatability TRD has been published, entitled "Management of Hazardous Waste Leachates" (SW-871; GPO Stock Number 055-000-00224-0). The construction manual effort is being pursued to provide inspection procedures and quality control criteria for the installation of flexible membrane liners at hazardous waste landfills.

Waste Modification

The objective of this activity is to evaluate the effectiveness of chemical stabilization and encapsulation processes as related to improving handling, reducing surface area, limiting solubility, detoxifying pollutants, and predicting performance with time. Validation efforts are being performed in the laboratory and field to correlate compatibility of the individual processes to specific waste types and to predict durability and leaching performance with time. Information produced will be published in a TRD.

The laboratory efforts are focusing on stabilization, encapsulation, and anaerobic degradation. *Stabilization* is defined as the chemical enhancement of a material to decrease or limit the solubility or to detoxify the waste constituents even though the physical properties may or may not be changed or improved. This usually involves addition of materials that ensure that the hazardous constituents are maintained in their least soluble form. The stabilization effort is assessing various stabilization processes to determine the effects of interfering materials (organic and inorganic) on process and product performance. Tests are being conducted on the fixed wastes to evaluate the degree of fixation, durability and strength, resistance to leaching, and similar factors. The encapsulation effort is categorizing general encapsulation processes according to organic and inorganic binders such as epoxides, polyesters, vinyls and cement, calcium, and silicates and evaluating their performance to the stated objectives. The anaerobic degradation effort is identified as a waste modification effort, although it focuses on the use of microorganisms of multiplasmid strains to enhance the degradation rate of organic compounds at hazardous waste sites.

The field efforts are investigating a portable overpack/drum ensemble and providing technical assistance as needed. The ensemble effort is a portable truck mounted unit which involves overpacking existing metal containers with rotamolded polyethylene (PE) overpacks and friction-welding a leak-tight PE cover onto the overpack. Limited-scale field demonstrations have been performed at two sites. The technical assistance effort will provide the capability to respond to EPA needs in assessing processes for encapsulating, solidifying, and fixing hazardous waste.

The TRD is focusing on the state of the art of stabilization/solidification of industrial waste as a pretreatment process for disposal of hazardous waste. The initial TRD has been published by the GPO in a document entitled "Guide to the Disposal of Chemically Stabilized and Solidified Wastes" (SW-872; GPO Stock Number 055-000-00226-6).

Surface Impoundments

The surface impoundment research program has been developed to provide a comprehensive understanding of the design, operation, and maintenance of surface impoundments as options for hazardous waste disposal. Information is being developed on the use of natural soils as liners and dikes. Also, the correlation of laboratory measurements with the construction standards achieveable in the field is being investigated. Of particular interest is the degree to which specification of construction techniques and inspection practice can influence uniformity and performance of the finished impoundment.

The components and unit operations of a surface impoundment to be addressed in this paper include:

- Assessment of design
- Volatile organic chemical (VOC) emissions
- Containment systems

Assessment of Design

The objective of this activity is to develop a comprehensive understanding of the design, operation, and maintenance of surface impoundments and to predict performance under varying designs. Information thus obtained would be modeled to predict the success or failure of future designs. Efforts are being pursued by literature reviews and laboratory and field studies. Information produced will be published in TRDs.

The literature review effort is assessing design and operating practices for surface impoundment technology. Surface areas versus depth, types of liners, construction practices, monitoring, and depth control are some factors being determined.

The laboratory efforts are investigating the concerns and benefits of locating disposal facilities in unsaturated low-permeability soils as compared with saturated low-permeability soils. Also, a technical assistance effort is being pursued to allow quick response to specific needs such as freeboard requirements and in-place closure of hazardous waste surface impoundments.

The field efforts are investigating success and failures of dike design, clay soil permeabilities and remedial action performance at surface impoundment sites. The success/failure effort is collecting data from specific owners/operators and contractors to identify mechanisms which contribute to a surface impoundment success or failure. The clay soil permeability effort is being conducted at an actual working site and is attempting to verify laboratory conclusions that organic chemicals affect the permeability of clay soils. The remedial action effort is evaluating the performance of various remedial actions at hazardous waste sites to compile and describe successful or unsuccessful methods that may be applicable to RCRA sites.

The TRDs are focusing on surface impoundment design, operation, maintenance, and closure. The initial effort has focused on the closure aspects of surface impoundment and a TRD has been published by the GPO entitled "Closure of Hazardous Waste Surface Impoundments" (SW-873; Stock Number 055-000-00227-4). Another TRD, entitled "Design, Construction, Maintenance and Evaluation of Clay Liners for Hazardous Waste Facilities," was completed in 1985.

VOC Emissions

The objective of this activity is to identify and assess VOCs from surface impoundments, determine pollutant release potential, and evaluate effectiveness of control technology techniques. Efforts are being pursued in the laboratory and the field to achieve this objective.

The laboratory effort is identifying volatilization mechanisms and parameters and developing techniques to measure these parameters. Models and predictive equations have been developed for these transport mechanisms and field validation studies are being pursued.

The field efforts are investigating wind dispersion and cover effectiveness. In the wind dispersal effort, assessment of gas volatilization is being investigated at actual disposal sites to determine the effects of wind velocity, temperature, and transfer coefficients. In the cover effectiveness effort, various control methods (synthetic membranes, soils, floatable materials) to minimize VOC emission release are being investigated at actual field sites by using previously developed measurement and modeling techniques.

The VOC emission area of activity for land disposal will be combined with VOC emission activities from incinerators and other disposal/treatment facility options to form a separate research category in FY 1986.

Containment Systems

The objective of this activity is to evaluate the effectiveness of soil dikes, liners, and other components with or without the incorporation of synthetic liners in containing hazardous sludges/liquids. Validation efforts are being performed in the laboratory and the field in relation to soil/waste interaction, hydraulic gradient, leak detection, and hydrostatic forces. Information produced will be published in a TRD and construction manuals.

The laboratory efforts are evaluating contaminant/soil interaction, containment times, and hydraulic conductivity. The contaminant/soil interaction effort is performing batch-type adsorption studies and using these data to estimate pollutant retention by earth material liners at hazardous waste disposal facilities. The containment times effort is comparing the accuracy and effort involved in using nonlinear forms of the unsaturated water flow equation to evaluate clay liner function by predicting leachate containment times. The hydraulic conductivity efforts are investigating the relationships of laboratory and field compacted clay soil hydraulic conductivity values for various clay soils of different location and composition. Also, the effects of hydraulic gradient on the hydraulic conductivity will be tested and measured on several soils and permeants.

The methodology efforts are those techniques that have applicability to locating liner leaks. One effort is investigating the use of electrical resistivity for determining liner failures. The other effort is investigating innovative acoustical emission and time domain reflectometry for monitoring early detection of leaks in surface impoundment liners.

The field efforts are investigating leak detection and upward hydrostatic forces. The leak detection effort has verified, at a 4-km^2 (1-acre) water-filled site, the usefulness of the electrical resistivity technique in locating leaks of 0.0254 m (1 in.), 0.127 m (5 in.), and 0.304 m (12 in.) diameter. The upward hydrostatic forces effort is evaluating the effects of hydrostatic pressure as exerted on various flexible membrane liners for a variety of point load conditions. Guidelines for installers will be developed from this effort. Technical assistance is also being pursued to allow for quick responses to specific needs as developed in the hazardous waste surface impoundment program.

The TRD which was identified under the previous surface impoundment section, entitled "Assessment of Design," also has application to this area of activity. Another effort which could develop into a TRD is investigating technical considerations in locating surface impoundments in low-permeability soils whereby specific site conditions will be modeled to determine leachate movement. The construction manual effort is being pursued to provide guidance to ensure proper control by use of quality assurance/quality control criteria during construction of surface impoundment dikes.

Underground Mines

The objective of this activity is to update the state-of-the-art technology on the use of underground mines for emplacement of hazardous waste. Efforts are being pursued both by literature review and planned field demonstration.

The literature review effort is updating the status of underground mine usage for hazardous waste emplacement from both the technical and economic feasibility viewpoints. Cost information is included to compare economics of various other disposal options.

The field effort was planned to participate with an owner of either an active or inactive mine in demonstrating on a limited scale the feasibility of this disposal concept in the United States. This effort has been delayed.

Dioxin Engineering

The objective of this activity is to provide a data base of information for input into the preparation and implementation of the National Dioxin Strategy to prevent further dioxin releases into the environment from contaminated sites. Validation studies are being performed in the laboratory and the field as relates to dioxin desorption, dioxin leachability, and fugitive dust emissions.

The laboratory efforts are concerned with 2,3,7,8-tetrachlorodibenzo-*p*-dioxin (TCDD); work will be expanded in the future to include other isomers as needed to support actions by the agency at specific locations. The laboratory studies of desorption and leachability of TCDD are investigating ten soils from TCDD-contaminated sites which are being extracted with water in batch and intact core studies, to determine the extent and rate of release and to identify mechanisms controlling this release. The stabilization study is investigating the effects of stabilizing dioxin contaminated soils with a bituminous and cement additive to minimize the leachability aspects as compared to nonstabilized soil.

The field effort is investigating a variety of dust suppressants to minimize fugitive dust emissions from dioxin-contaminated soils during application of various trafficking loads.

Conclusions

This paper has presented an overview of the hazardous waste land disposal research program as conducted by the U.S. EPA. Specific areas of research discussed were landfills, surface impoundment, underground mines, and dioxin engineering. This paper has attempted to mention all 75 research task activities being performed under this program. Consequently, more specificity has been introduced into this paper than is found in a previous paper by the author.[2]

All research data being produced by these research activities will be summarized for inclusion as best engineering judgment data into the TRDs. These documents are being utilized by regional personnel to evaluate the adequacy of the permit applications for new RCRA land disposal facilities. The plan is to update these documents routinely to reflect the latest research data being developed. More information about specific aspects of the hazardous waste land disposal research program may be obtained by contacting the author.

[2]Schomaker, N. B., "Hazardous Waste Landfill Research," *Hazardous and Industrial Waste Management and Testing: Third Symposium, ASTM STP 851,* American Society for Testing and Materials, Philadelphia, 1984, pp. 320-333.

Tommy E. Myers[1]

A Simple Procedure for Acceptance Testing of Freshly Prepared Solidified Waste

REFERENCE: Myers, T. E., "A Simple Procedure for Acceptance Testing of Freshly Prepared Solidified Waste," *Hazardous and Industrial Solid Waste Testing: Fourth Symposium, ASTM STP 886,* J. K. Petros, Jr., W. J. Lacy, and R. A. Conway, Eds., American Society for Testing and Materials, Philadelphia, 1986, pp. 263–272.

ABSTRACT: The U.S. Department of the Army (DA) is considering treatment and disposal options for hazardous wastes generated at DA facilities that involve solidification processing and landfilling of waste. A rapid test procedure that correlates to critical processing parameters, specifically process additive dosages, is needed for monitoring solidification processing. Candidate testing procedures were evaluated in bench-scale studies for their ability to distinguish between low-, intermediate-, and high-strength formulations of a solidification process. A simple, rapid, and inexpensive test procedure is described that showed promise in bench-scale studies of being a reliable means of real-time monitoring of solidification processing. The procedure uses the cone index (CI) of freshly prepared samples as measured with a cone penetrometer to provide data for a determination of the acceptance or rejection of material before it is actually landfilled. The test procedure involves molding and testing specimens of solidified waste immediately after solidification processing and consists of the following steps: (1) compaction using a standard compaction procedure; (2) determination of CI using standard Department of the Army procedures; and (3) comparison of results with previously established acceptance/rejection criteria. Establishment of acceptance/rejection criteria is based on empirical correlation of CI to specific process/waste formulations that are known to set and cure properly and are, therefore, acceptable for landfilling. Application of the recommended testing procedure to field acceptance testing is discussed. Data from the bench scale evaluation of various testing systems are also presented.

KEY WORDS: hazardous wastes, solidification, acceptance testing, cone penetrometer, concrete penetrometer, Vicat needle.

Hazardous wastes are generated at U.S. Department of the Army (DA) installations as by-products of manufacturing operations, as sludges from in-

[1]Ecologist, Environmental Laboratory, U.S. Army Engineer Waterways Experiment Station, Vicksburg, MS 39180.

dustrial wastewater treatment plants, by accidental spills, by demilitarization of obsolete weapons, and from past disposal practices. One treatment/disposal option that is sometimes considered for these wastes involves solidification processing of hazardous wastes followed by secure landfilling of the solidified waste. In the majority of the solidification processing systems available today, contaminant isolation and containment depend primarily on entrapment of the waste in a solid matrix formed by portland cement or a pozzolan substitute [1]. The typical product is a damp solid that is usually compacted when placed in a landfill. Waste solidification can provide three major advantages over raw waste management and disposal: (1) removal of free liquid, (2) development of structural integrity, and (3) reduction in leaching potential. The elimination of free liquid before disposal in a landfill is with a few exceptions a regulatory requirement. Structural integrity is important because the waste must have sufficient bearing capacity to support the overburden and the final cover material. Additionally, for solidification/disposal operations that use tracked or wheeled vehicles to move, place, and compact freshly solidified waste, the material must provide sufficient trafficability for these vehicles. Reduction of leaching potential provides direct benefits in terms of reduced environmental risks associated with a particular waste, and, in most instances, the trade-offs between costs of solidification processing and reduction in environmental risk is the basis of a decision to solidify.

Since the long-term environmental effectiveness of solidification/landfill projects is based in part on the projected physical-chemical properties of the landfilled wastes, a rapid field testing procedure is needed to confirm that the processed material will set as expected. This is in addition to simply monitoring the actual processing operation; that is, measuring additive feed rates, mixing time, or other material handling parameters. Lack of homogeneity in waste or additives can affect setting, and simple monitoring of the processing operation may not detect poorly setting batches that are caused by variability in waste characteristics or variability in additive reactivity. In addition to potential variability in waste characteristics, hazardous wastes often contain substances that interfere with setting reactions. The effect of these substances on solidification processing is generally unpredictable, and they can result in poorly setting batches of solidified waste. Therefore, field acceptance testing of freshly processed wastes should be a required element in quality control and quality assurance plans for major DA disposal projects.

A series of laboratory tests for assessing solidified waste has been used by several investigators [2-4]. In general, these assessment protocols have been used to provide background data on the environmental effectiveness and physical stability of solidified waste. However, none of these assessment protocols have been applied as field test procedures for acceptance testing of freshly processed waste. Assessment protocols typically involve the application of specific physical-chemical tests on specimens that have been cured for some specified time, usually 28 days. The physical-chemical properties tested include bulk density, specific gravity of the solids, wet-dry durability, freeze-thaw durability, unconfined compressive strength, permeability, and chemical leaching characteristics. At present there are no standardized rapid field test procedures available for monitoring solidification processing. Most of the standard soil and concrete test procedures either are not applicable to solidified waste or require too much time to be useful as rapid field procedures. Hence, there is a need for a rapid field test procedure that can be used for acceptance testing of fresh batches of solidified waste in order to determine if the fresh waste will set up and cure properly and thus if if is acceptable for landfilling.

This paper describes an evaluation of the unconfined compression test and three resistance-to-penetration tests for their potential as rapid field testing procedures for monitoring solidification processing of a DA hazardous waste. The objective of this evaluation was to find a rapid testing procedure that could reliably distinguish between a poor and a good set produced by different formulations of a specific solidification process and that could be incorporated into the framework of a proposed acceptance testing protocol. The proposed acceptance testing protocol involves conditional acceptance testing of freshly solidified waste, confirmed acceptance testing of partially set materials, and completed acceptance testing of cured materials.

Materials and Methods

Materials

A concentrated brine containing approximately 12 weight-percent chloride, 8 weight-percent organic carbon, 3 weight-percent ammonia-nitrogen, and various toxic substances such as aldrin, arsenic, and cyanide at parts per million levels was collected from a DA hazardous waste impoundment for study and testing. The liquid was solidified in 0.001-m^3 batches with a soil modification of a lime/fly ash solidification process (U.S. Patent RE 29 783). A native soil was added to the process formulation in order to absorb liquid and reduce the lime and fly ash needed to convert the liquid waste to solid form. Before solidification processing the waste was chemically pretreated in order to control ammonia off-gassing. Three formulations of the soil/fly ash/lime process were prepared in which the lime dosage was adjusted to produce three physically distinctive products: low-, medium-, and high-strength materials. The lime-to-waste weight ratios for the low-, medium-, and high-strength formulations were 0.5:1, 0.6:1, and 0.7:1, respectively. The soil-to-waste and the fly ash–to–waste weight ratios were 0.8:1 in each formulation. All materials were mixed in a 9.5×10^{-3} m^3 (2.5 gal) capacity grout mixer. The total mixing time was between 12 and 15 min. After a formulation was mixed, the freshly prepared solidified waste was allowed to set for 15 min in order to develop sufficient strength for the compaction step that followed.

After the waiting period, the freshly prepared solidified waste was compacted in standard compaction molds as specified in the U.S. Army Corps of Engineers Unconfined Compression Test (EM 1110-2-1906, Appendix VI). The compaction procedure used 25 blows of a 2.49-kg (5.5-lb) rammer and 304.8-mm (12-in.) drop. The compaction step required approximately 4 min. The strength of the solidified waste as indicated by resistance to penetration and by unconfined compressive strength was measured 15 min, 30 min, 60 min, 2 h, 4 h, and 24 h after preparation of the test specimens. Each process formulation was run in triplicate.

Methods

The testing systems evaluated were:

1. Vicat needle: ASTM Test for Time of Setting of Hydraulic Cement by Vicat Needle (C 191).
2. Concrete penetrometer: ASTM Test for Time of Setting of Concrete Mixtures by Penetration Resistance (C 403).
3. Unconfined compressive strength (UCS): U.S. Corps of Engineers Unconfined Compression Test EM 1110-2-1906, Appendix XI.
4. Cone penetrometer: U.S. Army Corps of Engineers Cone Index Test (Army TM 5-530), Chapter 1, Section XV.

The cone index (CI) is an index of soil bearing capacity or strength. Rohani and Baladi [5] have described the cone penetration process as the expansion of a series of spherical cavities and have shown that the forces resisting cone penetration of a soil can be described in terms of the fundamental properties of the soil (angle of internal friction, cohesion, stiffness, density, and so on). Theoretically, the cone penetrometer measures fundamental properties of the material; it is not simply an empirical tool without theoretical foundation for which useful correlations have been found. In practice the CI is the resistance of the test material to the penetration of a 30° right-circular cone. The CI is reported as the force per unit area of the cone base required to push the cone through the test material at a rate of 1.83 m (72 in.) per minute. The standard U.S. Army Corps of Engineers Waterways Experiment Station (WES) cone has a base area of 3.2 cm² (0.5 in.²); the airfield penetrometer has a base area of 1.3 cm² (0.2 in.²). It was convenient to use the standard WES cone on materials with a CI less than 690 kPa (100 psi) and to use the airfield penetrometer for materials with a CI greater than 2070 kPa (300 psi). For materials in between the above ranges both penetrometers generally give satisfactory results.

Results and Discussion

Evaluation of Test Procedures

The averaged data from each testing system for various cure times are presented in Table 1. The data show that the cone penetrometer measured a wider range of material strengths than either the Vicat needle or the concrete penetrometer. The Vicat needle and the concrete penetrometer did not have the sensitivity to measure the penetration resistance of solidified waste from the low- and medium-strength formulations with less than 30 min of curing. The penetration resistance of relatively hard materials developed by the high-strength formulation after 24 h curing were also beyond the measuring range of both the Vicat needle and the concrete penetrometer. In Fig. 1, CI is plotted versus cure time for the three formulations. These curves show that the cone penetrometer was able to distinguish among the low-, medium-, and high-strength formulations beginning at the 15-min mark and continuing throughout the 24-h testing period.

The data in the table also show that the cone penetrometer's ability to dis-

TABLE 1—*Strength of fresh solidified waste.*[a]

Process[b]	Cure Time, h	Vicat Needle, mm	Concrete Penetrometer, kPa[c]	Cone Index, kPa	Unconfined Compressive Strength, kPa[c]
Weak, $X = 0.5$	0.25	>50	<414	55	<1
	0.5	>50	<414	83	<1
	1	>50	<414	138	26
	2	41	690	290	ND[d]
	4	35	942	469	50
	24	<0.5	3785	1680	48
Medium, $X = 0.6$	0.25	<50	414	159	<1
	0.5	46	540	276	<1
	1	37	781	462	38
	2	35	919	607	ND[d]
	4	24	1220	904	43
	24	<0.5	4510	2140	65
Strong, $X = 0.7$	0.25	23	ND[d]	1470	<1
	0.5	20	2780	ND[d]	<1
	1	ND[d]	3100	1725	165
	2	13	3200	1790	131
	4	4	3760	2170	152
	24	<0.5	>4830	3470	145

[a]Average of three or more determinations.
[b]Process: soil/fly ash/lime/waste—0.8:0.8:X:1.
[c]No replication.
[d]ND = no data.

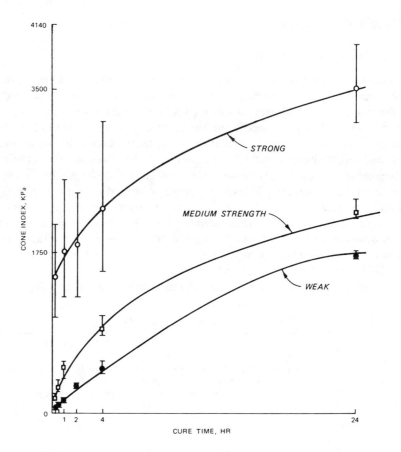

FIG. 1—*Cone index versus cure time for three solidification products.*

tinguish among different formulations of a soil/fly ash/lime process was superior to that of the Vicat needle and the concrete penetrometer. The differences in sensitivity to resistance to penetration were probably due to differences in the geometries of the shearing faces, differences in sizes of the shearing faces, and differences in the sensitivity of the force-measuring mechanism. All three measuring systems can be sized to meet a particular need; however, the Vicat needles and concrete penetrometers that are available as off-the-shelf items are not as versatile as available cone penetrometers.

Unconfined compressive strength was not measurable until cure times exceeded 30 min, and thereafter the UCS did not show marked change with cure time for the first 24 h. The UCS of the high-strength formulation was approximately twice that of the medium-strength formulation, which was only slightly higher than the UCS of the low-strength formulation.

Acceptance Testing

A proposed acceptance testing scheme for solidification processing of a waste that is then landfilled is shown in Fig. 2. In the first step in Fig. 2, a rapid field test is used to monitor the product delivered by solidification processing equipment before placement of the solidified waste in a landfill. If the specification for freshly prepared solidified waste is met, than the material is conditionally accepted for landfilling. Samples are collected for confirmation acceptance testing, and landfilling of the processed waste proceeds. If the conditional acceptance is confirmed in later laboratory or field tests, then placement of solidified waste on top of material that is already in place and that has been confirmed for acceptance can continue. Completed acceptance testing is performed on specimens cured for 28 days (or some other specified time) in order to determine if the landfilled solidified waste after curing meets the performance criteria on which the environmental impacts of the project were evaluated. In addition, tests can be conducted on cores taken from the disposal site.

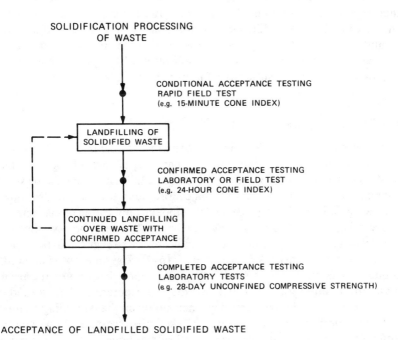

FIG. 2—*Proposed acceptance testing protocol.*

Conditional Acceptance

The objective of conditional acceptance testing is to determine as quickly as possible if the material being sent to the landfill will set properly, thereby avoiding the landfilling of any materials that do not have the required amount of solidification additive(s). The data collected in this study indicate that the 15-min cone index of compacted samples is a satisfactory test procedure for conditional acceptance testing of fresh solidified waste for processes that produce a damp solid that can be compacted. It can be directly related to critical additives such as the primary setting agent in a process formulation. For processes that produce a slurry that is pumped or poured into a disposal site the compaction step is not necessary, and CI could be determined on poured specimens after a sufficient curing time. The data collected in this study showed that the setting period before quantitative information can be obtained is less with the cone penetrometer than with other penetration tests.

The cone index–additive relationship is highly process- and waste-specific. The cone index of freshly prepared solidified wastes must be correlated to critical additive dosages in laboratory testing before one begins full-scale solidification and field testing. This would include the development of CI-versus–cure time curves and the selection of a specific CI value at a specific cure time as a conditional acceptance criterion. The actual value of the cone index chosen as a conditional acceptance criterion depends on the relationship of CI to the target additive/waste formulation. The target additive/waste formulation for full-scale processing is the one that produces a product in laboratory or pilot scale testing that will meet all the performance criteria required by regulatory authorities for landfill disposal.

Confirmed Acceptance

The primary purpose of confirmed acceptance testing is to indicate whether or not the material is setting properly before it is covered with additional waste. For a landfilling scenario in which the solidified waste must be spread, compacted, or otherwise worked, the confirmatory test should also indicate the trafficability of the in-place waste. Although application of the cone index to accessing the trafficability of solidified waste has never been verified, cone penetrometers have been used as an indicator of strength and/or vehicular mobility in a wide variety of environments [6-9]. The time period allowed for conducting confirmatory testing is generally short, because of the need to maintain a continuous solidification/disposal operation. This period can range from overnight to several days, depending on the landfilling technique used at a specific site. For example, the solidified waste could be placed in different cells on consecutive days to allow additional time for setting and testing before reentering the original cell.

Because of the short time period available for testing, it would be difficult to run conventional tests normally used for testing the physical properties of a material of this type. In the absence of an available alternative for conducting such tests, the cone penetrometer can be used for confirmed acceptance testing. In any case it will be necessary to establish the specific relationship between the results from the measuring system used in the confirmed acceptance tests and critical process additives for the process that is actually used and for whatever cure time is used.

Completed Acceptance

Performance criteria for the completed acceptance testing step are the basis for determining the environmental effectiveness and physical stability of landfilled solidified waste. For this reason there are no substitutes for the tests that must be run in order to provide the physical-chemical information required. Site-specific parameters influence which tests are required and which are weighed more heavily than others. Unconfined compressive strength and permeability are physical properties that are usually specified. Data from some type of chemical leach test (depending on the regulatory agency with primacy) is also required. Additional tests that are used include wet-dry durability, freeze-thaw durability, specific gravity of solids, bulk weight, percent moisture, and others that may be needed to evaluate special site-specific conditions. Whatever performance criteria are established for the landfilled waste in order to satisfy regulatory requirements (that is, to obtain required permits), the test methods used to test for compliance with the performance criteria constitute the completed acceptance test.

Conclusions

In the limited study reported here the cone index coupled with a standard compaction control procedure is a rapid and simple measuring system that (1) correlates to the amount of critical solidification reagent in a soil/fly ash/lime process formulated for a hazardous liquid waste; (2) is indicative of the strength of the material; and (3) follows a predictable increase that is a function of time. In comparison tests with the Vicat needle and the concrete penetrometer, the cone penetrometer was preferred on the basis of better correlation with critical additive dosages for the solidification process investigated. These conclusions are based on preliminary data and only one waste and one process were considered.

An acceptance testing scheme has been proposed that involves conditional acceptance testing within 15 min of solidification processing of the waste, confirmatory acceptance testing to verify the conditional acceptance, and completed acceptance testing to determine if performance criteria are met. Confirmation of these results and verification of the applicability of the proposed acceptance testing scheme will have to come from additional work.

Acknowledgments

The tests described and the resulting data presented, unless otherwise noted, were obtained from research conducted under the Department of Army's Installation Restoration Program. Mr. Donald L. Campbell, U.S. Army Toxic and Hazardous Materials Agency, Aberdeen Proving Ground, MD, was the project manager for the work unit under which this study was funded.

References

[1] Malone, P. G., Jones, L. W., and Larson, R. J., "Guide to the Disposal of Chemically Stabilized and Solidified Waste," Publication SW-872, U.S. Environmental Protection Agency, Cincinnati, OH, 1980.

[2] Bartos, M. J. and Palermo, M. R., "Physical and Engineering Properties of Hazardous Industrial Wastes and Sludges," Publication EPA 600/2-77-139, U.S. Environmental Protection Agency, Cincinnati, OH, 1977.

[3] Thompson, D. W. and Malone, P. G., in *Toxic and Hazardous Waste Disposal: Vol. 2, Options for Stabilization/Solidification,* R. B. Pojasek, ed., Ann Arbor Science, Ann Arbor, MI, 1979, pp. 35–50.

[4] Cote, P. L., "A Proposed Protocol for the Assessment of Solidified Wastes," presented to the Workshop on the Cooperative Project on Solidified Wastes Characterization, Vegreville, Alta, Canada, Sponsored by Environment/Canada, 21 and 22 Nov. 1983.

[5] Rohani, B. and Baladi, G. Y., "Correlation of Mobility Cone Index with Fundamental Engineering Properties of Soil," Miscellaneous Paper SL-81-4, U.S. Army Engineer Waterways Experiment Station, Vicksburg, MS, 1981.

[6] Hammitt, G. M., "Evaluation of Soil Strength of Unsurfaced Forward-Area Airfields by Use of Ground Vehicles," Miscellaneous Paper S-70-14, U.S. Army Engineer Waterways Experiment Station, Vicksburg, MS, 1970.

[7] Meyer, M. P., "Comparison of Engineering Properties of Selected Temperate and Tropical Surface Soils," Technical Report No. 3-732, U.S. Army Engineer Waterways Experiment Station, Vicksburg, MS, 1966.

[8] Meyer, M. P., "Naval Seafloor Soil Sampling and In-Place Test Equipment: A Performance Evaluation." Technical Report R-730, Naval Civil Engineering Laboratory, Port Hueneme, CA, 1971.

[9] Mitchell, J. K. and Houston, W. N., "Static Penetration Testing on the Moon," preprint for European Conference on Penetration Testing, Garston Engineering and Building Research Establishment, Stockholm, Sweden, 5–7 June 1974.

Philip G. Malone,[1] *Robert J. Larson,*[1] *James H. May,*[1] *and John A. Boa, Jr.*[2]

Test Methods for Injectable Barriers

REFERENCE: Malone, P. G., Larson, R. J., May, J. H., and Boa, J. A., Jr., "**Test Methods for Injectable Barriers,**" *Hazardous and Industrial Solid Waste Testing: Fourth Symposium, ASTM STP 886*, J. K. Petros, Jr., W. J. Lacy, and R. A. Conway, Eds., American Society for Testing and Materials, Philadelphia, 1986, pp. 273-284.

ABSTRACT: Grouts are becoming increasingly important in producing barriers to contaminated groundwater flow at hazardous waste sites. Grouted barriers can be used at depths and under conditions where slurry trenches are impractical. To employ grouts to advantage at waste sites it is necessary to select materials that solidify or gel in the presence of industrial waste compounds that may be present. Grouts must additionally be unaffected by aggressive chemical wastes and create impermeable barriers when injected into permeable units.

A wide variety of test procedures have been employed in bench and pilot-scale evaluation of grouts. Many testing systems do not duplicate field conditions closely enough to be useful. A standard approach using gelling time determinations, static chemical durability testing, and permeability determinations on cores of grouted material is suggested.

KEY WORDS: grout, gelling time, compatibility, permeability, hazardous wastes

Sealing materials pumped or injected into rock and soil have received widespread attention in the construction industry for control of groundwater and surface water infiltration. Grouting with chemical or particulate grout to control water movement has been routine procedure in remedial actions undertaken at dams or in tunnels [*1–3*]. Applications of grouts in waste control are becoming more common at sites where the depth to an impermeable horizon or the presence of coarse (boulder-sized) particulate material makes slurry trenching impractical. Large-scale vertical grout curtains have been employed in several waste control projects [*4,5*] and have been planned or evaluated at several others [*6,7*].

[1]Geologists, Geotechnical Laboratory, U.S. Army Engineer Waterways Experiment Station, Vicksburg, MS 39180.

[2]Geologist, Structures Laboratory, U.S. Army Engineer Waterways Experiment Station, Vicksburg, MS 39180.

As grouts are more commonly employed in hazardous waste containment, it is necessary to develop techniques for testing grouting material used at hazardous waste sites to determine:

(a) the ability of the grout to gel or set under adverse chemical conditions successfully,

(b) the ability of the gel to remain intact when exposed to aggressive chemical compounds, and

(c) the reduction in permeability that can be produced in grouting reactions.

The purpose of this paper is to examine techniques that have been used for testing grouts used for hazardous waste containment and to suggest how testing methods can be altered to better evaluate materials and methods used in producing injected barriers. This review concentrates on procedures used in laboratory evaluation of materials and small-scale field tests.

A summary of the major grout types that are available and their properties is given in Table 1. The materials vary from clay slurries that thicken or gel as water is lost to surrounding materials to low-viscosity organic materials that go through a complex series of linking reactions to form organic polymers. In all cases, the grouts become effective by forming a nonleaching solid or semi-solid that blocks the pore spaces in the permeable medium (for example, a sand or silt) in which it is injected. The successful use of grout may directly

TABLE 1—*Major grout types with potential application in waste containment.*

Type of Grout[a]	Major Components	Chemical Compatibility
Portland cement/clay (particulate)	bentonite and/or portland cement	resistant to most wastes except strong acids or organics
Sodium silicate	sodium silicate and organic acids or esters	resistant to most acids and organics; attacked by bases
Acrylates	acrylate monomer cross-linking agent and initiator and accelerator	resistant to most organics; attacked by acids, bases, and salts
Furan	furfuryl alcohol, filler and acid	resistant to everything but oxidizers
Lignosulfites	calcium lignosulfonate, acid and sodium dichromate	attacked by strong reducing agents
Urethanes	urethane monomer and water	attacked by aldehydes, ketones, acid, and bases
Phenoplasts	resorcinol, formaldehyde, and sodium hydroxide	attacked by hydrocarbons and bases
Aminoplasts	urea and formaldehyde	attacked by acids, bases, and ethers or epoxides

[a]Data are from Refs 8 and 9.

depend on the ability of a grout to penetrate the matrix material before gelling. In the case of particulate grouts such as cement or bentonite (clay), the pore size of the grouted unit must be relatively large (above that found in silt) in order for the grout grains to move between the matrix particles. Where solution grouts (usually referred to as chemical grouts) are used, the grout with the lowest viscosity has the best ability to penetrate. Because most groutable materials below a waste site contain water, viscosities near that of water are desirable. The grouts that have the widest application in grouting both coarse- and fine-grained sediments are chemical grouts. Portland cement and clay grout are limited to coarse sand or gravel unless a cavity is produced by such techniques as jetting or hydrofracting (splitting the substrate with pressurized grout).

Chemical Compatibility During Grout Setting

Most of the materials employed in grouts are organic polymers, such as acrylate or urethane, or soluble silicates, such as acidic or basic sodium silicates (Table 1). The gelling or polymerizing reactions in the various grouts can be slowed or stopped by a variety of chemical compounds found in industrial wastes. Many gelling reactions can also be accelerated by chemical compounds occurring in the permeable materials being grouted. These accelerated gelling or flash setting reactions can make it impossible to distribute grout in a permeable unit before hardening occurs. Table 2 summarizes some of the more common interfering reactions occurring in grouts. Not all of these reactions would occur in grouting chemically contaminated soil or rock units, but the high probability of some interaction of waste and grout necessitates laboratory testing for abnormal gelling reactions.

TABLE 2—*Common interfering reactions that occur in solidifying grouts.*

Type of Grout	Interfering Reaction
Portland cement/clay (particulate)	organics can attack clays; acids destroy lime in portland cement
Sodium silicate	acids and inorganics can cause flash sets; alkali can prevent gellation
Acrylates	strong oxidizers and reducers can interfere with gellation by destroying initiators
Furan	strong reducers can stop gellation
Lignosulfites	strong reducers can destroy the gelling initiator
Urethanes	ammonia compounds compete for the prepolymer and form urea compounds
Phenoplasts	strong acids can stop gellation
Aminoplasts	alkalis prevent gelling

Laboratory gel time determinations are approximate tests based on the time required for the grout to become too viscous to pour or too stiff to stir easily. Mechanical testing equipment to detect changes in viscosity is available but has limitations. Available equipment includes paddle gelometers [10,11], rotational viscometers [12], or penetrometer set measuring devices [10]. None of these measuring devices has proved completely satisfactory because of the gradual stiffening of chemical grouts and the lack of resistance to penetration observed in many chemical gels. Problems also arise from the shearing of the newly formed gel, which can reliquify part of the solid and give a false reading of the viscosity.

The simplest techniques for judging laboratory gel time for grouts are the glass rod method and the tilting tube method. The glass rod method involves drawing a glass rod across the surface of the liquid material and noting the time elapsed before the mark produced by the rod remains on the surface for several (two to three) seconds. The tilt tube method involves observing the time elapsed from grout mixing to the point at which the grout will no longer pour from a test tube [12]. Neither method is automated and constant attention is needed to obtain accurate results.

None of the commonly used gel time indicators accurately mimics the behavior of the grout in the field. Large-scale laboratory tests have shown that silicate grouts, for example, can be pumped into dry sand test beds for five to six times longer than the laboratory-determined gel times before viscosity increases become evident. The shearing of the grout in the pumps and hoses and in the sand test bed probably keeps the grout fluid longer than the typical gel time for stationary material [13].

Laboratory gel times are a valid technique for detecting problems related to unusual set times caused by background chemical interaction between the pore fluids and grout, but the use of laboratory set times in field operations may result in consistently underestimating the gel times.

Chemical wastes can be introduced into grout in a variety of ways. In field operations in sealing a waste site, the grout would be injected and would displace and partly mix with chemically contaminated water. A simple laboratory approach aimed at reproducing the field situation is to mix equal parts of the grout and the waste or simulated waste uniformly and to observe any changes in gelling time. This mixing procedure does not exactly duplicate field conditions where the advancing grout would not be well mixed with the chemical waste it encounters, but the test does provide a baseline condition for estimating interferences. The test program involving direct mixing of equal volumes of grout and waste has been effective in demonstrating that a variety of compounds potentially present in industrial waste can affect grout gel times even when present in pore water at concentrations below 10% by volume. Effects observed vary from a flash set of silicate grout with 10% copper sulfate solution to complete retardation of any gelling (10% ammonium hydroxide solution with urethane grout) [10]. Testing of retardants in portland cement is generally done in a similar way, by introducing the chemical of interest directly into the cement mix.

The concentration of contaminants in the grout can be varied to observe how the change in gel times is related to chemical interference during the setting reaction. Many waste materials will prevent gelling in grouts by destroying compounds added as initiators or accelerators. Other waste compounds can initiate competing reactions in grout that proceed faster than polymerization, so that reactive monomer is used up without any solid (or with a very weak, poorly polymerized solid) forming.

No standard methods exist for testing chemical interferences in grouts, but the general approach has been to determine gel times for grouts with the likely contaminants being mixed directly with the grout. More thorough approaches involving the systematic measurement of reduced strength or reduced permeability of grouted sand samples could also be used to produce realistic and applicable data.

Chemical Compatibility after Barrier Formation

Most grout barriers are put in place to last indefinitely. However, grouts are subject to chemical attack, just as flexible membrane liners or compacted clay liners would be. The question of chemical stability of grout materials is a serious one. Components in industrial waste can cause grout barriers to shrink, swell, dissolve, or embrittle. Grout layers can be leached away [14], or they may react with groundwater and wastes and cause precipitates to form at the grout boundary.

Two approaches have been commonly employed to test hardened solidified polymer durability in a reactive chemical environment. One approach has been to expose specimens of grout to wastes or simulated waste in a static test and to observe any changes, such as shrinking, swelling, cracking, or discoloration, that might indicate the grout polymer is breaking down. This type of static exposure test has been used to test grouts, but no standard test conditions are used.

The results of a 16-week test on acrylate gels exposed to 10% solutions of alcohols, ketones, hydrocarbons, acids, and metal salts were reported by Clarke [15]. A 20-day test on silicates, urethanes, acrylates, and portland cement with acids, bases, hydrocarbons, and metal salts was reported by Larson and May [10]. In the latter test, 10% (by volume) solutions or saturated solutions (for materials with water solubilities below 10%) were employed.

A 1- to 9-month durability test undertaken on sodium silicate–grouted sand specimens was reported by Lord et al [16]. The simulated wastes used were generally highly concentrated or saturated solutions (for example, 100% by volume sodium hydroxide and saturated sodium dichromate). The time for disintegration was arbitrarily taken as the time needed to observe an easily noticeable number of sand grains on the bottom of the test vessel.

Many important parameters—such as the volume–to–surface area ratio for the test specimens, presence or absence of supporting material (soil or rock), and the types of supporting material added—change from one reported procedure to another. The volumes and concentrations of waste or waste simulants used and the frequency with which the solutions are renewed are not standardized. At the present time no uniform protocol exists that can be used for static durability testing of grouts.

A second approach to testing grout durability involves the use of flow-through tests, where the test specimen of grouted material is placed in a permeameter and water containing the potentially deleterious waste is forced through the sample. The effect of the permeant is observed by noting changes in the permeability of the test specimens (although permeability changes may not always reflect grout deterioration).

Several significant problems are involved in using a flow-through test system. If a sample is prepared in such a way that the particulate material is saturated (or impregnated) with a grout and all the void space is filled, the permeability is extremely low and a true flow-through test becomes nearly impossible. For example, Doiron et al [17] prepared samples of sand grouted with acrylamide that had permeabilities of 0.2×10^{-9} cm/s. The sand was saturated with acrylamide and vibrated so that it would consolidate before gelling took place. Flow-through testing would be ineffective with such highly impermeable materials because only one surface would be effectively exposed to the permeant in any reasonable test period.

Other problems arise if the chemical permeant being used reacts with the grout to produce a precipitate or gas bubbles. The precipitate can clog the permeameter and prevent measurements from being taken [16]. Gas bubbles produce an unsaturated condition and cause inconsistent, erratic permeability measurements.

When flow-through chemical compatibility tests are used, the samples are typically saturated with water or a dilute calcium sulfate solution. After a reference hydraulic conductivity is established, the chemical permeant is introduced into the sample [16,18]. However, as the permeant changes the measured hydraulic conductivity will change because the kinematic viscosity of the new permeant is different from the baseline liquid (water or a dilute calcium sulfate solution). The higher the kinematic viscosity of the permeant the lower the permeability measured. To understand any permeability changes caused by chemical interaction between the grout and the permeant, it is necessary that any changes in the kinematic viscosity of the permeant be accurately determined.

The complications involved in flow-through compatibility tests make flow-through methods complex to use and limit the technique to the examination of materials (grouts and permeants) that do not react to produce gas or precipitate. Additionally, any changes in the viscosity of the permeant caused by dilution with interstitial water or interaction with grout must be determined independently so that the changes in the permeability of the grout sample can be documented. Static testing systems are less complex, but require consideration of a large number of variables related to sample geometry and the characteristics of test solutions. Standard test protocols for both static and flow-through tests are needed if useful data are to be developed.

Permeability Testing

Permeability tests on grouted material can produce widely different results when different sample preparation techniques are employed. In a field grouting situation, grout would be pumped out from a point or line source and would form a sphere or cylinder of grouted medium [13]. Grout cylinders or grout bulbs (Fig. 1) may have irregular shapes that reflect slight variations in the pump pressure, medium permeability, and grout viscosity. In many laboratory preparations, efforts are made to produce the most complete grout penetration possible, regardless of the limitations of field application. For example, in Lord et al [16], silicate grout was thoroughly mixed with Ottawa

FIG. 1—*Grout bulbs produced by injecting sodium silicate into a simulated water/waste mixture.*

sand and then compacted into molds with a Harvard compactor. As previously mentioned, Dorion et al [17] mixed a well-graded, washed, medium sand with acrylamide and mechanically vibrated the sample to ensure that maximum density was achieved. Spalding et al [7] prepared test samples by passing three pore volumes of grout through a packed soil test column; the column was then plugged before gelling of the grout occurred so that no grout could drain out. These techniques assured that all voids were filled. Samples prepared in the three studies referenced had extremely low permeabilities (usually at the limit of measurement of the test equipment), but these test results bear little relation to what will occur in the field.

Vinson [19] and Vinson and Mitchell [20] developed a preparation technique for grouted sand samples that closely approximates the behavior of grout in the field. The grout, a polyurethane in this report, was injected into a sand-filled mold and a sphere or cylinder of grouted material was produced. Both dry and saturated sands were used. The grouted sand cylinders were used for permeability testing. The limited space available for the expansion of the grout in the cylindrical molds did result in the specimens prepared as cylinders having slightly higher densities than the samples prepared as spheres, but the densities overlapped enough to indicate that the grout distribution in the cylinders was approximately what would be obtained if the grout had been allowed to flow outward with no cylinder walls present.

The bench-scale work was verified in a series of larger-scale tests where each test run involved injecting 9 kg of grout into 0.2 m³ of sand packed in a steel drum. A similar test system is shown in Fig. 2. Cores, 3.5 cm in diameter, were cut from the large grout bulbs produced in the tests. The permeabilities of the cores were comparable to those obtained from the injected cylinders. Of all the sample preparation systems employed to produce grouted sand test specimens, the cores from large grout bulbs represent the samples that most closely approach field conditions.

The laboratory procedures for determining permeability of grouted materials have generally followed standard triaxial permeability test protocols, where the sample is fully saturated and back pressure is employed as recommended by Zimmie [21]. Where fixed-wall permeameters have been employed [7,16], the samples were grouted directly in the test cell.

Field tests to determine permeability of grouted waste using monitoring wells established in the grout-modified soil horizon were reported by Spalding et al [7]. In this investigation, the wells were in place before grout injection and grout flowed through the screens into the wells. Slug tests on the wells showed that many of them were completely sealed, especially where the trench had been dry when the grout was injected. This procedure documented what happened to the wells, but did not directly measure the effect in the grouted wastes.

Alternative procedures that would provide more useful field verification would include coring and recovering a section of grouted material for perme-

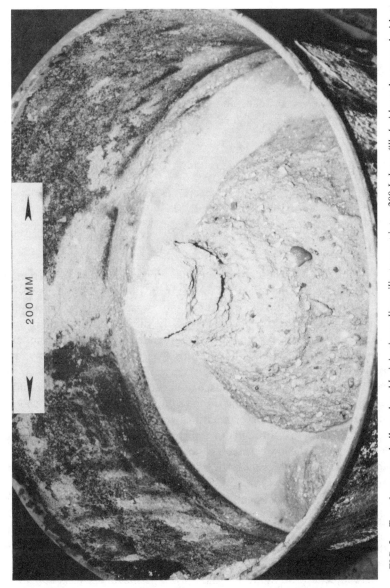

FIG. 2— *Top of grout test bulb produced by injecting sodium silicate grout into a 200-L drum filled with sand saturated with a simulated water/waste mixture.*

ability testing or drilling and setting casing in the grouted horizon for in-situ permeability tests [22]. Where conditions warrant, both types of tests could be performed.

Recommendations

Grouts are receiving wide attention in planning remedial action at hazardous waste sites because they can be used at greater depths and in geologic settings where slurry walls are not practical. In order to use grouts in waste containment it is necessary that standard approaches be developed for testing grouts. Such tests should be designed to permit the selection of materials that will:

(a) set or gel in the contaminated groundwater often encountered at a waste site,

(b) not disintegrate or dissolve under the chemical conditions at a waste site, and

(c) seal the treated layer sufficiently to assist in controlling groundwater flow.

The test approach that best answers the question of setting and variation in setting time caused by interaction with waste is the tilt tube method used with groundwater from a specific waste site or suitable waste simulants. The major disadvantage of using the tilt tube method is the difficulty in judging the gel point. Most results from gel time studies are intended to be used for comparison of one grout mixture to another. In this type of comparative study, approximate results are adequate.

Durability studies on grouts to be used at hazardous waste sites can best be conducted in static tests, where the samples of grouted materials are soaked in contaminated groundwater or solutions prepared to simulate contaminated groundwater. Flow-through durability tests are difficult to conduct and interpret because of the low permeability of many laboratory samples and the production of precipitates or gases that change the apparent permeability of the samples.

Determinations of the permeability of grouted material can be conducted on specimens prepared by using a variety of techniques. The most realistic permeability determinations are tests performed on cores prepared from large pods of grout injected into dry or saturated sand. The distribution of grout in pods or bulbs probably comes closest to the distribution occurring during field injection. Standard triaxial upflow permeameter tests appear to provide dependable data for grouted materials. Field permeability measurements should be performed by using an accepted slug or pump test with casing set to the grouted horizon.

Acknowledgments

The work reported in this paper was performed in the course of a larger investigation on the development of methods for in-situ hazardous waste stabilization by injection grout. The project is funded under Interagency Agreement D96930581-01-1 between the U.S. Army Engineer Waterways Experiment Station and the U.S. Environmental Protection Agency, Municipal Environmental Research Laboratory, Cincinnati, OH. Herbert R. Pahren is the project officer.

References

[1] Jiacai, L., Baochang, W., Wengguang, C., Yuhua, G., and Hesheng, C., "Polyurethane Grouting in Hydraulic Engineering," *Grouting in Geotechnical Engineering*, W. H. Baker, Ed., American Society of Civil Engineers, New York, 1982, pp. 403–417.

[2] Fox, R. C. and Jones, M. C., "Remedial Drilling and Grouting of Two Rockfill Dams, *Grouting in Geotechnical Engineering*, W. H. Baker, Ed., American Society of Civil Engineers, New York, 1982, pp. 136–151.

[3] Clarke, W. J., "Performance Characteristics of Acrylate Polymer Grout," *Grouting in Geotechnical Engineering*, W. H. Baker, Ed., American Society of Civil Engineers, New York, 1982, pp. 418–432.

[4] Henderson, J. K., "Guide to Alluvial, Rock, and Chemical Grouting," J. K. Henderson, Inc., Buffalo, NY, 1980.

[5] Montgomery, J. M., "Report on the Gel-injection Phase for Final Closure of the Stringfellow Class I Hazardous Waste Disposal Site," Consulting Engineers, Inc., Pasadena, CA, 1982.

[6] Tamura, T. and Boegly, W. J., Jr., "In-Situ Grouting of Uranium Mill Tailings Piles: An Assessment," ORNL/TM-8539, Oak Ridge National Laboratory, Oak Ridge, TN, 1983.

[7] Spalding, B. P., Hyder, L. K., and Munro, I. L., "Grouting as a Remedial Technique for Problem Shallow Land Burial Trenches of Low-Level Radioactive Solid Wastes," ORNL/NFW-83/14, Oak Ridge National Laboratory, Oak Ridge, TN, 1983.

[8] Karol, R. H., "Chemical Grouts and Their Properties," *Grouting in Geotechnical Engineering*, W. H. Baker, Ed., American Society of Civil Engineers, New York, 1982, pp. 359–377.

[9] Spooner, P. A., Hunt, G. E., Hodge, V. E., and Wagner, P. M., "Collection of Information on the Compatibility of Grouts with Hazardous Wastes," draft report, EPA Contract No. 68-03-3113, Task 40-3, JRB Associates, McLean, VA, 1983.

[10] Larson, R. J. and May, J. H., "Geotechnical Aspects of Bottom Sealing Existing Hazardous Waste Landfills by Injection Grouting (Phase I—Compatibility Studies), *Proceedings, First Annual Hazardous Materials Management Conference*, Tower Conference Management, Wheaton, IL, 1983, pp. 513–529.

[11] Malone, P. G., May, J. H., and Larson, R. J., "Development of Methods for In-Situ Hazardous Waste Stabilization by Injection Grouting," *Land Disposal of Hazardous Wastes*, EPA-600/9-84-007, U.S. Environmental Protection Agency, Cincinnati, OH, 1984, pp. 33–42.

[12] "Dynagrout T," product information sheet, Dynamit Nobel Aktiengesellschaft, Troisdorf, Federal Republic of Germany, 1982.

[13] Bader, T. A. and Krizek, R. J. "Injection and Distribution of Silicate Grout in Sand," *Grouting in Geotechnical Engineering*, W. H. Baker, Ed., American Society of Civil Engineers, New York, 1982, pp. 540–563.

[14] Petrousky, M. B., "Monitoring of Grout Leaching at Three Dam Curtains in Crystalline Rock Foundations," *Grouting in Geotechnical Engineering*, W. H. Baker, Ed., American Society of Civil Engineers, New York, 1982, pp. 105–120.

[15] Clarke, W. J., "Performance Characteristics of Acrylate Polymer Grout," *Grouting in*

Geotechnical Engineering. W. H. Baker, Ed., American Society of Civil Engineers, New York, 1982, pp. 418–432.

[16] Lord, A. E., Jr., Weist, F. C., Koerner, R. M., and Arland, F. J., "The Hydraulic Conductivity of Silicate Grouted Sands with Various Chemicals," *Management of Uncontrolled Hazardous Waste Sites, 1983,* Hazardous Materials Control Research Institute, Silver Spring, MD, 1983, pp. 175–178.

[17] Doiron, G. H., Burkhard, H., and White, M. L., *ASTM Bulletin,* Vol. 250, Dec. 1960, pp. 34–35.

[18] Haji-Djafari, S. and Wright, J. C., Jr., "Determining the Long-Term Effects of Interactions Between Waste Permeants and Porous Media," *Hazardous and Industrial Solid Waste Testing: Second Symposium, ASTM STP 805,* R. A. Conway and W. P. Gulledge, Eds., American Society for Testing and Materials, Philadelphia, 1983, pp. 246–264.

[19] Vinson, T. S., "The Application of Polyurethane Foamed Plastic in Soil Grouting," Ph.D. dissertation, University of California, Berkeley, 1970.

[20] Vinson, T. S. and Mitchell, J. K., *Journal of the Soil Mechanics and Foundation Engineering Division, American Society of Civil Engineers,* Vol. 98, No. SM6, June 1972, pp. 579–602.

[21] Zimmie, T. F., "Geotechnical Testing Considerations in the Determination of Laboratory Permeability for Hazardous Waste Disposal Siting," *Hazardous Solid Waste Testing: First Conference, ASTM STP 760,* R. A. Conway and B. C. Malloy, Eds., American Society for Testing and Materials, Philadelphia, 1981, pp. 293–304.

[22] Hvorslev, M. J., "Time Lag and Soil Permeability in Groundwater Observations," Bulletin 36, Waterways Experiment Station, U.S. Army Corps of Engineers, Vicksburg, MS, 1951.

Raymond C. Loehr,[1] John H. Martin, Jr.,[2] and
Edward F. Neuhauser[2]

Spatial Variation of Characteristics in the Zone of Incorporation at an Industrial Waste Land Treatment Site

REFERENCE: Loehr, R. C., Martin, J. H., Jr., and Neuhauser, E. F., "Spatial Variation of Characteristics in the Zone of Incorporation at an Industrial Solid Waste Land Treatment Site," *Hazardous and Industrial Solid Waste Testing: Fourth Symposium, ASTM STP 886,* J. K. Petros, Jr., W. J. Lacy, and R. A. Conway, Eds., American Society for Testing and Materials, Philadelphia, 1986, pp. 285–297.

ABSTRACT: The spatial variation in the characteristics of the zone of incorporation (ZOI) at an industrial waste land treatment site can be significant even when efforts are taken to achieve uniform distribution and incorporation. A single grab sample can be inappropriate for determining average ZOI characteristics.

The results of this study provide guidance in determining the number of samples necessary to provide reliable estimates of ZOI characteristics. The spatial variation at land treatment sites is a function of site-specific factors such as soil characteristics and methods of waste application and incorporation. A determination of specific site spatial variation is recommended as a prerequisite for developing an appropriate sampling program at land treatment sites.

KEY WORDS: land treatment, zone of incorporation, industrial wastes, spatial variation, oily wastes, hazardous wastes, sampling methods

Factors such as process simplicity and reasonable cost have made land treatment an increasingly attractive method for the treatment and disposal of industrial wastes. Land treatment uses the assimilative capacity of the soil to degrade and immobilize the applied wastes. The wastes are incorporated into the surface soil layer, usually the top 150 to 300 mm. This depth is known as

[1]Professor of civil engineering, University of Texas, Austin, TX 78712; formerly professor of engineering and environmental engineering, Cornell University, Ithaca, NY 14853.
[2]Research associates, Department of Agricultural Engineering, Cornell University, Ithaca, NY 14853.

the zone of incorporation (ZOI). This zone, in conjunction with underlying soils where additional treatment and immobilization of the applied waste constituents can occur, is referred to as the treatment zone. The treatment zone may be as much as 1.5 m deep. Soil conditions below this depth generally are not conducive to the degradation of the constituents in the applied waste. Most of the transformations, degradation, and immobilization will occur in the zone of incorporation.

To evaluate the performance of a land treatment site, samples of the soil in the ZOI are taken and analyzed for conventional, nonconventional, and toxic pollutants. Thus it is important that an adequate number of representative samples be taken to evaluate the degradation and immobilization that does occur at the site. It cannot be assumed that a single random grab sample will provide representative ZOI characteristics. Information about the representativeness of ZOI samples is key to both technical and regulatory decisions concerning a land treatment site.

Experience indicates that the spatial variation of the ZOI characteristics at a land treatment site can be substantial. The soil in the zone of incorporation at a land treatment site is not homogeneous and it rarely is possible to apply a waste and mix it with the soil in a completely uniform manner. The spatial variation of parameters in the ZOI at a land treatment site will affect the number of samples needed to achieve a desired level of accuracy when one is quantifying industrial waste land treatment site characteristics and performance.

This study was conducted as part of Cooperative Agreement CR-809285 between Cornell University and the Robert S. Kerr Environmental Research Laboratory (RSKERL) of the U.S. Environmental Protection Agency. A detailed description of this project is available elsewhere [1,2].

Objectives

The objectives of this study were:

1. To illustrate the spatial variation in the characteristics of the zone of incorporation that can occur at an industrial waste land treatment site.
2. To demonstrate how such information can be used to determine the number of samples necessary to achieve a desired level of accuracy in estimating the ZOI characteristics.

Experimental Approach

An oily waste was sprayed as uniformly as possible on two 16-m² field plots in June 1983. The application of the oily waste was 40 kg/m² (9.5 kg oil per square metre). Each plot was rototilled to a depth of 150 mm before, during, and after the application of the waste. There was no subsequent tillage of these plots.

Shortly after the oily waste was applied and the plots were rototilled, ten random samples were taken from the ZOI of each plot. Each sample was analyzed for oil and grease, volatile material, total Kjeldahl nitrogen, and pH. Ten random samples also were taken from the same plots in November 1983. The second series of samples helped identify the effects of time and degradative processes on the spatial variation. The November samples were analyzed for the four parameters noted above and also the moisture content.

Site

The two field plots are located adjacent to Cornell University in Ithaca, New York. The site is an old field that had not been used for agricultural purposes and had not received applications of lime, fertilizer, or pesticides for more than ten years. The site had been mowed annually to prevent establishment of woody plants.

The soil at this site is Rhinebeck silt loam, a deep, somewhat poorly drained, fine-textured soil that formed in clayey, calcareous lake deposits [*3*]. This soil has about 300 mm of moderately to slowly permeable heavy silt loam over slowly permeable silty clay loam or silty clay. The lower layer extends to a depth of 600 mm to 1 m and is underlain by layers of silty clay separated by thin layers of silt. Rhinebeck soils are moderately acidic.

Waste

The oily waste used in this study was obtained with the assistance of RSKERL personnel. The origin of the waste was unknown but it had physical and chemical characteristics, including odor, that were similar to those of oil refinery sludges. Analyses of this waste are summarized in Table 1.

Method of Application

Every effort was made to obtain a uniform soil-waste mixture in the zone of incorporation. Initially, each 16-m^2 plot was mowed and raked to remove the

TABLE 1—*Characteristics of the oily waste applied to the experimental field plots in June 1983.*

Parameter	$\bar{X} \pm SD^a$
Water, % WB^b	48.7 ± 0.2
Ash, % WB	15.5 ± 0.1
Oil and grease, % WB	23.9 ± 1.2
Total Kjeldahl nitrogen, g/kg	1.1 ± 0.1
pH	6.7 ± 0.2

$^a\bar{X} \pm SD$ = mean ± standard deviation, $n = 5$.
^{b}WB = wet basis.

existing vegetation. Both plots were then rototilled to an average depth of about 150 mm to facilitate subsequent waste incorporation. Before application the oily waste was thoroughly mixed to assure uniformly. The waste was applied to the plots by a hand-held sprayer in as uniform a manner as possible.

A split application approach was used. The total quantity of waste applied to each plot was divided into three equal parts. Following each waste application, the plots were rototilled immediately. Thus each plot was rototilled three times during the applications of the waste.

In addition, each plot was rototilled again two days after completion of the waste application process. At this time, the plots were rototilled twice in perpendicular directions. The plots received no further tilling for the duration of the study.

Sampling Protocol

The following summarizes the procedure used to sample each field plot randomly in June 1983 immediately after the application of the oily sludge and in November. Each 16-m² field plot was divided into 400 subplots measuring 0.04 m² each. No samples were taken from the edge subplots. Of the remaining 324 subplots, 20 were selected randomly without replacement by using a table of random numbers [4].

At each plot, the first ten subplots selected were sampled in June and the remaining subplots were sampled in November. The subplots sampled were located by placing a grid (Fig. 1) between permanent corner stakes. A soil core 7.6 mm in diameter and 150 mm deep was removed manually from each subplot for subsequent analysis.

Sample Preparation

Each soil sample was kept separate and was prepared for analysis in the following manner. First, each sample was mixed and hand-sorted to remove stones and large vegetative materials. After subsamples were taken for moisture and pH determinations, the remaining soil was air-dried at room temperature (about 20°C). When dry, the samples were pulverized and passed through a No. 16 U.S. standard sieve which had a mesh opening of 1.19 mm. The screened samples were then stored in screw-cap bottles at room temperature until analyzed.

Analytical Methods

The analytical methods used in this study are outlined in Table 2.

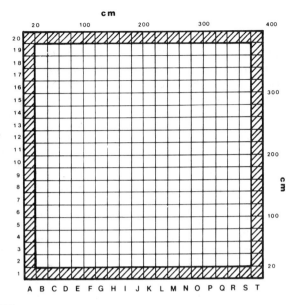

— INDICATES EDGE PLOTS NOT TO BE SAMPLED

FIG. 1—*Grid system used to locate subplots sampled.*

TABLE 2—*Methods used for analysis of the waste and subplot soil samples.*

		Sample	
Parameter	Method	Soil	Waste
Oil and grease	Extraction Method for Sludge Samples (APHA[a] 503D)	X	X
pH	pH Value—Glass Electrode (APHA 423) Method for Soil pH in 0.01 M CaCl$_2$ Solution (60-3.5) [5]		X
Total Kjeldahl nitrogen	Semimicro-Kjeldahl Method (APHA 420B and McKenzie and Wallace [6])	X	X
Moisture content	Total Residue Dried at 103–105°C (APHA 209A) and Gravimetry with Oven Drying (7-2.2) [5]	X	
Water	Test for Water in Petroleum Products and Bituminous Materials by Distillation (ASTM D 95)		X
Volatile material	Total Volatile and Fixed Residue at 550°C (APHA 209E)	X	
Ash	Test for Ash from Petroleum Products (ASTM D 482)		X

[a]American Public Health Association.

Results

The results of the June and November 1983 analyses are summarized in Tables 3 and 4, respectively. There were substantial variations in soil characteristics, particularly oil and grease, among each set of samples even though every effort was made to obtain the most uniform soil-waste mixture possible. Figures 2, 3, and 4 respectively illustrate the distributions of oil and grease, total Kjeldahl nitrogen, and volatile material in the subplot samples. Such normal distributions were typical for all four sets of subplot soil samples.

TABLE 3—*Summary of soil analyses—June 1983.*

Plot	Statistical Parameter	Oil and Grease, mg/g MFS[a]	Volatile Material, % MFS[a]	Total Kjeldahl Nitrogen, mg/g MFS[a]	pH
I	range	48.1–75.0	12.1–14.2	2.04–2.88	6.7–6.8
	$\bar{X} \pm SD$[b]	62.0 ± 7.1	13.1 ± 0.6	2.56 ± 0.25	6.7 ± 0.1
	CV,[c] %	11.4	4.6	9.8	...
	CI[d]	56.9–67.1	12.7–13.5	2.38–2.74	...
II	range	47.3–68.9	14.2–17.6	3.27–3.99	6.7–6.9
	$\bar{X} \pm SD$[b]	57.0 ± 7.2	16.2 ± 1.2	3.62 ± 0.26	6.8 ± 0.1
	CV,[c] %	12.6	7.4	7.2	...
	CI[d]	51.8–62.2	15.3–17.1	3.43–3.81	...

[a]Moisture-free soil, oven-dried at 103°C.
[b]Mean ± standard deviation, $n = 10$.
[c]Coefficient of variation.
[d]95% confidence interval estimate, $t_{0.5}$.

TABLE 4—*Summary of soil analyses—November 1983.*

Plot	Statistical Parameter	Oil and Grease, mg/g MFS[a]	Volatile Material, % MFS[a]	Total Kjeldahl Nitrogen, mg/g MFS[a]	pH	Moisture Content, % WB[b]
I	range	31.1–50.8	11.8–14.2	2.48–2.86	6.9	28.0–32.8
	$\bar{X} \pm SD$[c]	43.3 ± 5.6	13.1 ± 0.7	2.71 ± 0.13	6.9	31.0 ± 1.5
	CV,[d] %	12.9	5.3	4.8	...	4.8
	CI[e]	39.3–47.3	12.6–13.6	2.62–2.80	...	29.9–32.1
II	range	36.1–64.6	14.3–18.4	3.37–4.00	6.9–7.0	22.6–34.1
	$\bar{X} \pm SD$[c]	48.7 ± 7.8	16.1 ± 1.2	3.67 ± 0.22	7.0	28.4 ± 3.9
	CV,[d] %	16.0	7.4	6.0	...	13.7
	CI[e]	43.1 ± 54.3	15.2 ± 17.0	3.51 ± 3.83	...	25.6–31.2

[a]Moisture-free soil, oven-dried at 103°C.
[b]Wet basis.
[c]Mean ± standard deviation, $n = 10$.
[d]Coefficient of variation.
[e]95% confidence interval estimate, $t_{0.5}$.

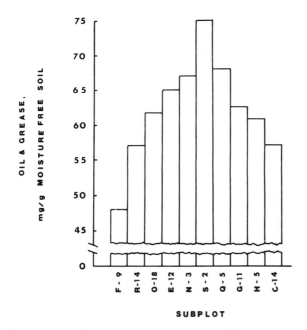

FIG. 2—*Distribution of soil oil and grease—Plot I, June 1983.*

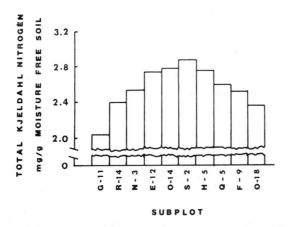

FIG. 3—*Distribution of total Kjeldahl nitrogen in soil—Plot I, June 1983.*

Variation Within Plots

There were substantial differences among each set of subplot samples for all of the parameters considered, with the exception of pH (Tables 3 and 4). The observed differences could be due to either spatial variation or random

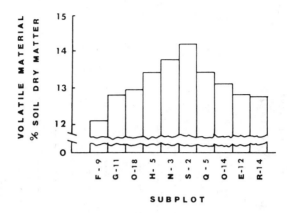

SUBPLOT

FIG. 4—*Distribution of volatile material in soil—Plot I, June 1983.*

analytical error. To separate these factors, the analytical results from each set of ten subplot samples, two replicates per sample for each parameter, were compared statistically.

The comparisons were made using a one-way analysis of variance [4] to test the null hypothesis that the subplot means for each parameter did not differ significantly. The results of the comparison (Table 5) indicated that the observed differences within each set of subplots were statistically significant ($P < .05$) and were due to spatial variation and not random error.

Variation Between Plots

The observed differences in mean values for oil and grease, total Kjeldahl nitrogen, and volatile material between plots and between sampling dates for each plot also were tested for statistical significance. These comparisons were

TABLE 5—*Results of analyses of variance[a] for differences among subplots.*

	Variance Ratio F		
Date and Plot	Oil and Grease	Total Kjeldahl Nitrogen	Volatile Material
June 1983			
Plot I	27.84	26.16	6.40
Plot II	44.74	9.54	21.24
November 1983			
Plot I	140.58	4.93	28.90
Plot II	153.34	12.45	126.95

[a]Subplots differ significantly ($P < .05$) if $F > 3.13$.

made using one-way analysis of variance [4], followed by Duncan's new multiple range test [7] to analyze differences between means.

The results of these statistical analyses are summarized in Table 6. Oil and grease concentrations in the two plots were not significantly different ($P <$ .05) in either June or November. However, the oil and grease concentrations were significantly different between June and November, indicating that reductions did occur in both plots between those months.

There were significant differences between plots in terms of total Kjeldahl nitrogen and volatile material concentrations in both June and November but no significant changes occurred in either plot with time. Although significant reductions in oil and grease occurred in both plots between June and November, the spatial variation as indicated by the coefficient of variation for each plot (Tables 3 and 4) remained essentially constant.

The differences noted in Table 6 for the specific parameters can be explained in terms of the amount of material that was added to the soil at each plot and by the natural variations between plots. Table 7 identifies the background characteristics as determined from control plots.

The oily waste added to the plots had a high oil and grease content but a

TABLE 6—*Results of analyses of variance for differences between plots and between sampling dates.*[a]

Date and Plot	Oil and Grease, mg/g MFS[b]	Total Kjeldahl Nitrogen, mg/g MFS[b]	Volatile Material % MFS[b]
June 1983			
Plot I	62.0 a	2.56 c	13.1 e
Plot II	57.0 a	3.62 d	16.2 f
November 1983			
Plot I	43.3 b	2.71 c	13.1 e
Plot II	48.7 b	3.67 d	16.1 f

[a]All values are means. Values in the same column with a common letter are not significantly different ($P <$ .05). $n = 10$.
[b]Moisture-free soil, oven-dried at 103°C.

TABLE 7—*Oily waste land treatment site—background soil characteristics.*[a]

Date	Oil and Grease, mg/g MFS[b]	Volatile Material, % MFS[b]	Total Kjeldahl Nitrogen, mg/g MFS[b]	pH
6/7/83	0.6 ± 0.2	8.8 ± 1.5	3.3 ± 0.7	5.8 ± 0.3
10/25/83	0.5 ± 0.1	9.4 ± 1.5	3.5 ± 0.8	5.8 ± 0.3

[a]Data from control plots; $n = 4$.
[b]Moisture-free soil.

low nitrogen content (Table 1). As a result of the waste application, the oil and grease and volatile material concentrations in the ZOI (Tables 3 and 4) were considerably greater than the background (Table 7). In contrast, the nitrogen additions were small and did not increase the ZOI concentrations measurably.

Discussion

These results demonstrate that the spatial variation in industrial waste land treatment site ZOI characteristics can be statistically significant even when efforts are made to ensure uniform waste distribution and incorporation. The coefficient of variation ranged from 11.4 to 16% for the oil and grease concentrations, 4.6 to 7.4% for the volatile material concentrations, 4.8 to 9.8% for the TKN concentrations, and 4.8 to 13.7% for moisture content (Tables 3 and 4). At an actual land treatment site, where less care to obtain uniform distribution and mixing may be exerted, the coefficient of variation may be greater.

These data help determine the number of random samples needed to obtain a statistically significant estimate. Assuming that there will be random sampling within a land treatment site, the number of samples required to provide a desired level of confidence that the sample mean $\bar{X}$ does not differ from the average ZOI characteristics by more than an identified acceptable error can be calculated by using the following relationship (4):

$$n = (Z^2 S^2)/L^2 \tag{1}$$

where

n = required number of samples
Z = normal deviate
S^2 = population variance
L = acceptable error

To use this relationship, it is first necessary to decide how accurate the sample estimate should be—that is, to identify the limits of error $\pm L$ that are acceptable. As shown in Fig. 5 using oil and grease data from Plot II, November 1983, the required sample size is inversely proportional to the square of the acceptable error. Clearly, the number of samples should provide an estimate that is useful, but the number should not be so large that the cost of sampling and analysis is excessive. The reason for obtaining the estimate of average site characteristics and the use of the results should be clearly identified before deciding how accurate the sample estimate should be.

If one wishes to express the acceptable error as a percentage of the sample mean—for instance, 30 mg/g $\pm$ 10%—the expected sample mean must be

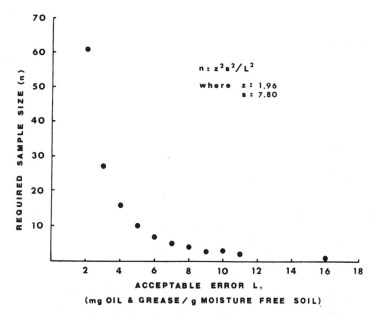

FIG. 5—*Illustration of relationship between acceptable error and sample size using oil and grease data from Plot II, November 1983.*

estimated before the required sample size can be calculated. The allowable error is now a function of the expected mean value and the required sample size, and the accuracy of the estimate increases as the expected mean value decreases.

It cannot be guaranteed that the sample mean $\bar{X}$ will fall within the limits of acceptable error $\pm L$, since the normal curve extends from minus infinity to plus infinity. However, the probability that this will occur can be specified. Assuming that a 5% chance that the acceptable error L will be exceeded is acceptable, the value of the normal deviation Z is 1.96. For 1%, the value of Z is 2.58. Values of Z for other probabilities can be obtained from appropriate statistical tables.

Finally, an estimate of the population variance (the standard deviation squared) is needed. It is preferable to rely on previous experience or results from similar industrial waste land treatment sites as the basis for the estimate. If such information is lacking, an educated guess may be necessary.

The results of this study indicate that spatial variations in ZOI characteristics vary little after waste is applied (Tables 3 and 4). Thus the results from an initial estimate of the needed number of samples can be used for subsequent sampling.

An example can indicate how the number of samples can be determined. Using data from Plot I, June 1983 (Table 3), Table 8 identifies the number of

TABLE 8—*Illustration of the number of random samples required for various levels of acceptable error and probability.*

Parameter[a]	Acceptable Error, %[b]	Number of Samples for Each Probability Level[c]		
		10%	5%	1%
Oil and Grease	20	1	2	3
(62.0 ± 7.1)	10	4	5	9
	5	14	20	35
Total Kjeldahl nitrogen	20	1	1	2
(2.56 ± 0.25)	10	3	4	7
	5	11	15	26
Volatile material	20	1	1	1
(13.1 ± 0.6)	10	1	1	2
	5	3	4	6

[a]Mean ± standard deviation for Plot I, June 1983 (Table 3).
[b]Percentage of sample mean that is the acceptable error.
[c]Probability that the acceptable error will be exceeded.

samples that are needed for different levels of acceptable error $\pm L$ and different probabilities that these limits will be exceeded. A large number of samples is required for a small acceptable error and small probabilities that the limits will be exceeded. As noted in Fig. 5, a larger number of samples will be required when the sample mean has smaller values.

Summary

The spatial variation of the ZOI characteristics at a land treatment site can be significant even when efforts are taken to obtain uniform application and incorporation. The coefficient of variation for the oil and grease analyses ranged from 11.4 to 16% and the coefficient of variation for volatile material, total Kjeldahl nitrogen, and moisture content generally were less than 10% at the site used in this evaluation.

These results provide guidance in determining the number of samples necessary to provide reliable estimates of average ZOI characteristics at industrial waste land treatment sites. However, the spatial variation at a land treatment site is a function of site-specific factors such as soil characteristics and methods of waste application and incorporation. Determination of specific site spatial variation is recommended as a prerequisite for identifying sampling requirements.

References

[1] Loehr, R. C., Martin, J. H., Jr., and Neuhauser, E. F., "Disposal of Oily Wastes By Land Treatment," *Proceedings, Purdue Industrial Waste Conference*, Ann Arbor Science Publishers, Ann Arbor, MI, 1983.

[2] Loehr, R. C., Martin, J. H., Neuhauser, E. F., Norton, R. A., and Malecki, M. R., "Land Treatment of an Oily Waste—Degradation, Immobilization, and Bioaccumulation," Final Report, Project CR-809285, Robert S. Kerr Environmental Research Laboratory, U.S. Environmental Protection Agency, Ada, OK, 1985.

[3] Soil Conservation Service, Soil Survey, Tompkins County, New York, Series 1961, No. 25, U.S. Department of Agriculture, Washington, DC, 1965.

[4] Snedecor, G. W. and Cochran, W. G., Statistical Methods, 7th ed., Iowa State University Press, Ames, 1980.

[5] Black, C. A., Evans, D. D., White, J. L., Ensminger, L. E., and Clark, T. E., Eds., Methods of Soil Analysis, American Society of Agronomy, Madison, WI, 1965.

[6] McKenzie, H. A. and Wallace, H. S., "The Kjeldahl Determination of Nitrogen: A Critical Study of Digestion Conditions—Temperature, Catalyst and Oxidizing Agent," Australian Journal of Chemistry, Vol. 7, 1954, pp. 55-71.

[7] Steel, R. G. D. and Torrie, J. H., Principles and Procedures of Statistics, McGraw Hill, New York, 1960.

Incineration

Kun-Chieh Lee,[1] James L. Hansen,[1] and Gary M. Whipple[1]

Characterizing Petrochemical Wastes for Combustion

REFERENCE: Lee, K.-C., Hansen, J. L., and Whipple, G. M., "**Characterizing Petrochemical Wastes for Combustion,**" *Hazardous and Industrial Solid Waste Testing: Fourth Symposium, ASTM STP 886,* J. K. Petros, Jr., W. J. Lacy, and R. A. Conway, Eds., American Society for Testing and Materials, Philadelphia, 1986, pp. 301–313.

ABSTRACT: Waste characterization is the first step for the design of an incinerator or for the prediction of a waste's combustion performance in an existing incinerator or other combustion device. Many petrochemical wastes are heterogeneous, so traditional analytical techniques that were developed to handle relatively pure or homogeneous coal or oil samples may not be efficacious. This paper will discuss the problems encountered in using traditional analytical techniques for waste characterization and some solutions to these problems. The analytical areas discussed include elemental composition, ash content, organic content, and metal content. A statistical analysis plan (not a sampling plan) is proposed for handling heterogeneous chemical waste.

KEY WORDS: incineration, wastes, combustion, characterization, petrochemical ash, elemental composition, metals, organic compounds, gas chromatography, chlorine, phosphorus, sulfur, sodium, heat of combustion, viscosity, hazardous wastes

In the design of a waste incinerator, waste characterization is the initial step of the project. Waste characterization data are first required to decide whether the waste can be incinerated. If it can, additional data are required for the design of the incinerator, design of the waste handling facility, and selection and design of the pollution control equipment.

The U.S. Environmental Protection Agency (EPA) regulates the generation, storage, transportation, and disposal of hazardous wastes. The regulations require that certain solid wastes be characterized; the required tests include flash point, extraction procedure (EP) toxicity test, pH, heating value, chlorine content, ash content, and hazardous organic content. Other tests, including viscosity, differential thermal analysis, differential scanning calo-

[1]Engineering scientist, research scientist, and product manager, respectively, Union Carbide Corp., South Charleston, WV 25303.

rimetry, and water content, may be required by the EPA administrator. The intent of these tests is (1) to decide whether a waste is a legally hazardous waste and should be regulated, and (2) to decide whether its characteristics are suitable for handling at a particular facility and within the permitted ranges of variation.

Although many standard and nonstandard methods have been developed to conduct the necessary tests, they were usually developed for the analysis of either pure chemicals or nearly homogeneous materials. Chemical wastes are typically highly heterogeneous and the sample matrix is almost always complex. Hence, many of the traditional methods or instruments have difficulty in handling a chemical waste sample. To make the situation worse, almost all state-of-the-art instruments analyze minute specimens. It is very unlikely that a few milligrams of a heterogeneous waste specimen are representative of the whole sample. Interference is another common problem in chemical waste analysis.

This paper discusses some of the problems encountered in characterizing chemical wastes and presents some ways of overcoming them. Topics discussed here are a summary of many experiences over the years on different projects. In the cases discussed, we did not expend a major effort to develop a standard method scientifically; since method development was not our prime function, few data on statistical precision were developed. As the properties of a chemical waste vary all the time, extremely precise data are of little use. In many cases, analyzing multiple samples generates more statistically valid results than replicating analyses of the same sample.

The discussion that follows will describe our experience with tests conducted on raw wastes and present a proposed statistical analysis plan (which is different from a sampling plan). All waste samples were obtained from various Union Carbide plants.

Tests Conducted on Raw Chemical Wastes

Organic Content Analysis

EPA publications [1,2] give detailed analytical methods to be used for hazardous components. A state-of-the-art gas chromatographic (GC) analysis typically uses about a 1-μL injection. Since chemical wastes are typically heterogeneous, it is important to find a good solvent to dissolve the samples first into a homogeneous matrix.

Most state-of-the-art gas chromatographs are equipped with a microprocessor to handle the data. Since chemical wastes are typically a mixture of many compounds with a wide range of molecular weights, the baseline of the chromatogram may not be a straight line. The software in the GC microprocessor may not be able to handle a chromatogram with a varying baseline, and the data printed out by the computer from an arbitrarily assigned baseline can be totally wrong. If the concentration of a particular hazardous organic compound is high, this may not cause serious problems, but if the concentration of the compound is on the order of 1 ppm, it can be very serious. For example, it could distort the results for a stack flue gas sample in an official Resource Conservation and Recovery Act (RCRA) Part B permit trial burn test. In a stack flue gas sample, the compounds of interest are always on the part-per-million level. The data printed out by the computer may give a false emission level that is several times higher than the actual one.

Figure 1 shows a gas chromatogram of a stack flue gas sample. Peak A is the compound of concern (the principal organic hazardous constituent). The "standard" baseline determined by the software of the microprocessor is shown in the figure as the line with long dashes. The correct baseline that should be used in calculating the area of Peak A is shown as the line with short dashes. As can be seen, the area reported by the computer is about five times larger than the true value. To overcome this problem, a data processing computer that is able to show precisely how the computer performed the calculation and to tell the computer to do the calculation in a precisely controlled way is very useful.

Other problems in the GC analysis of a complex organic sample include column contamination, interference, and retention time shifting. Heavy organics in the waste or the flue gas sample tend to be held in the GC column. The compounds may elute from the column very slowly and give a very erratic and nonreproducible baseline. Or, the compounds may coat the column packing and change the performance of the column. It is strongly suggested that the column be baked for a short period after each injection and baked overnight after every group of injections.

Since the sample matrix is complex, a GC peak may contain more than one compound. It is suggested that the identity(s) of the GC peak of interest be confirmed by gas chromatography/mass spectrometry. Another problem that results from complex sample matrices is retention time shifting. Because of the presence of other compounds in the waste, the retention time for the compound of interest may shift. Hence, the compound may have a retention time different from that determined by injection of a pure standard. It is important to spike the test specimen with the pure compound of interest and verify its retention time in that particular sample matrix.

Chlorine, Phosphorus, and Sulfur Contents

Data on chlorine, phosphorus, and sulfur contents are required in the design of air pollution control equipment and in the estimation of stack pollutant emission levels. Table 1 summarizes chlorine content measurement data for seven different wastes. Four different analytical approaches were used.

At first, a Dohrmann microcoulometer was used to measure chlorine content. Very quickly, it was realized that sample heterogeneity was making it

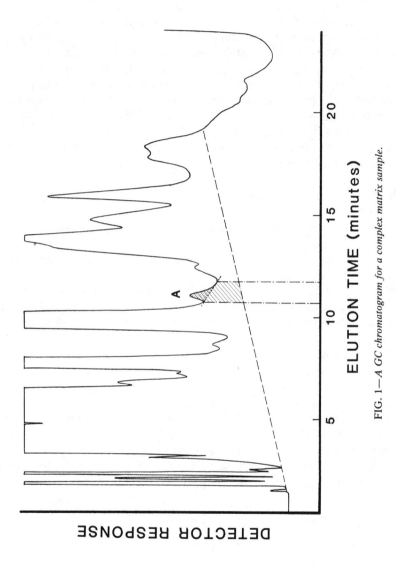

FIG. 1—A GC chromatogram for a complex matrix sample.

TABLE 1—*Chlorine content (weight percent) of liquid wastes.*[a]

	Waste						
Analytical Method	A	B	C	D	E	F	G
Theoretical	unknown	unknown	unknown	38.3	24	≈70	unknown
Dohrmann microcoulometer	0.31	1.06 0.043	1.25	...	...	...	...
X-ray fluorescence	0.49	0.55	1.01	...	...	...	...
Oxygen bomb and ion chromatography	...	0.48	1.73	13.5	23.3	28	20.4
Hydrolysis and titration	...	...	...	...	...	66.8	43.7

[a]See Table 2 for proposed peroxide bomb method.

difficult to obtain reliable data. This can be seen from the two measurements on the same Waste B sample. The method was abandoned.

Next, we tried X-ray fluorescence, a surface analysis technique. The sample is placed in a special cup and X-rays bombard only the part of the specimen right next to the bottom of the cup. Since waste normally contains suspended particles, different readings can be obtained even for the same specimen cup. This method was also abandoned.

The method that gave the most satisfactory results was to first digest the specimen in an oxygen bomb and then analyze the aqueous bomb wash by ion chromatography analysis. In this way, chlorine is converted into chloride. The sample matrix became very simple. The use of ion chromatography to analyze the chloride concentration minimized the possibility of interference from other known or unknown ions in the waste specimen. However, we realized later that there are problems with this approach as well. Depending on the chlorine content and the catalytic effect of the other elements in the waste, some of the chlorine may be converted into chlorine gas (Cl_2) (rather than chloride (Cl^-) during the oxygen bomb digestion. The chlorine gas will be lost to the air. For example, Waste D reported a chlorine content of 13.5%, while the correct value was known to be around 38%. On the other hand, Waste E (which had an even higher chlorine content and tended to produce more chlorine) reported a chlorine content of 23.3%, which was very close to the calculated value. A peroxide bomb should be used in the digestion step to make sure that all of the chlorine is converted into chloride. Additional data are discussed in Table 2.

Wastes F and G were reactive material. Upon being mixed with water, all chlorine will react with water and form hydrochloric acid (HCl). The reaction was so violent that the waste specimen had to be added to the water at a very

TABLE 2—*Chlorine, phosphorus, and sulfur contents of sludges.*[a]

	Waste		
Measurement	K	L	M
Chlorine			
Oxygen bomb	0.35	0.70	0.68
Peroxide bomb	0.54	0.70	0.87
Phosphorus			
Oxygen bomb	0.0	0.23	0.26
Peroxide bomb	1.03	0.77	0.90
Sulfur			
Oxygen bomb	1.5	2.2	1.3
Peroxide bomb	1.5	2.3	1.5

[a]All measurements are in weight percents.

slow rate to minimize the amount of HCl vapor lost to the air. The chloride content in the water was then measured by wet chemical titration using a silver nitrate standard solution. A much higher chlorine content, which was close to the expected level, was obtained.

Table 3 shows sulfur contents for three different types of waste. Basically it showed that data from X-ray fluorescence tests were nonreproducible and unreliable because of sample heterogeneity and possibly interference problems.

Table 2, in contrast, shows chlorine, phosphorus, and sulfur content data obtained simultaneously on sludge samples by using a bomb digestion followed by ion chromatography analysis. At first, an oxygen bomb was used to digest the sample; subsequently, the effort was repeated with a peroxide bomb, which gave a better chloride conversion than the oxygen bomb. The peroxide bomb was also more successful in converting phosphorus into a water-soluble form. There were no significant differences observed between the two bombs for sulfur content analysis.

TABLE 3—*Sulfur content (weight percent) of liquid wastes.*

	Waste		
Method	H	I	J
X-ray fluorescence	7.4	6.4	5.2
	9.2	4.5	...
Oxygen bomb and ion chromatography	no data	13.7	16.3

Water Content Analysis

There are basically two methods that may be used to measure the water content of a chemical waste. They are the ASTM Test for Water in Liquid Petroleum Products by Karl Fischer Reagent (D 1744) and the ASTM Test for Water in Petroleum Products and Bituminous Materials by Distillation (D 95). Unfortunately, neither method is suitable for complex wastes. In the Karl Fischer titration method, a list of known interferences are discussed. In our study, a waste sample with good combustion performance (the measured heating value was 8.3 kcal/g or 15 000 Btu/lb) was found to have a water content of 70% by Karl Fischer titration. From the process operating information, none of the interferences in the D 1744 list were known to be present. Since a chemical waste is a collection of unwanted material from one or more process units, it is difficult to pin down what caused the interferences.

The distillation method is also normally inapplicable, since a waste may contain material that has a boiling point close to that of water. Some method development is required in this area.

Ash Content

The ASTM Test for Ash from Petroleum Products (D 482) is typically used to measure the ash content of a chemical waste. If the waste contains organometals, the applicability of the method is questionable. During the ashing process, the organometals may be converted into submicron-sized oxides; the majority of those submicron particles are lost in the flue gas. Table 4 shows the test results for two organosilicone wastes. The results indicate that ASTM Test D 482 recovered only a very small fraction of the ash produced. The results improved after the sample was first hydrolyzed to convert the silicone into a more stable oxygenated compound.

The best results were obtained by combusting the sample in an oxygen bomb and then recovering the ash from the bomb wash. Since silica is insoluble in water, the bomb wash was filtered and the dried ash was weighed to

TABLE 4—*Ash content (weight percent) of organosilicone wastes.*

Method	Waste	
	N	O
ASTM D 482	1.45	11
Hydrolyze and ASTM D 482	26.3	26
Oxygen bomb + filtration	. . .	47
Theoretical	34.0	≈ 50

determine the ash content. If the ash is soluble, the ash content may be determined by evaporating the water from the bomb wash.

Elements of Special Interest

In some cases, it may be necessary to know the sodium, potassium, arsenic, or lead levels of a waste. Sodium exists in many chemical wastes. Almost all sodium salts have a low melting point that causes ash deposit and slagging problems in an incineration system. Sodium salts are also corrosive.

The sodium content may be measured directly by atomic absorption (AA) spectroscopy after the sample is diluted with a good solvent to a measurable range. The solvents normally used are methyl Cellosolve® solvent and dioxane.

Table 5 shows data on sodium content for Waste P. Three 20-mL specimens (identified as 1, 2, and 3) were obtained from a 1-gal jar after vigorous shaking. Each 20-mL specimen was divided into two portions (A and B). A total of six specimens (1-A through 3-B) were prepared from the same gallon jar. An 0.5-mL specimen was taken from each vial and diluted with methyl Cellosolve solvent by 25 000 times so that it could be measured by AA. Reasonably good data were obtained, as shown in Table 5. At least three readings were taken for the same diluted specimen and normally they were within 0.5% of one another.

Table 6 shows similar data for Waste Q, which had a relatively low sodium content. The sample was handled the same way as for Waste P, with the exception that in the final AA analyses, a 5-mL specimen was taken from each of the six vials for series dilution. The results also indicated a reasonable agreement among the specimens.

A quite different variation was observed for Waste R. Two series of tests (I and II) were conducted on the same gallon jar of sample. Three specimens (1, 2, and 3) were obtained in the Series I test. Specimen I-1 was collected from the jar with no stirring. Specimens I-2 and 3 were obtained after the jar was vigorously shaken. Each specimen was divided into two portions (A and B). A total of six specimens were prepared (I-1-A through I-3-B) for analysis.

TABLE 5—*Sodium content of Waste P.*

Specimen	Sodium Content, weight percent
1-A	14.1
1-B	14.5
2-A	14.0
2-B	14.5
3-A	14.1
3-B	14.0

TABLE 6—*Sodium content of Waste Q.*

Specimen	Sodium Content, ppm
1-A	53
1-B	45
2-A	50
2-B	46
3-A	45
3-B	53

A 5-mL portion was removed from each specimen vial and diluted to a level suitable for AA analysis. The results are shown in Table 7. The data were highly variable, ranging from 975 up to 9000 ppm. The data were reproducible for the same diluted specimen and the data reported were the average of at least three AA measurement readings. The readings were within 0.5% of one another for the same diluted specimen.

In the Series II test, four specimens (1, 2, 3, and 4) were obtained from the same gallon jar as the Series I test. Specimens 1 and 2 were obtained from the top of the jar after vigorously stirring the sample for 5 min. Specimens 3 and 4 were obtained near the bottom of the jar after stirring the sample for 10 min. Each specimen was then divided into two portions (A and B). A total of eight specimens (II-1-A through II-4-B) were prepared. The results are shown in Table 7. Again, the data were highly variable, ranging from 1340 to 10 900 ppm.

To determine whether suspended solid particles were responsible for the variations, a 10-mL specimen was obtained and separated by filtration into solid and liquid portions. The liquid was further divided into two equal portions and the sodium content of the three specimens was measured. The two liquid portions reported 8000 and 4800 ppm. The solid portion contained an equivalent of 500 ppm of sodium based on the original 10-mL specimen be-

TABLE 7—*Sodium content of Waste R.*

Series I Test		Series II Test	
Specimen	Sodium content, ppm	Specimen	Sodium content, ppm
I-1-A	975	II-1-A	10 900
I-1-B	6 200	II-1-B	4 200
I-2-A	3 500	II-2-A	8 560
I-2-B	5 500	II-2-B	3 900
I-3-A	9 000	II-3-A	1 800
I-3-B	3 100	II-3-B	1 340
		II-4-A	2 400
		II-4-B	1 180

fore filtration. It was concluded that suspended solids were not the source of the variations and that sample heterogeneity at the microscopic level explained the variable data.

In order to overcome the nonhomogeneity of the sample, a statistical analysis strategy is suggested. ASTM's Practice for Manual Sampling of Petroleum and Petroleum Products (D 4057) specifies sampling requirements. Once a statistically sound sample is collected, a statistical analysis plan is required. One possible statistical analysis plan is proposed later in this paper.

Heat of Combustion

The ASTM Test for Heat of Combustion of Liquid Hydrocarbon Fuels by Bomb Calorimeter (D 240) is typically used to measure heat of combustion. Modifications are required for handling highly volatile, toxic, or reactive chemical wastes. A small plastic bag (which has been preweighed and precalibrated for heating value) is an ideal way to contain the specimen before it is burned in the oxygen bomb.

TGA and DSC tests

For some solid wastes, thermal gravimetric analysis (TGA) and differential scanning calorimetry (DSC) may be required. Both tests use milligram quantities of specimen, which are unlikely to be representative of a highly heterogeneous solid waste. Unless a large-scale instrument is developed, a statistical analytical plan similar to that proposed in this paper will be required.

Viscosity

The performance of a fuel spray nozzle depends strongly on the viscosity of the fuel (chemical waste). If the viscosity of the fuel is higher than 20 cst, it is strongly suggested that the fuel be heated until the viscosity is lower than that level. Another approach is to dilute the fuel with a thin solvent. Hence, there is a need to obtain the viscosity-temperature curve for a viscous waste. The ASTM Test for kinematic viscosity of transparent and opaque liquids (and the Calculation of Dynamic Viscosity) (D 445) is normally used. When chemical waste contains suspended particles (they normally do) or some emulsion, D 445 does not work. A new method or procedure is required; a rotational type viscometer may have to be used.

Proposed Statistical Analysis Plan (Not a Sampling Plan)

As discussed previously, heterogeneous wastes are difficult to characterize. The standard procedures require analyses of a specimen when the error due to the analysis is less than 1%. The data in Tables 5, 6, and 7 indicate that the sampling errors caused by the heterogeneity of the material range from 2 to 59% of the sample mean. This indicates that from a statistical point of view it is better to do one analysis on each of several specimens than to do repeated analyses on a single specimen. The number of specimens to analyze can be calculated if these four items are known: the mean, the standard deviation, the desired level of accuracy, and the acceptable confidence level (Type I error). The number of specimens required to achieve a desired 95% confidence level (Type I error = .05) for various coefficients of variation (ratio of the standard deviation to the mean) and accuracy levels appear in Table 8. The entries in this table are calculated by using the formula [3]

$$n = \left(\frac{z \times CV}{d}\right)^2$$

where n is the number of specimens which need to be analyzed, $z = 1.96$ is the appropriate standard normal critical value, CV is the coefficient of variation, and d is the desired level of accuracy, expressed as a percentage of the mean.

The required number of specimens increases as the heterogeneity increases or the level of accuracy decreases for a fixed confidence level. A 95% confidence level was chosen for the selection of the standard critical value ($z = 1.96$) because it represents a reasonable trade-off between correctly characterizing the waste with a very high confidence level—which is unfeasible in practice—and a lower level which incorrectly characterizes the waste more often. The number of specimens required (Table 8) is economically unfeasible in several instances, but those cases are ones in which the standard deviation is very high relative to the average to be estimated.

An efficient procedure for dealing with heterogeneous wastes is sequential. First four specimens are analyzed. Then, based on the results and the desired

TABLE 8—*Specimens necessary to characterize heterogeneous wastes.*

Coefficient of Variation[a]	Desired Accuracy		
	5%	10%	20%
0.05	4[b]	1	1
0.10	16	4	1
0.20	64	16	4
0.30	144	36	9
0.40	256	64	16
0.50	400	100	25

[a]Ratio of the standard deviation to the mean.
[b]Calculated assuming a 95% confidence level.

accuracy, a decision is made on whether or not to analyze more specimens. For example, if it is desirable to characterize a waste component to within 10% and the initial results give a coefficient of variation of 0.20, then an additional twelve specimens will need to be analyzed. When the required number of specimens is too large, then trade-offs between desired accuracy and the required number are necessary. For the previous example, as can be noted from the table, the initial four samples give an estimate that is within 20% of the mean. This accuracy level may be acceptable. For heterogeneous wastes this procedure can be summarized as follows. First, analyze at least four specimens. Then calculate the 95% confidence interval for the average of the four analyses. If the average and the interval are acceptable, further specimen analysis is unnecessary. If they are not acceptable, then analyze more specimens to achieve the desired accuracy. The procedure identifies difficulties involved in heterogeneous waste analysis and quantifies the risk of using small numbers of specimens.

Conclusions

Almost all the traditional analytical methods are developed for the analysis of homogeneous samples or a sample with relatively simple matrix. In the analysis of chemical waste, many of those methods are not applicable or experience problems because of sample heterogeneity. Interference resulting from a complex sample matrix is another common problem.

Special problems experienced and overcome in GC analysis of chemical waste include column contamination by heavy organics, inability of the state-of-the-art GC computer to analyze data for trace chemicals, and interference caused by complex sample matrices. For analysis of chlorine, phosphorus, and sulfur contents, the waste sample should be digested in an oxygen bomb first so that a simple sample matrix can be obtained. In some cases, a peroxide bomb may have to be used. The current methods for determining water and ash contents are inadequate and development of improved methods is required. New methods or procedures are also needed for heat of combustion and viscosity measurements for certain special chemical wastes. Some approaches have been suggested.

Much discussion has been devoted to the importance of a statistically sound sampling plan in collecting a representative sample for a chemical waste. Because of sample heterogeneity and the use of microscopic quantities of specimens in the state-of-the-art analytical instruments, a statistically designed analysis plan is also needed. This paper has proposed one approach for such a plan.

Acknowledgment

The authors wish to thank the analytical personnel at the Union Carbide South Charleston Technical Center for their help in conducting the analytical work discussed here.

References

[1] "Test Methods for Evaluating Solid Waste; Physical/Chemical Methods," SW-846, 2nd ed., U.S. Environmental Protection Agency, Washington, DC, July 1982.
[2] Harris, J. D., Larsen, D. J., Rechsteiner, C. E., and Thrun, K. E., "Sampling and Analysis Methods for Hazardous Waste Combustion," EPA 600/8-84-002, NTIS PB84-155845, National Technical Information Service, Springfield, VA, Feb. 1984.
[3] Dixon, W. J. and Massey, F. J., "Introduction to Statistical Analysis," McGraw-Hill, New York, 1969, p. 80.

Lawrence G. Doucet[1]

Diagnostic Test Burns and Considerations for Selecting Hazardous Waste Incineration Systems

REFERENCE: Doucet, L. G., **"Diagnostic Test Burns and Considerations for Selecting Hazardous Waste Incineration Systems,"** *Hazardous and Industrial Solid Waste Testing: Fourth Symposium, ASTM STP 886,* J. K. Petros, Jr., W. J. Lacy, and R. A. Conway, Eds., American Society for Testing and Materials, Philadelphia, 1986, pp. 314–320.

ABSTRACT: It is difficult to procure hazardous waste incineration systems that will meet demanding performance and regulatory requirements and still be cost-effective. Test burns have proven to be of great value in this regard. These consist of burning representative waste samples under controlled conditions to observe and measure reactions. Test burn data are useful for evaluating technologies and designing systems and equipment. Despite the many benefits, test burns are not frequently used. There are difficulties associated with conducting test burns, and very little has been publicized concerning test burn procedures, performance, and applications. Standards and guidelines are needed for test burns, possibly through a technical organization such as ASTM.

KEY WORDS: hazardous waste, incineration systems, test burns, combustion, incinerability, burning characteristics, waste characterization, calorimetry, proximate analysis, ultimate analysis evaluation

It is usually a difficult task to evaluate, select, and design hazardous waste incineration systems that will meet demanding performance and regulatory requirements and still be cost-effective. The difficulty is compounded by unknown factors and uncertainties relating to compliance with permit conditions, as well as associated financial risks should such compliance prove unobtainable. Published data and other sources of information and guidance do not adequately cover the detailed incineration system requirements for specific wastes generated by most industries.

An approach that has proven of great value in the procurement of hazard-

[1]Principal, Doucet & Mainka, P.C., Peekskill, NY 10566.

ous waste incineration systems is the performance of diagnostic test burns to identify incinerator operating requirements and develop design and construction criteria. Test burning generally involves the combustion of representative wastes under controlled conditions in order to observe, monitor, sample, and analyze reactions and the formation of by-products. Test burn programs can range in complexity from batch burning tests intended merely to observe reactions to highly sophisticated programs involving a matrix of testing using different incinerator designs, waste feeds, and combustion conditions. Test burns are normally conducted before selecting and designing incineration systems, as opposed to trial burns and performance tests, which are conducted following installation and start-up.

Despite the many potential benefits of test burns, they are not frequently used. Some test burns are poorly planned and improperly organized so that costs are excessive and the data are of little value. Also, test burn data may be misinterpreted, misapplied, or ineffectively used. These shortcomings are largely the result of difficulties associated with conducting test burns, as well as insufficient publications concerning test burn procedures, performance, and applications. The discussion that follows will highlight some of these issues.

Benefits

Test burns provide data that may not be available from other sources or obtainable by other means. First of all, they provide specific, qualitative information concerning the burnability or incinerability and the burning characteristics of wastes. This includes indications of how well the wastes will burn under varying incinerator loading and operating conditions, as well as identification of such factors as ignition and reaction rates, flame patterns and impingement, ash insulation effects and slagging tendencies, and the formation of undesirable by-products. These indications and factors are not predictable from observing or examining the wastes, and they cannot be derived from engineering calculations. Furthermore, waste characterization data, including calorimetry and proximate and ultimate analysis, provide minimal information concerning waste incinerability and reactions during actual incineration conditions.

Test burns also provide quantitative data that are representative of the actual planned incineration operations. These data include combustion air and supplemental fuel requirements; combustion temperatures; flue gas volumes and constituents, including wet and dry gas compositions; particulate emission rates and size distributions; and concentrations of hazardous organic compounds. Most of these parameters can be evaluated from combustion calculations and heat and mass balances. However, such calculations are typically based on "average" waste characterization data, loading rates, and combustion conditions. In actuality, their values may deviate widely, frequently, and unpredictably from these averages. For example, flue gas volumes may rapidly surge to more than 300% of average hourly rates immediately upon loading certain wastes into an incinerator.

Some parameters, such as emissions of specific hazardous organic compounds, cannot be calculated or predicted with any precision. Test burn data may be necessary to predetermine what incinerator operating conditions will limit such emissions to levels permitted by hazardous waste regulations. These levels are typically expressed in terms of hazardous waste destruction (and removal) efficiencies.

Test burns are initially useful for evaluating alternative incineration technologies and for determining the feasibility of innovative technologies. However, and more importantly, test burn observations and data are useful for selecting, sizing, and establishing criteria for constructing and operating all major components and equipment, including combustion chambers, waste and residue handling systems, flue gas handling systems, and heat recovery and air pollution control equipment. Test burns provide a data base for obtaining guaranteed performance from equipment suppliers, as well as assurance of successful compliance with regulations. Test burn data may not only greatly affect overall system performance and efficiencies, but could also result in major capital and operating cost savings.

Test Burn Facilities

An ideal test burn facility would be similar in type and capacity to the system under consideration, have a permit for burning the selected hazardous wastes, be operable under widely varying conditions, and be readily available for use. Unfortunately, such ideal facilities are extremely rare. In fact, it may be difficult to locate a test facility that satisfies any of these desirable conditions. One exception to this circumstance is the use of an existing incineration facility to make test burns with different wastes or operational changes considered for that particular system.

Sometimes incineration systems owned and operated by other firms may be identified as possible candidates for conducting test burns. However, these systems usually have limitations that restrict or prohibit their use. One major drawback is that normal operating schedules for such facilities typically make it difficult, if not impossible, to obtain sufficient time for conducting test burns. In addition, there are risks and liabilities associated with using another firm's incineration system for test burns, including the possibility of equipment damage or environmental endangerment, since the system may not be adequately designed or constructed for burning specific test materials.

Test burns are most commonly conducted at facilities owned or made available by incineration equipment manufacturers. Nearly all of the manufacturers have some type of test unit or facility available. Most of these are very basic designs, with minimal accessories and limited flexibility. Such test facilities are only suitable for accommodating very simple testing programs. Several incinerator manufacturers, however, offer highly sophisticated testing facilities, with provisions for combining numerous types and configurations of incinerator furnaces, heat recovery systems, quenchers, and air pollution control devices. The components typically are matched to fit together as a system.

Manufacturers of specialized and innovative incineration systems usually have specially designed facilities and provisions to test and demonstrate the capabilities of their equipment. These facilities are often used in conjunction with in-house design development operations.

There are several private companies, typically associated with stack testing laboratories, that offer test burn services in their own facilities. These firms are reportedly independent of incineration equipment manufacturers.

Some manufacturers of incineration system components, such as burners, nozzles, waste handling devices, and air pollution control systems, have equipment available for testing purposes. These can usually be adapted for testing in conjunction with other systems and equipment. For example, at least one flue gas scrubbing system manufacturer has a truck-mounted, small-scale system that can be connected to a slip stream of almost any incineration system for testing.

Most test burn facilities have a relatively small capacity. The majority are estimated to be less than about 1.06 GJ (10^6 Btu) per hour. A few manufacturers offer testing on bench-scale equipment suitable for burning only a few kilograms per hour. The larger, more sophisticated test burn facilities have equipment in the range of about 3.16 GJ to about 5.27 GJ (3 to 5 million Btu) per hour.

As indicated, test burn programs often involve several different testing facilities. For example, for evaluating technologies, testing may be conducted in both a rotary kiln and a fixed-hearth test facility. Also, for evaluating operating criteria, testing may be conducted in both a low-temperature pyrolyzing kiln and a high-temperature oxidating kiln.

Test Burn Programs

There are no standard procedures for conducting test burn programs. Instead, most test burns are conducted on a case-by-case basis, tailoring various standardized procedures for burning specific wastes in a selected test facility. Careful planning, organization, and coordination are essential for obtaining needed data at least cost and within an acceptable time period. Conversely, inadequate planning or inattention to details could result in the compilation of useless data at a great expenditure.

Selecting Waste Materials

The first step is to identify and select wastes for test burning. This choice is particularly complex for industries that generate a wide variety of waste types and mixtures. Although it may be desirable to obtain test burn data on every individual waste and possible combination of wastes, this would most likely be prohibitively expensive. On the other hand, the omission of even a few difficult-to-burn materials could leave significant data gaps.

Wastes are usually selected for test burning because they appear to have poor incinerability or adverse burning characteristics. They are sometimes selected because other data and analysis indicate possible burning reactions and formations for which additional data may be required. However, as noted above, these factors are not always obvious. In fact, they may only become evident through the test burns themselves. Thus it is easy to overlook difficult wastes in the selection process.

Establishing Procedures

The second step in a test burn program is to establish actual data collection procedures. There are literally dozens of parameters that can be sampled and analyzed for each waste tested. For each of these parameters there are at least several sampling techniques available. These methods typically cover a wide range of sophistication, precision, and cost. In addition, sampling duration and frequency can range from continuous to instantaneous, one-shot grab samples.

It is usually prohibitive in cost—and seldom necessary—to sample every possible parameter for each waste. Specific choices, including sampling method, frequency, and duration, must be carefully evaluated in terms of data requirements, budgetary constraints, and regulatory demands.

Qualitative data collection also requires careful planning and coordination. Observations and monitoring of factors such as flame appearance, color changes, volume changes, explosive reactions, smoldering, and stack appearance, must be documented in an organized manner. Such data must also be correlated with sampling and analysis data.

Additional Considerations

Besides the difficulties often associated with locating a test facility, selecting representative wastes, and establishing testing procedures, there may be other relevant considerations. Some of these are listed in the following paragraphs.

Facility Availability

The better, more sophisticated test burn facilities are in relatively high demand. Some of these may have previous commitments so that their equipment is unavailable for as long as six months or more.

Data Restrictions

Several equipment manufacturers may attempt to restrict the use of data obtained at their facilities. In fact, a signed affidavit may be required that prohibits using the data for design development and competitive bidding.

High Costs

Most of the simple, basic test burn units may be leased for a nominal charge of about $1000 per day. The better, more sophisticated facilities may be leased at a nominal charge upwards of about $10 000 per day, plus a separate, additional charge for equipment setup. Several highly specialized systems have been offered for testing at even higher costs. However, some manufacturers may deduct all or portions of the test facility usage costs should an order be signed to purchase equipment within a certain period after testing.

Costs for sampling and analysis are separate and additional to facility usage costs. These are highly variable and dependent on the type and number of parameters sampled, methods used, numbers of wastes and test runs, and test facility provisions and limitations. These costs can easily be at least several times facility usage costs. Sampling and analysis costs for a nominal two-day test for a few wastes at a single facility could easily range upwards of $50 000.

Data Interpretation

Proper interpretation and application of test burn data are difficult. Typically, specialized expertise is required, and the data should be analyzed by those responsible for ultimate system performance. Data obtained from typical test burn facilities must be scaled upwards many times in relation to planned incinerator operations. Also, there may be little similarity between the test facility and the planned installation. Similarity and scale-up data are sometimes difficult to extrapolate and may be misleading.

Future Needs

Test burns are an important adjunct in the overall field of hazardous waste incineration. However, in light of the above considerations, a lot of work is needed to encourage their use and improve their benefits. The major effort appears to be needed in the development of standards and guidelines for the selection of wastes, testing facilities, and procedures. Guidelines are also needed for interpreting and applying test burn data.

Test burn standards and guidelines might best be developed under the aegis of a technical organization such as the ASTM. This organization could solicit, coordinate, and consolidate the necessary support and inputs from all sectors, including manufacturers, designers, and owners of hazardous waste incineration systems. In addition, it could serve as the clearinghouse for the documentation, publication, and dissemination of information on test burn activities, innovations, and developments.

Joseph J. Santoleri[1]

Test Burns: A Critical Step in the Design of Hazardous Waste Incineration Systems

REFERENCE: Santoleri, J. J., "**Test Burns: A Critical Step in the Design of Hazardous Waste Incineration Systems,**" *Hazardous and Industrial Solid Waste Testing: Fourth Symposium, ASTM STP 886*, J. K. Petros, Jr., W. J. Lacy, and R. A. Conway, Eds., American Society for Testing and Materials, Philadelphia, 1986, pp. 321-334.

ABSTRACT: With the final implementation of the Resource Conservation and Recovery Act covering the disposal of hazardous wastes by incineration, certain basic criteria have been established. Most critical is the ability to reach levels of 99.99% destruction and removal efficiencies (DRE) of the principal organic hazardous constituents (POHC). In order to be assured that this is possible, a pilot test burn is necessary. If these wastes are presently being burned in a similar incineration system, the data can be used in the permit application.

However, there are many wastes which have been disposed of by landfill, land treatment, and chemical treatment that still present a hazard to our environment, and high-temperature oxidation is the only sure method of reaching destruction to four nines.

Test burns provide information for the proper design of the oxidizer related to air flow, fuel flow, oxidation temperature, mixing technique to achieve turbulence, atomizer type, residence time, and resultant emissions. Other problems inherent in handling the waste—such as pumping, piping, valving and control, atomization fluid and pressure, and atomizer location—are also encountered in the test burn and provide insight in the design of the full-scale system.

Certain control techniques which may be necessary for handling a variety of wastes are encountered in the test burn and provide the opportunity to check the effects of the unit operation.

The need for proper design and operation of a test burn facility will be discussed. Problem test burns will be reviewed and related to the effect on scale-up of full scale operating units. When a process changes, resulting in a different waste product, the pilot unit will permit testing and optimization of design without interruption of the full-scale system.

KEY WORDS: incineration, hazardous wastes, test burns, trial burns, turbulence, oxides of nitrogen, carbon monoxide, combustion, atomization, high-intensity combustion

Before entering into a design and construction contract for a hazardous waste incinerator facility, the generator or waste disposal firm must be aware

[1]Principal consultant, Four Nines, Inc., Plymouth Meeting PA 19462.

of the most recent legislation covering the requirements of owning and operating a treatment, storage, or disposal facility. The Resource Conservation and Recovery Act (RCRA) of July 1, 1982 [1], gives the detailed requirements for the design and operation of the facility. It is very important that the waste characteristics be well defined, as outlined in Table 1. These data will enable the designer, as well as the operator of the equipment, to understand better the problems that need to be overcome in meeting all of the requirements of RCRA.

In many cases the data indicate that the design of the system can be based on past experience with a similar waste material. However, with many of the wastes listed in Part 261, Appendix VIII of RCRA, there may be first-time applications. These would include the problems of material handling, flow control, atomization, combustion, and total oxidation. Many of these problems can be solved in bench-scale tests, which provide a method for determining individual solutions to the problems listed above and others identified by the process engineer, design engineer, or operator. However, when one is attempting to integrate these individual solutions into a working system, one proven solution is to set up and operate a complete pilot unit. Such a unit will bring to light problems that might never be considered on paper but could cause emergency situations in a full-scale system.

Pilot Plant Equipment Requirements

A pilot plant should be designed to provide flexibility in all areas of system operation. A typical liquid waste incineration system consists of the major components listed in Table 2.

During operation, a pilot system will point out problems that can be expected. Modifications can be made to a pilot system more economically than with a full-scale commercial industrial system. Modifications can be made to the materials of construction, piping design, pump selection and speed, atomizers, burner location, and the like. Critical to the total operation of the

TABLE 1—*Waste liquid fuel data.*

Chemical composition
Density
Heat of combustion
Viscosity
Corrosivity
Chemical reactions
Polymerization
Solids content
Ash reaction—refractories
Slag formation
Combustion gas analysis
Nitrogen composition (NO_x)

TABLE 2—*Major components of typical liquid waste incineration system.*

Storage—tanks or drums	Combustion air systems
Pumps	Incinerator configuration
Filtration	Slag collection and disposal
Piping	Refractory design
Control valves	Heat recovery
Metering	Quench—cooling
Atomization	Scrubbers
	Stack

pilot facility is the requirement of RCRA for the destruction and removal efficiency (DRE) of the principal organic hazardous constituents (POHC). As we all know, with hazardous wastes the EPA, as well as most states, requires 99.99% DRE of the POHC. In order to ensure that this requirement is met, methods must be available to analyze the waste for its POHCs. There are many articles describing methods; the U.S. Environmental Protection Agency (EPA) has also provided "Sampling and Analysis Methods for Hazardous Waste Incinerations, issued in February 1984. This allows measurement of not only the waste but also the emissions in the quench zone, the slag discharge, and the stack emissions. A proper analysis of POHC in waste-in and all detectable POHC in all emission points will determine the ability of the system to meet 99.99% DRE.

Pilot Test Design Information

A pilot test burn allows plant operators to determine what is needed in the design stage to minimize downtime and reduce maintenance costs. It can also indicate when certain wastes should be separated to minimize slagging or emission problems. Table 3 lists key items in a system that will point out where these critical conditions will occur.

Why a Pilot Test Burn?

A pilot test permits a small quantity of the waste to be tested. Test engineers recognize the disadvantages of using small samples. Long-term effects on the refractories, nozzles, and pumps cannot be determined. Scale-up to full size is also of major concern. A pilot test will indicate that total auxiliary heat input–to–waste ratio is much higher than the heat and material balance calculations for the full-scale system. This is due to a greater heat setting loss in a pilot unit in relation to total heat released. However, experienced operators of pilot plant facilities accustomed to the results of a scale-up design obtained from the pilot test are able to predict the needs of a full-scale unit. Pilot test burns are needed to:

1. Obtain heat release data to determine auxiliary fuel requirements.
2. Obtain effect of temperature, residence time, and mixing on DRE.
3. Determine effect of excess air rate on DRE.
4. Compare DRE to emissions data on carbon monoxide and hydrocarbons.
5. Quantify stack emission rate through particulate sampling (air pollution control equipment selection and permit data).
6. Determine emissions into quench system, as well as stack gases.
7. Determine stack appearance (opacity).

These tests provide management information as to the combustibility of the waste, the composition of ash emissions into the quench system and stack, and the practicality of heat recovery, as well as potential by-product recovery.

In some cases, a pilot test burn is also used to determine the problems that may arise if a new process waste is introduced into an existing incinerator. Plant operations may elect not to attempt the test burn in an existing full-scale incinerator. If the test burn is not properly controlled, the operating unit may be damaged. The plant may create a nuisance or serious emissions problem which leads to shutdown of the operation. A pilot test burn in a similarly designed unit will provide the data and the conditions necessary for safe and legal operation of the full-scale system.

Test Burn Combustion Data

When one is conducting test burns on a hazardous waste, it is extremely important to understand the critical data needed to determine the overall effectiveness of the combustion system. It is also important to project the operating costs of the full-scale system, which will include fuel costs, horsepower, and utilities (water, compressed air, steam). A second purpose of the test is to determine the minimum operating cost that will permit the system to reach the required destruction efficiencies. A pilot system will permit operation with different injector designs, pressures, and atomizing fluids; minimum to maximum excess air levels; a range of incinerator temperatures; and a turndown range to provide varying residence times will permit the test engineer to establish the effects each will have on the overall destruction and emission levels [2].

A variation from 10 to 50% excess air will have an effect on the minimum heating value of the waste fuel that can be burned directly in the combustor [3] (see Fig. 1). It also has a direct effect on the horsepower requirements of the combustion air fan because of the increased volume of air, as well as the increased pressure drop across the system. If a venturi scrubber is needed to minimize particulate emissions, and the pressure drop necessary to reach stack grain loading is determined to be 127 mm (50 in.), the increased volume of excess air will force a larger flow area for the scrubber, resulting in in-

TABLE 3—*Key items in a pilot test burn system.*

Component	Key Items to Consider
Storage tank	mixing (internal mixers, external recycle pumps, air or steam spargers), venting, solids buildup, bottom discharge and design (cone or dished head)
Pumping and transport	pump type (gear, screw, positive displacement, progressive cavity), pump speeds, pump materials, operating pressure and temperature, filtration, pipe size (minimum diameter to avoid bridging or blockage), elbow design (long radius, 90° bends, or tees), trap location
Valves (control and block/ safety shutoff)	ball, plug, etc.; materials (erosion, corrosion), necessity of control
Metering equipment	flow meter (magnetic, bellows type, venturi, orifice), minimum flow volume due to settling of solids, rotameters—very viscous and unclear liquids prevent their use
Nozzles	type (hydraulic, two-fluid); internal or external atomizer; steam, air, or other gas; materials of construction (corrosion, erosion); ability to handle solids; droplet characteristics; position in burner or incinerator chamber; maximum flow rate
Burner	type (high-intensity), flame characteristics (laminar, turbulent), position in chamber
Secondary air	volume, inlets to incinerator, velocity and swirl (effect on combustion characteristics)
Incinerator chamber	volume, velocity, refractory construction, orientation (vertical or horizontal), injector locations, burner location, secondary air inlets, sample connections, thermocouple (locations and type), maximum temperature
Heat recovery	temperature limits (maximum, minimum), gas composition, solid content ash (dry, sticky, molten), materials of construction, dew point of gases, gas-to-gas or gas-to-liquid heat exchanger
Scrubber (type: packed column, spray quench, venturi, Sub-X)	composition of products (gaseous and particulate), method of quench (spray, flooded, or submerged), materials of construction (refractory lining—hot/cold interface, fiber-reinforced plastics, rubber-lined steel, alloys—corrosion resistance), scrubber solution (water, caustic, lime), reaction products (neutral—salts; reactive—sodium hypochlorite), efficiency, recycle and bleed or once-through, total organic carbon (TOC) or quench discharge, chemical oxygen demand (COD) of quench discharge
Stack	overall height, materials of construction, sample ports

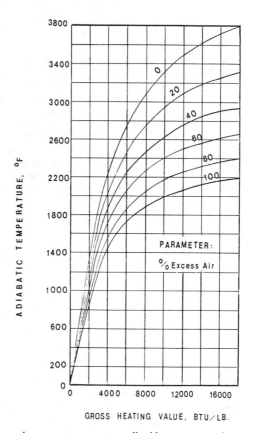

GROSS HEATING VALUE, BTU/LB.

FIG. 1—*Heating value versus temperature—liquid wastes at varying excess air contents.*

creased capital and operating costs. This applies to all of the major items of equipment in the system (see Fig. 2).

Table 4 shows recent test burn conducted in the Trane Thermal laboratory facility. Major objectives of the test burn were to determine the effects of temperature, excess air, and residence time on the stack emissions and scrubber effluent during burning of the subject waste. It was recognized that the waste had a high percentage of fuel-bound nitrogen which would result in very high levels of emissions of oxides of nitrogen (NO_x) under normal incineration conditions. The test program agenda was established to reach the following objectives:

1. Evaluate the system under normal incineration conditions.
2. Demonstrate that NO_x can be reduced to an acceptable level.
3. Determine carbon monoxide (CO), hydrocarbon (HC), and NO_x emissions as a function of excess air and incineration temperature (temperature variation controlled by injecting quench water).

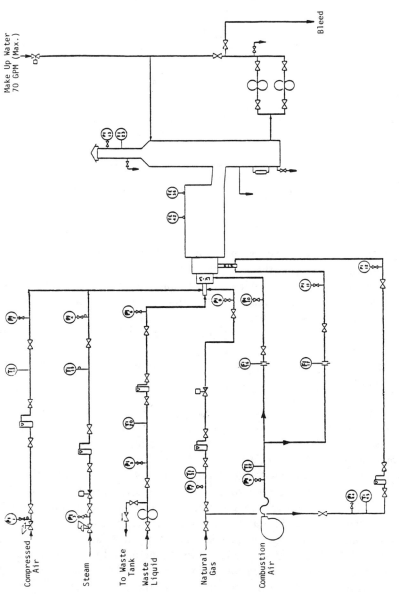

FIG. 2—*LV-3 test incineration system (Trane Thermal Laboratory).*

TABLE 4—*Components of waste used in test burn.*

Component	Percent by Weight	EPA Hazardous Waste Designation[a]	Heating Value, kcal/g
Hexane	39.2	UN 1208 (I)	...
Formamide	22.3	...	...
Ethanol	7.3	U 173	...
MEK (methyl ethyl ketone)	6.8	U 159 (I, T)	8.07
Toluene	5.0	U 220	10.14
Pyridine intermediates	4.7	U 196	7.83
Aluminum chloride	3.8	NA 9085	...
Adenine	2.9	...	...
Water	2.5	...	...
DCFT (Dichloro fluorotoluene)	1.5	...	...
Aniline	1.5	U 012 (I, T)	8.73
Aliquat 336	1.4	...	...
Other organic compounds	1.1	...	...

[a]Listings are those in DOT/EPA Hazardous Materials (CFR49 Part 172) Subpart B, and EPA identification and listing on Hazardous Waste (CFR 40 Part 261); I indicates ignitable, T is for toxic.

4. Determine emissions of specific compounds (aniline and dichlorofluorotoluene [DCFT]) in the stack gases.

5. Determine the quality of the scrubber water; analyze for aniline, DCFT, chemical oxygen demand (COD) and total organic carbon (TOC).

6. Determine the best injector for use with this waste.

7. Determine areas of operating problems in handling, pumping and atomizing waste.

8. Determine the higher heating value of the waste using the bomb calometer.

The pilot used for the test burn was the Model LV-3 incinerator with a refractory lined packed tower scrubber (Fig. 3). Proper instrumentation was selected and calibrated for measuring flows, temperatures, and pressures throughout the system (Fig. 2).

Stack gases were sampled through proper sample conditioning to provide a direct readout of CO, carbon dioxide (CO_2), oxygen (O_2), NO_x, and HC. The EPA Method 5 for Particulate Sampling was conducted by an independent laboratory and analyzed for aniline and DCFT. Quench scrubber water samples were also taken by an independent laboratory and analyzed for aniline, DCFT, COD, TOC, and suspended solids (mineral and volatile). The data obtained from these tests are shown on the curves in Figs. 4 to 9.

The triggering level for principal source discharge (PSD) of NO_x by present U.S. EPA regulations from a specific incinerator system is 36.3 metric tons a year (40 tons a year). In Fig. 4 this level is indicated by a conversion rate of 3.8% of fuel nitrogen in the waste sample. Note that this occurred at the

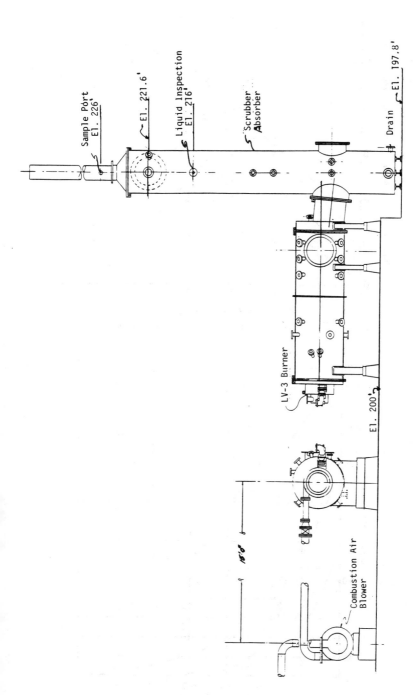

FIG. 3—*LV-3 test incinerator (Trane Thermal Laboratory).*

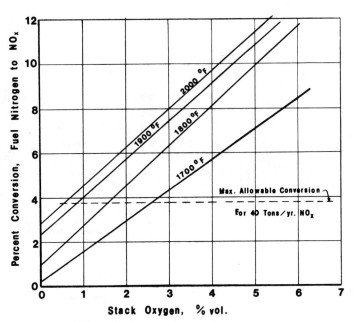

FIG. 4—*Fuel nitrogen conversion versus oxygen.*

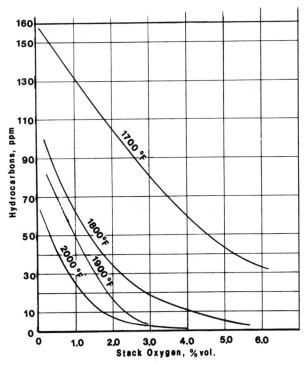

FIG. 5—*Hydrocarbons versus oxygen.*

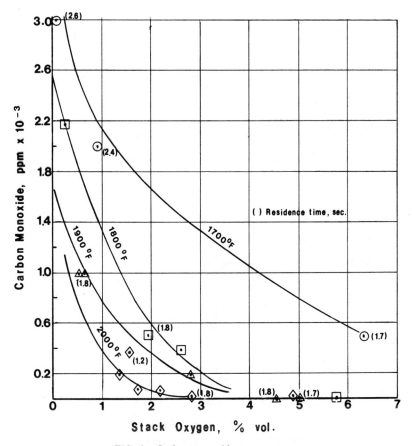

FIG. 6—*Carbon monoxide versus oxygen.*

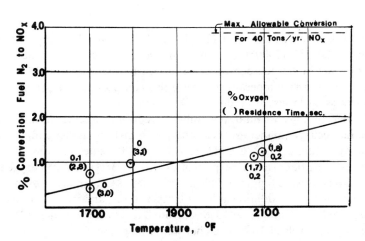

FIG. 7—*Fuel nitrogen conversion versus temperature.*

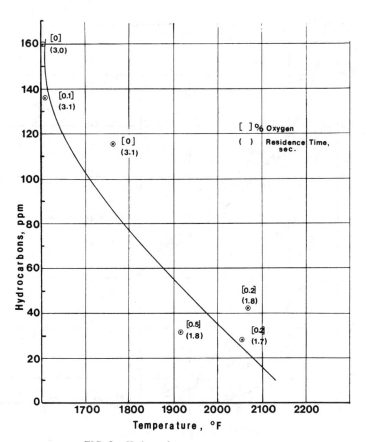

FIG. 8—*Hydrocarbons versus temperature.*

lower levels of oxygen content (2.5% and below). As the temperature increased to 1093°C (2000°F), the oxygen level also had to be reduced. This indicates that higher temperatures will increase the conversion of fuel nitrogen to NO_x. Lower temperatures reduce this conversion. Operation at minimum excess air levels also reduces NO_x formation.

In reviewing Figs. 5 and 6, the ability to reduce hydrocarbons and carbon monoxide occurs at higher excess air (oxygen levels) and higher temperatures. At a temperature of 1093°C and 0.35% oxygen, the stack emissions were 50 ppm of HC (methane) and 1000 ppm of CO. At 1093°C and 2.0% oxygen, the stack emissions were 8 ppm of HC and 60 ppm of CO. In order to maintain these same levels of oxidation at lower temperatures, an increase in excess air was required; for example, at 1038°C, from 2.5% to 3.3% oxygen. At 927°C excess oxygen levels of 10% or higher would be required.

Figure 7 shows the effect of the operating temperatures at excess oxygen

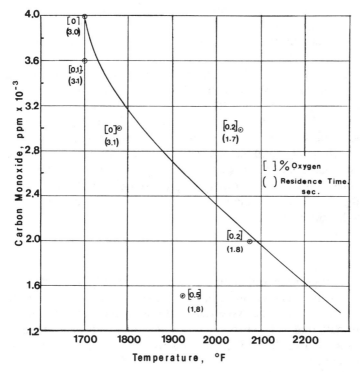

FIG. 9—*Carbon monoxide versus temperature.*

levels of 0 to 0.5% on fuel nitrogen conversion. Again note the increase in fuel nitrogen conversion as temperature levels are increased.

Figures 8 and 9 indicate the high levels of hydrocarbons and CO that exist at these low levels of excess air (oxygen).

The results of these tests point out that the full-scale system will require operation in the primary chamber at very low levels of excess air at temperatures in the range of 1093°C or below. Secondary air may then be introduced to increase the oxygen level in the system and permit oxidation of the hydrocarbons and carbon monoxide to CO_2 and H_2O. Controlling NO_x requires this type of operation.

The data indicated the amounts of aniline and DCFT in the stack sample to be <6 μg/L and <1 μg/L, respectively. The scrubber waste sample indicated <3 μg/L and <0.5 μg/L. Because of the method of analysis used, this was the lowest detection limit of the instrument. The test results indicate the DRE of the POHC was better than 99.99%.

In this particular test, problems occurred with plugging of the control valves in the waste piping system. The nozzle selected had no internal orifices and used external atomization with steam. This ran smoothly with no plugging problems. The heating value of the waste was determined to be 5133 kcal/kg (9240 Btu/lb) by the ASTM Test for Gross Calorific Value of Solid Fuel by the Isothermal-Jacket Bomb Calorimeter (D 3286-82). This was checked by mass balance of the combustion air used for oxidizing the waste and found to be within 5%.

Scale-up to a commercial-size unit was made possible as a result of the pilot test. Care was taken in the waste handling section design to prevent any plugging or blockage of the lines. The full-scale system was designed with two-stage combustion with secondary air introduced downstream of the primary chamber. The instrumentation supplied with the full-scale system permits control of the oxygen and temperature in the primary zone to minimize NO_x emissions. Secondary air was controlled to permit sufficient oxygen at the proper incineration temperature to keep the stack emissions at a level of less than 50 ppm CO and 15 ppm HC. At these levels, tests indicated better than 99.99% DRE.

Summary

A pilot test program will minimize the investment required in selection of a full-scale incineration system. It will provide the operator and owner with sufficient data to design the system better to meet final requirements. It also provides data that will be necessary for discussions with state environmental agencies as well as the U.S. EPA. This will be the basis of design for the construction permit.

Pilot tests also provide the basis for the trial burn, which will have to be conducted before final approval of the operating permit. If an existing unit is unavailable to provide this information, a pilot test is the next best answer. Data resulting from the pilot test burns, as well as trial burns from full-scale units, are being compiled by the U.S. EPA. Hopefully, within a ten-year period, enough information will be available for full-scale systems to be designed in confidence that they will meet all regulations. We are not yet at that point.

Pilot units are also valuable in determining the effect on the incinerator of a change in process conditions, which may produce a new waste stream. It is much safer to test with the pilot system than to risk a problem on the full-scale unit.

References

[1] Resource Conservation and Recovery Act, Standards for Owners/Operators of Waste Facilities: Incinerators, 40 CFR 264, RCRA 3004, *Federal Register*, Vol. 47, No. 122, 25 Jan. 1981, revised 1 July 1982.
[2] Santoleri, J. J., "Hazardous Liquid Waste Disposal: Incineration—The Ultimate Solution," National Conference on Hazardous Wastes and Environmental Emergencies, Hazardous Materials Control Institute, Houston, TX, 12–14 March 1984.
[3] Kiang, Y. H., "Incineration of Hazardous Organic Wastes," National Waste Processing Conference, American Society of Mechanical Engineers, New York, 1980.

Larry D. Johnson[1]

Development of the Volatile Organic Sampling Train for Use in Determining Incinerator Efficiency

REFERENCE: Johnson, L. D., "Development of the Volatile Organic Sampling Train for Use in Determining Incinerator Efficiency," *Hazardous and Industrial Solid Waste Testing: Fourth Symposium, ASTM STP 886,* J. K. Petros, Jr., W. J. Lacy, and R. A. Conway, Eds., American Society for Testing and Materials, Philadelphia, 1986, pp. 335–343.

ABSTRACT: Development and field application are described for the volatile organic sampling train (VOST) for testing incinerator efficiency. The concepts behind the VOST, several stages in its development, and the current status of this technology are discussed. Early experience indicates that the VOST works well in the field and yields good results when used with adequate care and precautions.

KEY WORDS: hazardous wastes, volatile organic compounds, stack sampling, sample contamination, emissions testing, purge-and-trap tests

The development and field application of the volatile organic sampling train (VOST) has progressed unusually quickly—to address an unusually difficult task. The VOST has generated a high degree of interest in the industrial and contractor sectors, primarily because of its application to incinerator trial burns, but also because it fills a void in the stack sampler's arsenal.

This paper presents the concepts behind the VOST, describes several stages in its development, and offers a summary of the current status of VOST technology.

Background

Although an improved stack sampling procedure for volatile organics has been needed for years, the situation only became critical in relation to hazard-

[1]Research chemist, Air and Energy Engineering Research Laboratory, U.S. Environmental Protection Agency, Research Triangle Park, NC 27711.

ous waste incineration. A reference document designed to assist in planning or review of sampling and analysis programs for hazardous waste incinerators or related combustors has been produced by the U.S. Environmental Protection Agency (EPA) [1]. That document recommends the Modified Method 5 (MM5) train or Source Assessment Sampling System (SASS) for collection of organic compounds with boiling points greater than 100°C. The sorbent breakthrough characteristics associated with lower boiling organic compounds make the use of these trains inadvisable for quantitative collection of compounds boiling below 100°C. Early drafts of Ref 1 recommended the use of either plastic sampling bags or glass sample bulbs for collection of such compounds. These recommendations were based on the assumption that at least 1000 μg/g of the material was present in the waste feed, and that the destruction efficiency of interest was 99.99%. In early 1982, an urgent need was identified by EPA R&D engineering programs and the Office of Solid Waste. A number of projects had been identified where volatile principal organic hazardous constituents (POHC) were of interest in waste feed at 100 μg/g concentrations, and it was necessary to characterize destruction efficiencies of the order of 99.999%. At such low concentrations of POHC in the stack, glass sampling bulbs are useless and plastic bags are only effective if combined with a concentration technique. Although it is possible to concentrate organic compounds collected in bags by the use of sorbent tubes or other techniques such as cold trapping, it is not recommended. There are drawbacks to any sampling technique; when two techniques (such as bag sampling of the source and sorbent sampling of the bag) are applied sequentially, the problems are likely to be compounded.

Since the need for a new sampling approach was apparent, and time was an important factor, a meeting of EPA and contractor scientists and engineers was held in April 1982 to discuss the problem. As a result of that meeting, the basic concepts and train arrangement for the VOST were agreed upon.

Basic Concepts

The schematic diagram of the VOST is shown in Fig. 1. The active elements for collection of volatile organic compounds are the two sorbent cartridges, although some compounds may also be collected in the condensate under certain circumstances. The first sorbent tube is filled with Tenax® porous polymer as the primary collection agent, while the second tube contains Tenax and activated charcoal as an extra precaution against breakthrough of the more volatile materials. The glass sorbent tubes are 100 by 16 mm and hold 1.6 g Tenax or 1 g Tenax plus 1 g charcoal. These tubes were chosen because they were the largest of the commercially available ambient air sampling tubes and were compatible with existing desorption equipment. Larger tubes allow a larger sample volume before breakthrough occurs. Tubes much

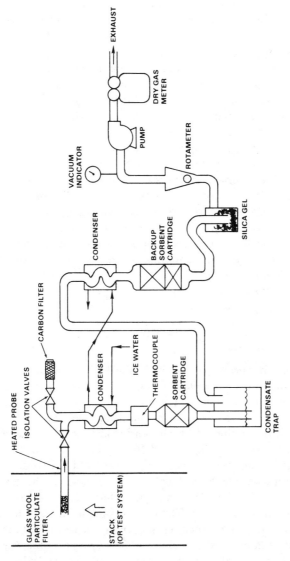

FIG. 1—*Schematic drawing of the VOST.*

larger than those chosen, however, become difficult to heat-desorb in an effective manner, and would require a new desorber design.

The VOST operates at 1 L/min while the sorbent tubes are maintained at 20°C or less during sampling. It is important that this temperature be maintained at all times, since higher temperatures may cause breakthrough of the more volatile organic compounds. Six sets of 20-min (also 20-L) samples are collected over a 2-h sampling period. The first set is analyzed as a "range finder." If adequate amounts of the material of interest are found, then the other five sets are also analyzed in an identical manner. In some instances only three sets of tubes are analyzed, and others are retained temporarily as possible backup samples. In the event that the first analysis detects no compound of interest, the remaining five sets are desorbed and recollected onto one Tenax tube. Subsequent analysis of that tube gives an effective detection limit five times as low as the single set analysis.

The analysis procedure used with the VOST consists of heat desorption of the Tenax tubes into a small water vessel, followed by analysis of the water trap according to EPA Method 624—Purgeables. EPA Method 624 is a purge-and-trap procedure followed by gas chromatography/mass spectrometry. The water trap is included since the Tenax cartridges are often very wet and cause problems if desorbed directly into the gas chromatography column. Several attempts have been made to improve this part of the procedure, but none have succeeded as yet.

For further details concerning operation of the VOST or the associated analysis system, see Refs 2, 3, and 4.

Initial Laboratory Studies

The first laboratory evaluation of the VOST system was assigned to Midwest Research Institute. A synthetic gas generation system was constructed that was capable of producing gas streams fortified with appropriate levels of test compounds. The compounds chosen for evaluation were vinyl chloride, carbon tetrachloride, trichloroethylene, and chlorobenzene. The concentration levels tested were 0.1, 1.0, 10, and 100 ng/L. A series of ten test runs were made. The test matrix included runs at each of the four concentration levels, replicate tests at one level, blanks, and a run where the gas included hydrochloric acid (HCl). The HCl run was included since many incinerator emissions contain this compound. Three trains sampled the gas manifold system during each run. The average results from three trains from the experimental series are summarized in Table 1. The single-pair data were derived from analysis of individual pairs of cartridges, one Tenax and one Tenax/charcoal. The data for combined pairs result from desorption, recollection, and subsequent analysis of multiple pairs following the procedure described earlier. No combined-pair experiments were carried out at the higher concentration levels, since that mode of operation was designed for low concentra-

TABLE 1— *VOST recovery efficiencies, initial studies.*

Expected Concentration, ng/L	Recovery, %			
	Vinyl Chloride	Carbon Tetrachloride	Trichloroethylene	Chlorobenzene
Combined Pairs				
0.1	111	176	97	104
1.0	48	88	108	131
1.0	146	113	105	115
1.0[a]	79	110	105	96
Single Pairs				
0.1	79	221	79	95
1.0	63	47	116	153
1.0	85	55	115	105
1.0[a]	95	40	95	90
10	142	70	109	106
100	43	108	132	101

[a]Included HCl.

tions. The combined-pair data appear to exhibit somewhat less deviation from the desired results, but most of the recoveries fall within ±50%. It is important to note that the presence of HCl did not produce a significant effect in the recovery levels of the organic compounds tested. Some of the variability in the carbon tetrachloride results is believed to result from blank variability, which was improved for later applications by better cleaning, storage, and handling procedures. Vinyl chloride is quite difficult to handle because of its very low boiling point, and some of the data presented in Ref *3* indicate breakthrough of the compound at higher concentration levels. The major conclusion from the first VOST laboratory study was that the concept was sound and that the train and subsequent analysis procedures were capable of producing the data for which they had been designed. Further details of the laboratory study are presented in Refs *3* and *4*.

Initial Field Application

Immediately upon completion of the laboratory study, Midwest Research Institute was faced with field application of the train as part of an engineering study for EPA's Hazardous Waste Engineering Research Laboratory in Cincinnati in support of the Office of Solid Waste. The train was repackaged into a more compact and rugged arrangement and taken to the field. Difficulties were encountered on the earliest jobs in this series, primarily from high blanks and field contamination. These were soon brought under control, but certainly emphasized that sampling low concentrations of volatile organics in contaminated surroundings can be very difficult indeed. Severe contamination was encountered with bag samples as well; clearly, adequate field blanks are just as necessary with bag sampling as with VOST.

During the first series of field tests of the VOST, the equipment was operated at two different sampling rates and two different total volumes were collected. The low flow rate operation, designated slow VOST, collected 5 L of gas at 250 mL/min. The stack concentrations obtained by the two methods compared very favorably for carbon tetrachloride, chloroform, and tetrachloroethylene. The results for trichloroethylene were less definitive [4]. Comparability of the slow VOST and original VOST operating procedures is important because it shows that breakthrough is generally not occurring and that the equipment may be operated successfully in more than one mode.

Further Evaluation of the VOST

The next major project using the VOST was emissions testing for the incinerator ship Vulcanus II [5,6]. The VOST used in this project employed the inside-inside cartridge design, and was constructed for Radian Corp. (formerly TRW Corp.), Research Triangle Park (RTP), NC, by Nutech Corp. (RTP). The inside-inside (I/I) cartridge differs from the inside-outside (I/O) cartridge in that the ends of the glass tube are drawn down to 6.4 mm ($\frac{1}{4}$ in.) in order to accommodate gas chromatography ferrules. During the heat desorption phase of the analysis, purge gases travel only through the inside of the I/I cartridge but flow over as well as through the I/O tube.

Before the application of the VOST to the Vulcanus II project, Radian Corp. carried out a laboratory evaluation not unlike the earlier one performed by Midwest Research Institute. A synthetic gas stream was generated and sampled by the I/I-style VOST. The data in Table 2 are derived from more detailed results given in Ref 5. It appears that the precision, and probably the accuracy, of these data are improved over those in Table 1. The two reasons that seem to best explain the improvement are that much had been learned about control of contamination and blanks, and perhaps that the I/I design was inherently more resistant to contamination. At any rate, the data in Table 2 strongly support the conclusion of the initial laboratory study: that the VOST concept is sound and will perform as originally intended.

The first known application of the VOST to sampling highly water-soluble compounds was carried out by EPA's Air and Energy Engineering Research Laboratory at RTP. The VOST was used to sample methyl vinyl ketone and tetrahydrofuran emissions from a pilot-scale fluidized bed combustor. This particular VOST was of the I/O type and was constructed for EPA by Envirodyne Engineers following the Midwest Research Institute design. In theory, highly water-soluble compounds are more likely to penetrate the first sorbent cartridge and be found in either the condensate or the second sorbent cartridge. It is necessary to perform spiking and recovery experiments in order to determine whether the compounds of interest may be analyzed by the purge-

TABLE 2— *VOST recovery efficiencies, further studies.*

Compound	Expected Value, ng/L	Recovery, %
1,1-Dichlorethane	54	105
	77	90
Chloroform	69	147
	118	98
1,2-Dichlorethane	77	95
	80	98
Carbon tetrachloride	73	91
	74	70
Trichloroethylene	67	99
	96	97
1,1,2-Trichloroethane	66	78
	94	68
	75	120
Tetrachloroethylene	107	105

and-trap method. In the event they are not, some other method (such as direct injection or high-performance liquid chromatography) will be necessary for analysis of the condensate.

In the project described above, the two compounds of interest were both quantitatively recovered by a slightly modified version of the purge procedure. Longer purge times were used but total volume was kept the same in order to avoid breakthrough. The two compounds were found primarily on the first sorbent cartridge in spite of their relatively high water solubility. All equipment and procedures appeared to work well even though this was the EPA crew's first experience with them.[2]

Current Status

The VOST is now available commercially, and ownership and field experience have become more widespread. Most of the experience has still been with respect to the more "popular" incineration-related compounds, such as those listed in Tables 1 and 2. The consensus from users of the train is that it works well in the field and yields good results when used with adequate care and precautions.

Reference 2 is now available to users of the train and those reviewing incinerator sampling plans.

Collection of compounds with boiling points between 100 and 30°C is considered more or less routine for the VOST. The train may be quite effective

[2]Personal communication from R. G. Merrill, Air and Energy Engineering Research Laboratory, U.S. Environmental Protection Agency, Research Triangle Park, NC.

for certain compounds in the 100 to 150°C boiling point range and for many with boiling points considerably below 30°C, but certain precautions and preliminary laboratory checks are necessary. Vinyl chloride, for example, boils at −14°C and was sampled adequately in the studies cited. Further discussion and recommendations concerning this subject are given in Ref 2.

A validation program for VOST is under way and is being managed by EPA's Environmental Monitoring Systems Laboratory at RTP. Audit gas cylinders have been prepared and are available from the same laboratory. A number of other VOST-related research programs are near completion and should provide additional insight into the limitations of the train.

Summary

Laboratory experiments and field experience to date have led to the following conclusions:

1. The VOST concept is sound—the train performs well in the field and yields good results in the hands of careful and experienced operators.
2. Sampling and analysis of low concentrations of organics in the presence of high levels of contamination is very difficult. Strong and well-planned quality control is essential.
3. Both the I/O and I/I designs are capable of producing good results if proper precautions are taken.
4. Sampling and analysis of water-soluble organics may not be as difficult as feared, but this area stills needs exploration.

It appears that VOST is a generally useful and flexible piece of equipment that will be valuable for other sources in addition to incinerators.

References

[1] Harris, J. C., Larsen, D. J., Rechsteiner, C. E., and Thrun, K. E., "Sampling and Analysis Methods for Hazardous Waste Combustion," EPA-600/8-84-002, PB84-155845, U.S. Environmental Protection Agency, Washington, DC, Feb. 1984.
[2] Hansen, E. M., "Protocol for the Collection and Analysis of Volatile POHCs Using VOST," EPA-600/8-84-007, PB84-170042, U.S. Environmental Protection Agency, Washington, DC, March 1984.
[3] Jungclaus, G. A., Gorman, P. G., Vaughn, G., Scheil, G. W., Bergman, F. J., Johnson, L. D., and Friedman, D., "Development of a Volatile Organic Sampling Train (VOST)," presented at Ninth Annual Research Symposium on Land Disposal, Incineration, and Treatment of Hazardous Waste, U.S. Environmental Protection Agency, Washington, DC, May 1983.
[4] Jungclaus, G. A., Gorman, P. G., and Bergman, F. J., "Sampling and Analysis of Incineration Effluents with the Volatile Organic Sampling Train (VOST)," Proceedings, National Symposium on Recent Advances in Pollutant Monitoring of Ambient Air and Stationary Sources, EPA-600/9-84-001, U.S. Environmental Protection Agency, Washington, DC, Jan. 1984.
[5] Ackerman, D. G., Beimer, R. G., and McGaughey, J. F., "Incineration of Volatile Organic Compounds on the M/T Vulcanus II," TRW Inc., Energy and Environmental Division,

Redondo Beach, CA; report to Chemical Waste Management, Inc., Oak Brook, IL, April 1983.

[6] Ackerman, D. G., McGaughey, J. F., Wagoner, D. E., and VanderVelde, G., "Emissions Testing Onboard the Incinerator Ship Vulcanus II Using a Volatile Organic Sampling Train," presented at Symposium on Organic Emissions from Combustion, 187th National Meeting, American Chemical Society, New York, April 1984.

Summary

Summary

This volume is the fourth in a series of special technical publications (STPs) sponsored by ASTM Committee D-34 on Waste Disposal. As noted in the Introduction to this volume, the prior volumes[1,2,3] covered a wide variety of topics in solid and hazardous waste management. This STP, and the symposium on which it was based, were intended to provide even greater insight into a number of waste characterization methods and to present state-of-the-art information on land treatment and disposal, incineration, and risk assessment technologies. A special effort was made to include papers in all of these areas which described ongoing research by the U.S. Environmental Protection Agency (EPA).

The papers included herein have undergone peer review and often extensive revision since their presentation at the symposium in Arlington, Virginia, in May 1984. The information discussed can be used to assist the solid waste community in developing better testing and handling methods for industrial and hazardous solid wastes.

Analysis and Characterization of Wastes

The most important factor in determining the proper handling methods for a solid waste is the actual character of the waste. Methods that define the chemical composition and physical characteristics of a waste are essential to ensuring that such materials are treated or disposed of in a manner that is protective of human health and the environment.

The precision and accuracy of the standard Soxhlet extraction method for the determination of oil and grease content in soils was evaluated by *Martin and Loehr*. They found that the coefficient of variation for five different oil/soil mixtures never exceeded 4%. The authors state that careful adherence to procedural details allows this method to be used to quantity the oil and grease content in soil samples accurately.

Francis et al conducted extensive leaching studies of four different industrial wastes, using batch and column leaching methods and five different

[1] *Hazardous Solid Waste Testing: First Conference, ASTM STP 760,* American Society for Testing and Materials, Philadelphia, 1982.

[2] *Hazardous and Industrial Solid Waste Testing: Second Symposium, ASTM STP 805,* American Society for Testing and Materials, Philadelphia, 1983.

[3] *Hazardous and Industrial Waste Management and Testing: Third Symposium, ASTM STP 851,* American Society for Testing and Materials, Philadelphia, 1984.

leaching media. This work was performed in part to provide a data base for further development of regulatory testing programs to determine the hazard potential of solid wastes disposed in landfills. Data from large-scale lysimeter tests were compared with the results of 32 different laboratory leaching tests. The authors conclude that a batch leaching test using carbonic acid as the extraction fluid may provide a suitable test for further evaluation, but caution that the selection of the final test conditions may be dependent on the intended use of the test results.

Current EPA plans to use the data presented in the paper by *Francis et al* are the topic of the next paper by *Kimmell and Friedman*. The authors describe the testing needs of the agency, explaining how the development of a new generation batch extraction method will be applied in the regulatory framework. The advantages and disadvantages of a batch extraction method are discussed, as well as future EPA research activities in this area. Significant discussion of the technical validity of a short-term extraction procedure versus long-term real-world behavior of landfilled wastes is also included.

The concentrations of specific compounds or classes of chemicals in waste samples are often critical in determining ultimate disposal techniques. *Barsotti and Palmer* compare the results of two different analytical methods, manual distillation and an automated technique, on the cyanide content of an aqueous waste stream. While both methods show acceptable precision, the automated technique was easier and had a larger sample throughput. Depending on the types of cyanide species present in the sample, the manual distillation method could yield erroneous results when used to measure free cyanides.

Ziegler et al describe a rapid method for determining total organic halogens in industrial wastes. The procedure combusts a waste sample in a Parr combustion bomb, then measures the converted inorganic halides by any of three different methods. A detection limit of 3.5 ppm can be obtained. Actual precision data on a large number of halogenated organics are presented.

Friedman discusses ongoing EPA research activities for a number of specific analytical techniques. The single and multilaboratory programs to evaluate 21 analytical methods included or considered for inclusion in the EPA's *Solid Waste Analytical Manual* (SW-846) are described. Other primary areas of focus included two different analytical schemes for analyzing hazardous constituents in groundwater; monitoring methods for volatile and nonvolatile organic compounds from incinerators or boilers, as well as volatile compounds from land treatment and disposal facilities; and leaching potential (mobility) of hazardous constituents from solidified wastes and wastes that contain pyritic sulfur.

At times, the ability to characterize a waste properly is limited by the lack of acceptable test methods. *Michael et al* describe ongoing research to identify a test suitable for quantitatively measuring the ignitability potential of a solid waste. Three different tests—the radiant heat ignition, rate of flame propagation, and difficulty of extinguishing the waste with water—were evaluated on 26 different waste samples and commonly found materials. The test results were used to determine an "ignitability factor" for each of the materials; the factor takes all of the data into account for determining the overall tendency of a waste to ignite and cause a hazard. The authors note that the results are not conclusive enough to classify materials qualitatively as hazardous or nonhazardous based on the calculated ignitability factor.

The difficulty of determining the amount of hydrogen cyanide or hydrogen sulfide evolved from a waste is discussed in the paper by *Handy et al.* Their attempts to analyze 16 selected waste streams as part of an interlaboratory testing program show that further work is needed before reliable results can be obtained with the type of testing procedures they evaluated. The procedure showed poor precision for hydrogen cyanide, and the authors recommend that it not be considered further. The hydrogen sulfide results are more encouraging but demonstrate a need for optimizing testing conditions to improve the precision of the method.

Problems associated with the analysis of lead in two different types of soil are presented by *Perket and Barsotti.* A multilaboratory analysis has revealed that several laboratories consistently generated aberrant data. Statistical evaluation of all laboratories suggests that problems exist with sample preparation as well as actual laboratory analysis. Recommendations for better quality assessment programs for remedial site investigations are presented.

Gurka et al propose the use of a sophisticated instrumental technique for identifying a complex array of organic constituents in environmental samples. The combination of Fourier-transform infrared spectrometry and the more common gas chromatography/mass spectrometry (GC/MS) technique allows a more detailed composition of samples to be obtained. Data on two specific samples are presented to show the utility of such an analysis. The needs for lower-cost, better computer software and for improvements in the infrared techniques are suggested as areas for further evaluation.

Risk Assessment/Biological Test Methods

An area that has been receiving an increasing level of attention in both the public and private sectors is the concept of risk assessment of chemical wastes and various waste management methods. This volume includes a special section on risk assessment. In addition, several papers address the more specific area of biological testing of wastes in order to ascertain the level of risk involved with various management techniques.

Human exposure assessment is discussed in the paper by *Dragun and Erler.* They describe the various types of environmental releases and exposure routes for chemicals and provide a real-world example calculation for determining exposures during cleanup of a site contaminated by polychlorinated biphenyls (PCBs). In this case, the assessment reveals that greater human exposure to PCBs would occur if contaminated soil were disposed off-site than if it were simply covered and left at the site.

Matthews and Bulich discuss a new method for assessing the suitability of organic wastes for land treatment. The test uses marine luminescent bacteria to measure differences in toxicity before and after a waste is subjected to land treatment. The reduction of the toxicity is regarded as a measure of treatability of that waste. The data can also be used to determine optimum operating criteria (for example, the waste application rate or moisture/nutrient effects).

Newhauser et al describe an innovative test using earthworms to determine the biological impact of wastes applied to soil. Data are presented comparing the relative toxicities of a variety of organic chemicals and metals to earthworms under two different test conditions—one using filter paper impregnated with a test solution, the other using an artificial soil mixture. The toxicity results from the earthworm tests were qualitatively similar to published results on rat oral median lethal dose (LD_{50}) values for a number of the organic chemicals tested.

Andon et al describe several tests using chemical extracts to determine the level of biological activity in a number of industrial waste samples. Samples were extracted with ethanol, dichloromethane, and dimethyl sulfoxide (DMSO); the extracts were tested for toxicity and mutagenicity by using the Chinese hamster ovary and Ames tests. A new procedure utilizing thin-layer chromatography of the pure waste streams is also described. All of the waste samples showed some biological activity, although the tests differed in their ability to detect the toxic or mutagenic properties.

A full toxicological screening program is proposed by *Lewtas and Andon.* Their protocol involves feeding hazardous wastes orally to rats for ten days, followed by a comprehensive battery of tests to identify potential toxic effects of the waste. The testing program is currently being validated by using known toxic compounds. A prescreen test sequence is also being developed to allow prioritization of waste samples for full toxicological screen testing.

Land Treatment and Disposal Test Methods

The historical practice of disposing of solid and hazardous wastes in landfills or other land disposal systems has been seriously questioned over the past ten years because of the environmental problems caused by this waste management method. A number of papers address research and testing efforts designed to provide a better technical basis for determining what types of wastes may be best suited for land disposal and how to design such facilities better.

The paper by *Bowders et al* provides an up-to-date description of the types of tests available for predicting leachate effects on natural soil liners. A thorough discussion of the advantages and disadvantages of using different permeability tests is presented. A number of recommendations regarding permeameter selection, graphic presentation of the results, and the use of index tests and batch equilibrium tests are presented. Actual test results are included showing the effects of landfill leachates and pure chemicals on clay soils when using the recommended tests.

The current EPA research program in land disposal is described in the paper by *Schomaker*. Specific programs addressing landfills, surface impoundments, underground storage in mines, and dioxin engineering are discussed as to general approach, study content, and application of the information within the regulatory framework. Most research efforts culminate in the issuance of Technical Resource Documents (TRDs).

Myers describes a test program utilized by the U.S. Department of the Army for determining the adequacy of solidified waste for landfilling. A 15-minute cone index test on compacted samples of freshly solidified waste was shown to be a good indicator of the suitability of the material for landfilling.

The need for proper testing of chemical grouts used as subsurface barriers at hazardous waste sites is described by *Malone et al*. The standard tests currently in use are discussed, and specific deficiencies in these tests are noted. Procedures for judging the compatability of grouts with specific groundwater environments and test methods for determining the permeability of the in-situ material are two areas where standardized test methods are needed.

Loehr et al discuss the need for a statistical sampling program to obtain valid data from a land treatment facility. Because of such variables as soil nonhomogeneity, differences in waste application, and uneven mixing characteristics, the coefficients of variation for a number of parameters ranged from 5 to 16%, even for a small pilot-scale land treatment plot. A proposed sampling plan, based on acceptable error in measuring critical parameters, is presented for obtaining valid data on waste constituent concentrations in the zone of incorporation.

Incineration

As regulations and public concern over the effects of landfilling continue to reduce dependency on this method of disposal, particularly for hazardous wastes, an increased emphasis on incineration technology has resulted. Although incineration is a far more desirable technology for managing most organic wastes, there are a number of problems that are deserving of study, including improvement in analytical techniques. This volume has sought to emphasize the importance of incineration by having a special section for this technology.

Lee et al discuss a variety of problems encountered in analyzing complex chemical wastes to determine the characteristics that are important for incineration. Topics covered include GC/MS analysis and data interpretation; chlorine, phosphorus, sulfur, sodium, water, and ash content; and tests for heat of combustion and viscosity. Solutions are offered to a number of the analytical problems discussed and several areas are suggested for further research. As a general suggestion, a statistical plan for obtaining valid analytical data from complex, nonhomogeneous chemical wastes is proposed that will help minimize the problem of nonrepresentative analysis.

Doucet describes the importance of test burns to assist in the proper design and selection of waste incineration systems. Such testing programs allow the owner to determine the incinerability of a waste under actual operating conditions. Numerous qualitative and quantitative benefits, such as observing flame patterns, ash slagging tendencies, air and fuel requirements, and organic destruction efficiency, are outlined for this approach. The problems of identifying proper facilities and specifying the waste and operating criteria to be used in the test are also discussed.

Santoleri presents a similar paper on the importance of pilot test burns. Data from a real test burn are also presented to demonstrate the value in this approach; an operational method for controlling emissions of oxides of nitrogen while achieving the necessary efficiency in organic destruction was determined as a result of information collected during the test.

As indicated in the previous two papers, a critical aspect in assessing the suitability of an incineration system is organic destruction efficiency. The paper by *Johnson* describes EPA efforts at developing a volatile organic sampling train (VOST) to use for stack sampling. A summary of some of the laboratory evaluation data using various chlorinated organics is presented. Some of the potential problem areas, such as sample breakthrough and the need for extensive quality assurance/quality control data to validate the data, are reviewed.

James K. Petros, Jr.

Union Carbide Corp., South Charleston, WV 25303; symposium cochairman and coeditor.

William J. Lacy

U.S. Environmental Protection Agency, Washington, DC 20460; symposium cochairman and coeditor.

Richard A. Conway

Union Carbide Corp., South Charleston, WV 25303; coeditor.

Indexes

Subject Index

355

Author Index

Cumulative Index

The roman numerals printed in boldface type refer to the particular symposium in this series of solid waste testing symposia. The first symposium (**I**) is reported in *ASTM STP 760*,[1] the second (**II**) in *STP 805*,[2] the third (**III**) in *STP 851*,[3] and the fourth (**IV**) in *STP 886*.[4]

Subject Index

[1]*Hazardous Solid Waste Testing: First Conference, ASTM STP 760*, R. A. Conway and B. C. Malloy, Eds., American Society for Testing and Materials, Philadelphia, 1981.

[2]*Hazardous and Industrial Solid Waste Testing: Second Symposium, ASTM STP 805*, R. A. Conway and W. P. Gulledge, Eds., American Society for Testing and Materials, Philadelphia, 1983.

[3]*Hazardous and Industrial Waste Management and Testing: Third Symposium, ASTM STP 851*, L. P. Jackson, A. R. Rohlik, and R. A. Conway, Eds., American Society for Testing and Materials, Philadelphia, 1984.

[4]*Hazardous and Industrial Solid Waste Testing: Fourth Symposium, ASTM STP 886*, J. K. Petros, Jr., W. J. Lacy, and R. A. Conway, Eds., American Society for Testing and Materials, Philadelphia, 1986.

Author Index

369